Basic
Crystallography

Basic Crystallography

J.-J. Rousseau
Department of Physics, University of Maine, Le Mans, France

Translated from the French by A. James
University of Picardie, France

JOHN WILEY & SONS

Chichester · New York · Weinheim · Brisbane · Singapore · Toronto

National 01243 779777
International (+44) 1243 779777
e-mail (for orders and customer service enquiries): cs-books@wiley.co.uk
Visit our Home Page on http://www.wiley.co.uk
or http://www.wiley.com

Reprinted February 1999

Originally published in French as "Cristallographie géométrique et radiocristallographie."
©Masson Editeur, Paris 1995

Published with the help of the Ministère de la Culture

Other Wiley Editorial Offices

John Wiley & Sons, Inc., 605 Third Avenue,
New York, NY 10158-0012, USA

WILEY-VCH Verlag GmbH, Pappelallee 3,
D-69469 Weinheim, Germany

Jacaranda Wiley Ltd, 33 Park Road, Milton,
Queensland 4064, Australia

John Wiley & Sons (Asia) Pte Ltd, 2 Clementi Loop #02-01,
Jin Xing Distripark, Singapore 129809

John Wiley & Sons (Canada) Ltd, 22 Worcester Road,
Rexdale, Ontario M9W 1L1, Canada

Library of Congress Cataloging-in-Publication Data
Rousseau, J.-J.
[Cristallographie géométrique et radiocristallographie. English]
Basic crystallography/J.-J. Rousseau; translated from the French by A. James.
p. cm.
Includes bibliographical references (p. —) and index.
ISBN 0-471-97048-4 (cloth: alk. paper). —ISBN 0-471-97049-2
(pbk.:alk.paper)
1. Crystallography. I. Title.
QD905.2.R6813 1998
548'.81–dc21
97-44185
CIP

British Library Cataloguing in Publication Data

A catalogue record for this book is available from the British Library

ISBN 0 471 97048 4 cloth
0 471 97049 2 paperback

Typeset in 10/12pt Times from the author's disks by Dobbie Typesetting, Tavistock, Devon
Printed and bound in Great Britain by Biddles Ltd, Guildford, Surrey
This book is printed on acid-free paper responsibly manufactured from sustainable forestry,
in which at least two trees are planted for each one used for paper production.

Contents

Preface

This book is the English translation of a teaching manual for French undergraduates.

In it, I have used my teaching experience to give the reader as straightforward an account as possible of the general principles of crystallography using only mathematical tools accessible to the average science undergraduate.

During my years of teaching I have realised that many students have difficulty in visualising objects in space; for this reason I have given considerable importance to geometrical crystallography, and I describe in detail the stereographic projection as a means of tackling this problem. The more difficult parts of the course are illustrated with the help of exercises of various lengths and complexities. Solutions are given, but these have deliberately been kept brief in order to encourage the reader to think for himself or herself.

Some changes have been made since the first edition: I have corrected a number of typographical errors and added several more illustrations and exercises. There is now a chapter on the numerical calculations used in crystallography, and a PC disc containing programs accompanying and illustrating the course is included with the book. Many of the illustrations were in fact produced from these programs, which, I believe, will be of help to students who are studying the subject on their own. Throughout the book vectors are printed in bold characters and the letter j is used for complex numbers.

I would like to thank most sincerely my colleagues, Professors A. Bulou and G. Courbion for their views and comments, and Mr. A. James for his kind collaboration in producing this translation.

Le Mans, October 1997

Historical background

Crystallography is the science of crystals. It concerns the external shape, the internal structure, the growth and the physical properties of crystals.

The word 'crystal' is of Greek origin (krustallas) and means 'solidified by cooling'; the Greeks thought that rock crystal (quartz) was formed through a transformation of ice by cooling.

Originally, crystallography was purely descriptive and was considered a branch of mineralogy. Later it was realised that the crystalline state is not confined to mineral salts alone and that it is actually a very common state of matter; by the middle of the nineteenth century crystallography had become a science in its own right.

For a long time it has been thought that the external appearance of crystals reflects some regular internal ordering of matter. The first references to this are to be found in the work of Johannes Kepler (1619), Robert Hooke (1665) and Christian Huyghens (1690). Studying the birefringence of calcite, Huyghens suggested that its optical properties could be explained by the rules governing the internal arrangement within the crystal.

The first quantitative law of crystallography (the law of constant angles) was foreseen in 1669 by Nils Steensen from measurements of the angles between the faces of a quartz crystal. The law was formally expressed in 1772 by Jean-Baptiste Romé de l'Isle in his 'Essai de Cristallographie'.

The second law (the law of rational indices or simple truncations) was stated in 1774 by the Abbé Réné-Just Haüy. He had noticed that when a crystal of calcite cleaved, the pieces obtained had shapes identical to that of the original crystal. He assumed that crystals are made up of identical parallelepipeds which he called 'molécules intégrantes'. This proposition leads to the fact that the position in space of each face of a crystal can be described by three whole numbers.

The ideas of Haüy were refined by W.H.Miller who introduced the methods of analytical geometry into crystallography and who proposed the notation system which is still in use.

The contribution of Auguste Bravais to crystallography is particularly important. In his 1849 work 'The Lattice Structure of Crystals', he stated the following postulate, which has become the basis of crystallography:

THE BRAVAIS POSTULATE:

Given any point P in a crystal, there exists an infinite number of discrete points, unlimited in the three directions of space, around each of which the arrangement of matter is the same as it is around the point P.

From this postulate there comes the notion of the *three-dimensional crystal lattice* and all the symmetry considerations involved. Bravais also introduced into crystallography the fundamental notion of the reciprocal lattice (the dual space of mathematicians).

Following the work of Bravais, numerous studies were carried out on the problems of crystal symmetry, aided by the development by mathematicians of group theory. In particular, the problem of enumerating and classifying space groups was solved by Schönflies and Fedorov.

This theoretical work could not have been carried out without the work of the engineers who developed crystallographers' measuring instruments. The first goniometer (application goniometer) was made in 1782 by Carangeot. Babinet and Wollaston designed the first one-circle goniometers in 1810, and Wulff devised his net and developed the first two-circle goniometers; these were perfected by Fedorov (1853–1919).

Until the beginning of the twentieth century, crystallography was purely axiomatic. The first X-ray diffraction experiments were performed in 1912 by W.Friedrich and P.Knipping, following the ideas of M.von Laue; the work of W. and L.Bragg then confirming the accuracy of the Bravais postulate. Diffraction measurements gave direct experimental proof of the regular arrangement within crystals.

The introduction of new experimental diffraction techniques enabled the rapid development of X-ray crystallography. Finally, since 1960, computer tools have been used systematically for processing the data from diffraction experiments on crystals.

Today, in a well equipped laboratory, a few days are sufficient for the absolute structure determination of a newly synthesised inorganic crystal.

Part 1
GEOMETRICAL CRYSTALLOGRAPHY

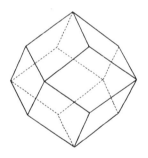

Chapter 1

The Postulates of Crystallography

One of the principal features of the crystalline state is anisotropy of physical properties, and this anisotropy can actually be seen in a crystal's external appearance, with its natural flat surfaces.

Before stating the postulates of crystallography we shall briefly review the two experimental laws relating to the shapes of crystals which led to the formulation of these postulates: the law of constancy of angles and the law of rational indices.

1 THE LAW OF CONSTANCY OF ANGLES

Certain crystals will cleave perfectly in definite directions. When cleavage occurs, the *position* of the crystal face changes, but not its *orientation*.

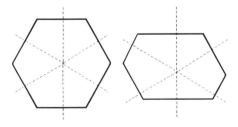

Figure 1.1

Quartz crystals are in the form of a right prism of hexagonal section with pyramidal ends. Figure 1.1 shows right sections of two quartz crystals with the normals to the faces of the prism.

For every sample of quartz ever examined, the same dihedral angle has been found between adjacent faces: exactly 120°.

For a given crystalline species, the dihedral angles between the faces are constant. On the other hand, the relative position in space of the faces can vary from one sample to another. The faces of a crystal are thus defined in orientation but not position, and this leads to the law of constancy of angles:

> *The bundle of lines originating in any point in a crystal and normal to the faces of the crystal is an invariable characteristic of the crystalline species.*

NOTE: The position—and in some cases the number—of faces on a crystal depends on the conditions in which the crystal grew; these conditions are nearly always anisotropic (e.g. gravity, impossibility of matter being transported to the face on which the crystal is lying). Note that the faces we observe are those whose growth rate was small, since rapidly growing faces are eliminated during growth. Figure 1.2 shows the appearance of a crystal at different stages of growth, with identical and non-identical rates; the higher the growth rate, the smaller the face.

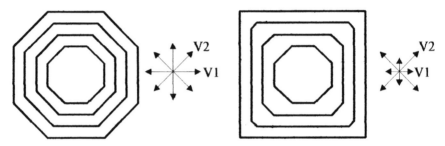

Figure 1.2

2   THE LAW OF RATIONAL INDICES

The faces of a crystal are not formed of arbitrary polyhedra. Suppose we decide to study a crystal; we take a suitable system of coordinates and we choose three non-coplanar axes **a**, **b** and **c**. If we imagine a plane intersecting these three axes, we can define the lengths a/b, b/c and c/a. All we are interested in here is the direction of the faces, not their position, and so the absolute values of a, b and c are irrelevant.

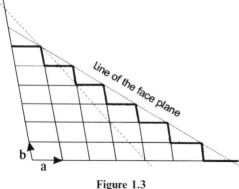

Figure 1.3

Any face of the crystal will intersect the axes to give lengths pa, qb and rc. From our preceding remark, only the ratios *pa/qb*, *qb/rc* and *rc/pa* concern us.

Figure 1.3 shows as an example a section of a crystal through a plane **a,b** with lines representing two faces. (continuous line: $p = 1$, $q = 1$) (dotted line: $p = 1$, $q = 2$).

THE LAW OF RATIONAL INDICES: *The numbers p, q and r that characterise a face are small integers which are relatively prime.*

If the three numbers are not relatively prime, then they have a common denominator n. The face defined by $p' = p/n$, $q' = q/n$ and $r' = r/n$ is a face parallel to the face defined by p, q and r. Since we are only interested in the orientation of the faces we can impose the condition that the indices be relatively prime. As a consequence of the law, the crystal must be made up of a regular three-dimensional stack of identical parallelepipeds. The basic parallelepiped is constructed along the three vectors **a**, **b** and **c**, and the stacking of elementary cells leads us to the notion of the **lattice**.Thus on the microscopic scale, the majority of the faces of a crystal have a step structure, but on the macroscopic scale appear to be smooth. We also note that this law implies the law of constancy of angles.

3   THE POSTULATES OF CRYSTALLOGRAPHY

The law of rational indices was formalised by Bravais in a much more general form as follows:

THE BRAVAIS POSTULATE: *Given any point P in a crystal, there exists an infinite array of points in the three directions of space, around which the arrangement of matter is the same as around the point P, and with the same orientation.*

At the end of the nineteenth century, this postulate was expanded and reformulated virtually simultaneously and independently by Schönflies and Fedorov:

THE SCHÖNFLIES–FEDOROV POSTULATE: *Given any point P in a crystal, there exists an infinite array of points in the three directions of space, around which the arrangement of matter is either the same as around the point P, or is an image of that arrangement.*

The difference between this postulate and that of Bravais is that it no longer requires an identical orientation of the environment around equivalent points, and the notion of image (symmetry with respect to a point) is introduced. This leads us to distinguish proper operations, which leave the orientation of space unchanged, and improper operations which modify the orientation. The consequences of this postulate are both numerous and important: the set of homologous points in a crystal constitutes a *periodic space lattice*, characterised by three basic translations. Any given lattice is characterised by a *set of symmetry operations* or translation operations defining displacements in space that leave the lattice invariant overall. The periodicity of the lattice greatly limits the number and nature of symmetry operations that leave the lattice unchanged. For a given crystal, the set of translation operations is mathematically known as '*an orientation symmetry group*' or '*a space group*' or '*a Schönflies–Fedorov group*'.

4   LATTICES, PATTERNS AND STRUCTURES

An ideal crystal is made up of a regular repetitive arrangement of atoms. Knowledge of the whole crystal only requires knowledge of the three vectors which define the lattice and the arrangement of atoms within the unit cells. The latter constitutes the basic repeating unit of a pattern, and a crystalline structure is the periodic repetition of this unit by lattice translations.

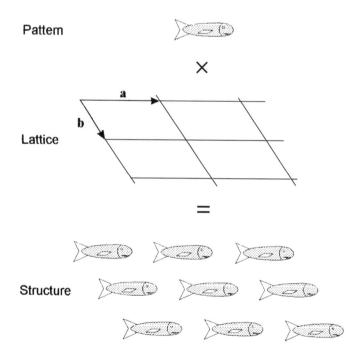

Figure 1.4

Two-dimensional equivalents of crystal structures can be found in wallpaper, pavings and tilings.

NOTE: The lattice only describes the periodicity of the structure and hence only the symmetry properties. The nodes of a lattice do not correspond to any physical entity and must not be confused with the atoms themselves. In particular, the lattice origin is completely arbitrary and may be chosen to lie at any point in a unit. In figure 1.4, we pass from one point to another analogous point—for example from one fish eye to another fish eye, by a lattice translation equal to $n.\mathbf{a} + m.\mathbf{b}$ (n, m are integers).

5   SYMMETRIES OF POSITION AND ORIENTATION

The symmetry operations which, as far as macroscopically observable properties are concerned, bring the medium to a position indistinguishable from the initial position, form a group which mathematicians refer to as a '*point group*'. The symmetry operations concerned (orientation symmetries) are also those which leave unchanged a bundle of lines emanating from any arbitrary point O in a crystal.

The relation between the symmetries of orientation and position of a crystal is simple: the former is on the macroscopic scale, while the latter is on the microscopic scale.

The orientation symmetries only involve changes in spatial orientation, because in symmetry operations translation takes place at the atomic level, and is undetectable at the macroscopic level.

Point groups describe the symmetry of finite objects whereas space groups describe the symmetry of infinitely repeating structures.

6   THE CRYSTALLINE STATE

A perfect crystal is totally ordered over a long distance. Real crystals are never perfectly ordered; their structures are all more or less disordered, but certain types of disorder allow us to define a perfectly ordered *average structure*. In particular, in a real crystal there is thermal vibration of the atoms about a mean position: translational symmetry exists only over a time-averaged structure. There is also chemical disorder: although the sites available to the atoms form a periodically repeating pattern, actual occupation of the sites by various types of atom may be more or less random. Finally, there are point defects (holes and interstitials), dislocations, grain boundaries (interfaces between two crystalline regions with different orientations), all of which perturb the crystalline order. However, when the number of defects is fairly small compared with the total number of atoms, we can use the ideal crystal model.

With advances in the techniques of solid state physics and X-ray crystallography, came the discovery, around 1980, of structures which have long-distance order but which are not strictly periodic: the incommensurables and the quasi-crystals.

In incommensurables, certain atoms are displaced from their ideal positions according to a modulating wave whose wavelength λ is incommensurable with the lattice translation T in the same direction (λ/T is an irrational number).

The first reported example of a quasi-crystal was in 1984 by Shetchtman (by rapid quenching of the alloy $Al_{86}Mn_{14}$). Quasi-crystals have symmetries (particularly five-fold) which are incompatible with lattice symmetries. These structures are now known to result from the aperiodic tiling of space by several types of unit cell.

Recent mathematical work shows that incommensurables and quasi-crystals can be studied using crystallographic systems in spaces of more than three dimensions.

Chapter 2

Point Lattices

1 THE DIRECT LATTICE

1.1 Definitions

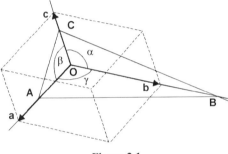

Figure 2.1

Consider three vectors defining a right-handed trihedral which can be oblique: **a**, **b**, **c**.

Let α, β, γ be the angles between these vectors, where:

$$\alpha = \{\mathbf{b},\mathbf{c}\},\ \beta = \{\mathbf{a},\mathbf{c}\},\ \gamma = \{\mathbf{a},\mathbf{b}\}$$

The vectors **a**, **b**, **c** are the **base vectors**.

The parallelepiped constructed on these three vectors is the **unit-cell**.

Let the vector $\mathbf{OP} = \mathbf{r} = u.\mathbf{a} + v.\mathbf{b} + w.\mathbf{c}$

If u, v, w are *integers*, **r** is said to be a **row** and the point P a **node**. The infinite set of nodes forms the **lattice**.

In the case of a *crystal*, this lattice describes the periodicity of the structure and it makes up the *crystal lattice*.

The base vectors, which are in general arbitrarily chosen, form an oblique coordinate system. For a given lattice, there exists more than one possible choice of base vectors and hence of unit-cell; this is illustrated in figure 2, which shows a two-dimensional lattice.

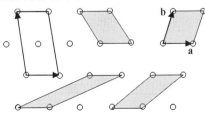

Figure 2.2. Primitive unit-cells in grey

A unit-cell is said to be **primitive** if it only has nodes at the vertices of the corresponding parallelogram (in two-dimensional lattices) or prism (in three-dimensional lattices). A primitive unit-cell is the smallest entity enabling all the nodes to be generated by whole translations of the lattice.

If further nodes exist (inside the unit-cell, on the faces or on the edges), the cell is said to be **multiple**.

In a two-dimensional lattice, the area of every primitive unit-cell is identical. Similarly, for a three-dimensional lattice, the volume of a primitive unit-cell is constant and corresponds to the volume around each node.

Using matrices, we can represent a row by:

$$\mathbf{r} = u.\mathbf{a} + v.\mathbf{b} + w.\mathbf{c} = (u,v,w) \begin{pmatrix} \mathbf{a} \\ \mathbf{b} \\ \mathbf{c} \end{pmatrix} = (\mathbf{a},\mathbf{b},\mathbf{c}). \begin{pmatrix} u \\ v \\ w \end{pmatrix}$$

The scalar product of the two vectors is:

$$\mathbf{r_1}.\mathbf{r_2} = (u_1.\mathbf{a} + v_1.\mathbf{b} + w_1.\mathbf{c}).(u_2.\mathbf{a} + v_2.\mathbf{b} + w_2.\mathbf{c})$$

which can then be expressed in the form:

$$\mathbf{r_1}.\mathbf{r_2} = (u_1,v_1,w_1). \begin{pmatrix} \mathbf{a}^2 & \mathbf{a}.\mathbf{b} & \mathbf{a}.\mathbf{c} \\ \mathbf{a}.\mathbf{b} & \mathbf{b}^2 & \mathbf{b}.\mathbf{c} \\ \mathbf{a}.\mathbf{c} & \mathbf{b}.\mathbf{c} & \mathbf{c}^2 \end{pmatrix} . \begin{pmatrix} u_2 \\ v_2 \\ w_2 \end{pmatrix} = \mathbf{u}_1^T.M.\mathbf{u}_2$$

The row matrix $\mathbf{u}^T$ is the transpose of the column vector $\mathbf{u}$, and the matrix M represents a tensor called the 'metric tensor'.

1.2   Double Vector Product

The following vector products hold:

$$\mathbf{a} \wedge (\mathbf{b} \wedge \mathbf{c}) = \mathbf{b}.(\mathbf{a}.\mathbf{c}) - \mathbf{c}.(\mathbf{a}.\mathbf{b}) \tag{1}$$

$$(\mathbf{a} \wedge \mathbf{b}).(\mathbf{c} \wedge \mathbf{d}) = (\mathbf{a}.\mathbf{c}).(\mathbf{b}.\mathbf{d}) - (\mathbf{a}.\mathbf{d}).(\mathbf{b}.\mathbf{c}) \tag{2}$$

1.3   Volume of the Unit-cell

It can be shown, for example by expressing the base vectors in an orthonormal frame, that the determinant of the matrix M is equal to the triple scalar product (**a**, **b**, **c**) and hence to the square of the volume of the cell. We then deduce:

$$V = abc[1 - \cos^2 \alpha - \cos^2 \beta - \cos^2 \gamma + 2\cos \alpha.\cos \beta.\cos \gamma]^{1/2} \tag{3}$$

We can also consider the identity:

$$(||\mathbf{a}||.||\mathbf{b} \wedge \mathbf{c}||)^2.\cos^2 \theta \equiv (||\mathbf{a}||.||\mathbf{b} \wedge \mathbf{c}||)^2(1 - \sin^2\theta)$$

$$(\mathbf{a}.(\mathbf{b} \wedge \mathbf{c}))^2 = (\mathbf{a},\mathbf{b},\mathbf{c})^2 \equiv \mathbf{a}^2.||\mathbf{b} \wedge \mathbf{c}||^2 - ||\mathbf{a} \wedge (\mathbf{b} \wedge \mathbf{c})||^2$$

and from this directly deduce the cell volume:

$$a^2.||\mathbf{b} \wedge \mathbf{c}||^2 = a^2b^2c^2 \sin^2 \alpha = a^2b^2c^2(1 - \cos^2 \alpha)$$

$$||\mathbf{a} \wedge (\mathbf{b} \wedge \mathbf{c})||^2 = ((\mathbf{a}.\mathbf{c}).\mathbf{b} - (\mathbf{a}.\mathbf{b}).\mathbf{c})^2 = a^2b^2c^2(\cos^2 \beta + \cos^2 \gamma - 2 \cos\alpha.\cos\beta.\cos\gamma)$$

1.4 Planes in the Direct Lattice

Consider a plane having the equation:

$$h\frac{x}{a} + k\frac{y}{b} + l\frac{z}{c} = 1 \qquad (4)$$

For $y = z = 0$ (Figure 1), we obtain the intersection A of this plane with the axis Ox. From equation (4) we have:

$$\mathbf{OA} = \frac{a}{h} \quad \mathbf{OB} = \frac{b}{k} \quad \mathbf{OC} = \frac{c}{l}$$

$$\text{and } \mathbf{AB} = \frac{b}{k} - \frac{a}{h}; \quad \mathbf{AC} = \frac{c}{l} - \frac{a}{h}; \quad \mathbf{BC} = \frac{c}{l} - \frac{b}{k} \qquad (5)$$

In a crystal lattice, a plane passing through three nodes and hence containing an infinite number of nodes is a **lattice plane**. The set of parallel lattice planes is a **family of planes** containing all the lattice nodes. If the points A, B and C are nodes, then:

$$\mathbf{OA} = x = u.a, \ \mathbf{OB} = y = v.b, \ \mathbf{OC} = z = w.c, \text{ where } u, v, w \text{ are } integers.$$

The general equation of the lattice planes in a family h, k, l is therefore, from equation (4), of the form:

$$h.u + k.v + l.w = n$$

The first plane in the family which does not contain the origin has the equation:

$$h.u + k.v + l.w = 1$$

h, k and l are the reciprocals of the lengths of the intercepts of this plane on the axes.

Each node of the lattice belongs to a lattice plane and as a result, for a crystal lattice, h, k and l are integers. These three indices define a family of lattice planes, and it is always possible to choose them to be relatively prime, since we do not distinguish parallel planes characterised by h, k, l and by $H = nh$, $K = nk$, $L = nl$.

1.5   Nomenclature

According to international convention, we call a **row** $\mathbf{r} = u.\mathbf{a} + v.\mathbf{b} + w.\mathbf{c}$ of a crystal lattice [u v w]. (*Indices between square brackets without separating commas*). Negative indices have bars: $\bar{\mathbf{u}}, \bar{\mathbf{v}}, \bar{\mathbf{w}}$.

$$\text{e.g.: } [1\,\bar{3}\,2], [1\,0\,0], [1\,0\,\bar{1}]$$

The family of **lattice planes** with the equation $h.u + k.v + l.w = \mathrm{n}$ is denoted (h k l). (Indices between brackets without separating commas).

$$\text{e.g.: } (2\,\bar{3}\,4), (0\,1\,0), (1\,0\,\bar{1})$$

The indices u, v, w for the rows and h, k, l for the planes are the *Miller indices*.

2   THE RECIPROCAL LATTICE

The reciprocal lattice[1] may appear rather artificial, and is by no means essential in geometrical crystallography; however, it does very often simplify calculations. In addition, the reciprocal lattice makes a natural appearance in the study of diffraction of periodic structures.

2.1   Definition

The reciprocal lattice has its base vectors defined from those of the direct lattice, together with the unit-cell volume, by the following relations:

$$\mathbf{A}^* = \frac{\mathbf{b} \wedge \mathbf{c}}{V} \quad \mathbf{B}^* = \frac{\mathbf{c} \wedge \mathbf{a}}{V} \quad \mathbf{C}^* = \frac{\mathbf{a} \wedge \mathbf{b}}{V} \tag{6}$$

---

[1]Geometrically, the direct and reciprocal lattices are interrelated by a polar reciprocal transform, and analytically by a Fourier transform.

We can also use the equivalent formulation, based on the scalar product:

$$\mathbf{A}^*.\mathbf{a} = \mathbf{B}^*.\mathbf{b} = \mathbf{C}^*.\mathbf{c} = 1$$

$$\mathbf{A}^*.\mathbf{a} = \mathbf{A}^*.\mathbf{c} = \mathbf{B}^*.\mathbf{a} = \mathbf{B}^*.\mathbf{c} = \mathbf{C}^*.\mathbf{a} = \mathbf{C}^*.\mathbf{b} = 0$$

These relations can be abbreviated as:

$$\mathbf{a}_i.\mathbf{A}_j^* = \delta_{ij} \begin{cases} \delta_{ij} = 1 & i = j \\ \delta_{ij} = 0 & i \neq j \end{cases} \tag{7}$$

For the reciprocal lattice, just as for the direct lattice, we can define nodes, rows and families of lattice planes.

NOTATION: Throughout this book, all reciprocal values are given an asterisk (*).

2.2 An Example of a Reciprocal Lattice

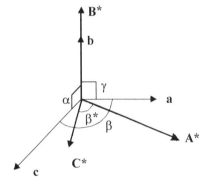

Figure 2.3

Figure 3 shows the direct and reciprocal base vectors of a mono-clinic lattice with:

$$\alpha = \gamma = \pi/2; \ \beta > \pi/2; \ a \neq b \neq c.$$

$$\mathbf{A}^* \perp \mathbf{b}, \ \mathbf{A}^* \perp \mathbf{c}$$
$$\mathbf{C}^* \perp \mathbf{b}, \ \mathbf{C}^* \perp \mathbf{a}$$
$$\mathbf{B}^* \perp \mathbf{a}, \ \mathbf{B}^* \perp \mathbf{c}$$

In this example, vectors **b** and **B*** are colinear.

$$(\alpha = \alpha^* = \gamma = \gamma^* = \pi/2)$$

In tri-orthogonal lattices ($\alpha = \beta = \gamma = \pi/2$), all the base vectors of the direct and reciprocal lattices are colinear. The lengths of the reciprocal axes are the reciprocals of those of the direct axes (hence the term reciprocal!).

2.3 Calculation of Reciprocal Values

□ ANGLES BETWEEN BASE VECTORS

The angles α^*, β^* and γ^* between the base vectors of the reciprocal lattice can be expressed in terms of the angles α, β and γ through the scalar product **A***.**B***.

α is the angle between **b** and **c**, β is the angle between **a** and **c**, γ^* is the angle between **A*** and **B***. According to the relations in definition (6), $||\mathbf{A}^*|| = A^* = b.c.\sin\alpha.V^{-1}$.

Using relation (2), we have:

$$\mathbf{A}^*.\mathbf{B}^* = \frac{(\mathbf{b}\wedge\mathbf{c}).(\mathbf{c}\wedge\mathbf{a})}{V^2} = \frac{(\mathbf{b}.\mathbf{c}).(\mathbf{a}.\mathbf{c}) - c^2.(\mathbf{a}.\mathbf{b})}{V^2} = \frac{b.c.\cos\alpha.a.c.\cos\beta - a.b.c^2.\cos\gamma}{V^2}$$

Direct calculation of the scalar product gives:

$$\mathbf{A}^*.\mathbf{B}^* = A^*.B^*.\cos\gamma^* = \frac{b.c.\sin\alpha.a.c.\sin\beta.\cos\gamma^*}{V^2}$$

Comparing the two expressions we have:

$$\cos\gamma^* = \frac{\cos\alpha.\cos\beta - \cos\gamma}{\sin\alpha.\sin\beta}$$

By cyclic permutation we obtain:

$$\cos\beta^* = \frac{\cos\alpha.\cos\gamma - \cos\beta}{\sin\alpha.\sin\gamma}$$

$$\cos\alpha^* = \frac{\cos\gamma.\cos\beta - \cos\alpha}{\sin\gamma.\sin\beta}$$

Similarly, the angles of the direct lattice are deduced from those of the reciprocal lattice by relations of the form:

$$\cos\alpha = \frac{\cos\gamma^*.\cos\beta^* - \cos\alpha^*}{\sin\gamma^*.\sin\beta^*}$$

❒   MODULUS OF THE BASE VECTORS

Taking the vector product of the reciprocal lattice base vectors, we obtain through relations (1) and (2):

$$\mathbf{A}^*\wedge\mathbf{B}^* = \frac{(\mathbf{b}\wedge\mathbf{c})\wedge(\mathbf{c}\wedge\mathbf{a})}{V^2} = \frac{\mathbf{c}.(\mathbf{b},\mathbf{c},\mathbf{a})}{V^2}$$

Calculation of the moduli of the first two terms gives:

$$||\mathbf{A}^*\wedge\mathbf{B}^*|| = \frac{b.c.\sin\alpha.c.a.\sin\beta.\sin\gamma^*}{V^2} = \frac{c.V}{V^2}$$

Hence: $V = a.b.c.\sin\alpha.\sin\beta.\sin\gamma^* = a.b.c.\sin\alpha^*.\sin\beta.\sin\gamma = a.b.c.\sin\alpha.\sin\beta^*.\sin\gamma$

$$||\mathbf{A}^*|| = \left\| \frac{\mathbf{b} \wedge \mathbf{c}}{V} \right\| = \frac{b.c.\sin\alpha}{a.b.c.\sin\alpha.\sin\beta.\sin\gamma^*} = \frac{1}{a.\sin\beta.\sin\gamma^*} = \frac{1}{a.\sin\beta^*.\sin\gamma}$$

2.4 Properties of the Rows of a Reciprocal Lattice

❏ ORIENTATION

Consider the reciprocal vector $\mathbf{N}^*_{hkl} = h.\mathbf{A}^* + k.\mathbf{B}^* + l.\mathbf{C}^*$ and the plane Π of the direct lattice which we denote $(h\ k\ l)$, whose equation is:

$$h\frac{x}{a} + k\frac{y}{b} + l\frac{z}{c} = 1$$

Since this plane intersects the direct axes at A, B and C, the vectors $\mathbf{AB}$ and $\mathbf{BC}$ lie in the plane Π. From relations (5) and (7) we have:

$$\mathbf{N}^*_{hkl}.\mathbf{AB} = (h\mathbf{A}^* + k\mathbf{B}^* + l\mathbf{C}^*).\left(\frac{\mathbf{b}}{k} - \frac{\mathbf{a}}{h}\right) = 0$$

The scalar products $\mathbf{N}^*_{hkl}.\mathbf{AB}$ and $\mathbf{N}^*_{hkl}.\mathbf{BC}$ are null and it follows that:

$$\boxed{\mathbf{N}^*_{hkl} \perp (h\ k\ l)}$$

The reciprocal row $\mathbf{N}^*{}_{hkl}$ is normal to the planes $(h\ k\ l)$ of the direct lattice.

❏ MODULUS OF THE RECIPROCAL ROWS IN A CRYSTAL LATTICE

If the plane Π is a lattice plane, then it belongs to a family of equidistant parallel planes denoted $(h\ k\ l)$. Let d_{hkl} be the distance between any two planes in the family. This distance is the projection of the vector $\mathbf{OA}$ onto the normal to the plane, the normal being in the direction of the vector $\mathbf{N}^*{}_{hkl}$:

$$d_{hkl} = \frac{\mathbf{N}^*{}_{hkl}.\mathbf{OA}}{||\mathbf{N}^*{}_{hkl}||} = \frac{(h\mathbf{A}^* + k\mathbf{B}^* + l\mathbf{C}^*)}{||\mathbf{N}^*_{hkl}||}.\frac{\mathbf{a}}{h} = \frac{1}{||\mathbf{N}^*_{hkl}||}$$

$$\boxed{d_{hkl}.||\mathbf{N}^*_{hkl}|| = 1} \tag{8}$$

☞ *Any family (h k l) of planes in the direct lattice can be associated with the reciprocal row [h k l]* which is orthogonal to it.*

2.5   A Property of Reciprocal Planes

Relation (7) defining the reciprocal lattice is symmetrical in $\mathbf{a}_i$ and $\mathbf{A}_j^*$. *The reciprocal lattice of a reciprocal lattice is the initial direct lattice.* Any family $(u\ v\ w)^*$ of reciprocal lattice planes can be associated with the direct row, denoted $[u\ v\ w]$, which is orthogonal to it. Let D_{uvw}^* be the distance between two planes in the family, and $\mathbf{n}_{uvw}$ the normal direct row. From relation (8) we have:

$$\boxed{D_{uvw}^* \cdot ||\mathbf{n}_{uvw}|| = 1}$$

3   MILLER INDICES

Many notation systems for lattice planes have been suggested in the past (Lévy–Des Cloizeaux, Weiss–Roze, Nauman, Goldschmidt) but the system which finally won the day was that of Miller in 1839.

A family of lattice planes which has as its normal the reciprocal row of indices $[h\ k\ l]^*$ is denoted $(h\ k\ l)$. This new definition of Miller indices is equivalent to the one given in paragraph 1.4: the Miller indices of a family of lattice planes are the reciprocals of the intercepts on the axes by the first plane of the family (which is the plane of equation $h.u + k.v + l.w = 1$).

The main advantage of Miller indices is that the notations of a family of lattice planes in the direct lattice (reciprocals of the intercepts) and reciprocal lattice (indices of the normal) are identical.

SPECIAL CASE: If a plane is parallel to an axis, it intercepts it at infinity and the corresponding Miller index is therefore null. Consequently, planes containing base vectors have the notations:

$$xOy \Rightarrow (001) \quad yOz \Rightarrow (100) \quad xOz \Rightarrow (010)$$

In the example shown in figures 2.4 and 2.5, the planes (102) have been drawn in a lattice in which $\alpha = \beta = \gamma = \pi/2$ and $\beta > \pi/2$ (monoclinic lattice).

The first plane in the family intercepts the Ox axis at a distance a, the Oy axis at infinity and the Oz axis at $c/2$.

In figure 2.5, which is drawn in the xOz or (010) plane, lattice nodes and several planes in the family (102) are shown, together with their normal N^*_{102} which enables the constant spacing of the $d_{102} = 1/||N^*_{102}||$ planes to be determined.

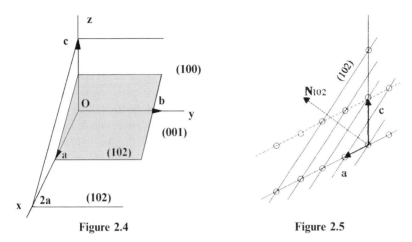

| Figure 2.4 | Figure 2.5 |

Figure 2.6 is for an orthorhombic lattice ($a \neq b \neq c$, $\alpha = \beta = \gamma = \pi/2$) where the lattice planes of the families (001), (101) and ($1\bar{1}1$) have been drawn.

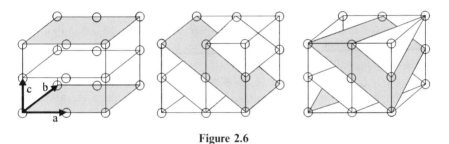

Figure 2.6

NOTE: Weiss indices are the reciprocals of Miller indices and correspond to the intercepts on the axes by the first plane in the family.

4 TRANSFORMATIONS OF THE COORDINATE SYSTEM

4.1 Covariance of the Miller Indices of Planes

Consider two direct coordinates **a**, **b**, **c** and **a′**, **b′**, **c′**, in a lattice, such that:

$$\mathbf{a}' = \alpha_{11}.\mathbf{a} + \alpha_{12}.\mathbf{b} + \alpha_{13}.\mathbf{c}$$

$$\mathbf{b}' = \alpha_{21}.\mathbf{a} + \alpha_{22}.\mathbf{b} + \alpha_{23}.\mathbf{c} \qquad (9)$$

$$\mathbf{c}' = \alpha_{31}.\mathbf{a} + \alpha_{32}.\mathbf{b} + \alpha_{33}.\mathbf{c}$$

We can associate these with the reciprocal coordinates $\mathbf{A}^*$, $\mathbf{B}^*$, $\mathbf{C}^*$ and $\mathbf{A}'^*$, $\mathbf{B}'^*$, $\mathbf{C}'^*$.

Consider a reciprocal row $[h\ k\ l]^*$. This will be an *invariant* in the coordinate transformation:

$$\mathbf{N}^*_{hkl} = h.\mathbf{A}^* + k.\mathbf{B}^* + l.\mathbf{C}^*$$

In the new coordinate system, this row becomes $[h'\ k'\ l']^*$:

$$\mathbf{N}^*_{hkl} = h'.\mathbf{A}'^* + k'.\mathbf{B}'^* + l'.\mathbf{C}'^*$$
$$h'.\mathbf{A}'^* + k'.\mathbf{B}'^* + l'.\mathbf{C}'^* = h.\mathbf{A}^* + k.\mathbf{B}^* + l.\mathbf{C}^* \tag{10}$$

Scalar multiplication of both sides of (10) by the vector $\mathbf{a}'$ (9) gives:

$$(h'.\mathbf{A}'^* + k'\mathbf{B}'^* + l'.\mathbf{C}'^*).\mathbf{a}' =$$
$$(h.\mathbf{A}^* + k.\mathbf{B}^* + l.\mathbf{C}^*).(\alpha_{11}.\mathbf{a} + \alpha_{12}.\mathbf{b} + \alpha_{13}.\mathbf{c})$$

Now: $\mathbf{A}'^*.\mathbf{a}' = \mathbf{A}^*.\mathbf{a} = 1 \ldots$ and $\mathbf{B}'^*.\mathbf{a}' = \mathbf{B}^*.\mathbf{a} = \mathbf{C}'^*.\mathbf{a}' = \mathbf{C}^*.\mathbf{a} = 0 \ldots$ From which we deduce that:

$$h' = \alpha_{11}.h + \alpha_{12}.k + \alpha_{13}.l$$

We can similarly show that:

$$k' = \alpha_{21}.h + \alpha_{22}.k + \alpha_{23}.l$$

$$l' = \alpha_{31}.h + \alpha_{32}.k + \alpha_{33}.l$$

☞ | *In a coordinate transformation, the Miller indices of reciprocal rows (or planes in the direct lattice) are transformed as base vectors of the direct lattice.*

EXERCISE:
Establish the relations between the Miller indices of a direct plane, expressed in the new coordinates, in terms of the plane indices in the old coordinates. Show that the transformation matrix is the reciprocal of the transpose of the matrix which relates the base vectors. Hint: the row $\mathbf{r} = u.\mathbf{a} + v.\mathbf{b} + w.\mathbf{c}$ is an invariant in the transformation.

4.2 Generalisation

Consider a transformation from coordinates $\mathbf{a}$, $\mathbf{b}$, $\mathbf{c}$, to coordinates $\mathbf{a}_1$, $\mathbf{b}_1$, $\mathbf{c}_1$. The relations between the base vectors, the reciprocal vectors, the direct rows and the reciprocal rows are written in the following matrix forms:

$$\begin{pmatrix} \mathbf{a}_1 \\ \mathbf{b}_1 \\ \mathbf{c}_1 \end{pmatrix} = (\mathbf{A}) \begin{pmatrix} \mathbf{a} \\ \mathbf{b} \\ \mathbf{c} \end{pmatrix}, \begin{pmatrix} \mathbf{A}_1^* \\ \mathbf{B}_1^* \\ \mathbf{C}_1^* \end{pmatrix} = (\mathbf{A}^*) \begin{pmatrix} \mathbf{A}^* \\ \mathbf{B}^* \\ \mathbf{C}^* \end{pmatrix}, \begin{pmatrix} u_1 \\ v_1 \\ w_1 \end{pmatrix} = (\mathbf{U}) \begin{pmatrix} u \\ v \\ w \end{pmatrix}, \begin{pmatrix} h_1 \\ k_1 \\ l_1 \end{pmatrix} = (\mathbf{H}) \begin{pmatrix} h \\ k \\ l \end{pmatrix}$$

With $(\mathbf{U}^T)$ denoting the transposed matrix of $(\mathbf{U})$, we also have:

$$(u_1, v_1, w_1) = (u, v, w).(\mathbf{U}^T)$$

The direct row $\mathbf{r}$, the reciprocal row $\mathbf{R}^*$ and their scalar product $\mathbf{r}.\mathbf{R}^*$ are the invariants of this transformation:

$$\mathbf{r} = (u, v, w) \begin{pmatrix} \mathbf{a} \\ \mathbf{b} \\ \mathbf{c} \end{pmatrix} = (u_1, v_1, w_1) \begin{pmatrix} \mathbf{a}_1 \\ \mathbf{b}_1 \\ \mathbf{c}_1 \end{pmatrix} = (u, v, w)(\mathbf{U}^T)(\mathbf{A}) \begin{pmatrix} \mathbf{a} \\ \mathbf{b} \\ \mathbf{c} \end{pmatrix} \Rightarrow (\mathbf{A}) = (\mathbf{U}^T)^{-1}$$

$$\mathbf{R}^* = (h, k, l) \begin{pmatrix} \mathbf{A}^* \\ \mathbf{B}^* \\ \mathbf{C}^* \end{pmatrix} = (h_1, k_1, l_1) \begin{pmatrix} \mathbf{A}_1^* \\ \mathbf{B}_1^* \\ \mathbf{C}_1^* \end{pmatrix} =$$

$$(h, k, l)(\mathbf{H}^T)(\mathbf{A}^*) \begin{pmatrix} \mathbf{A}^* \\ \mathbf{B}^* \\ \mathbf{C}^* \end{pmatrix} \Rightarrow (\mathbf{A}^*) = (\mathbf{H}^T)^{-1}$$

$$\mathbf{r}.\mathbf{R}^* = (u, v, w) \begin{pmatrix} h \\ k \\ l \end{pmatrix} = (u_1, v_1, w_1) \begin{pmatrix} h_1 \\ k_1 \\ l_1 \end{pmatrix} =$$

$$(u, v, w)(\mathbf{U}^T)(\mathbf{H}) \begin{pmatrix} h \\ k \\ l \end{pmatrix} \Rightarrow (\mathbf{H}) = (\mathbf{U}^T)^{-1}$$

We also have: $(\mathbf{A}^T)^{-1} = (\mathbf{U})$ and $(\mathbf{H}^T)^{-1} = (\mathbf{U}) = (\mathbf{A}^*)$. From this we deduce the relations:

$$(\mathbf{A}^*) = (\mathbf{A}^T)^{-1} = (\mathbf{U})$$
$$(\mathbf{H}) = (\mathbf{A})$$

☞ | *The base vectors and the indices of the planes (h k l) transform covariantly, whereas the reciprocal base vectors and the indices of the rows [u v w] transform contravariantly.*

5   LATTICE CALCULATIONS

These calculations are often made simpler by use of the reciprocal lattice.

❑   ZONES AND ZONE AXES

DEFINITION: A **zone** is formed by the set of direct lattice planes which intersect each other along parallel lines. The common direction of these lines is the **zone axis**. In a crystal they correspond to the edges of the faces.

The reciprocal row $[hkl]^*$ ($\equiv \mathbf{N}^*_{hkl}$) is perpendicular to the plane $(h\ k\ l)$, and is therefore perpendicular to all rows $[u\ v\ w]$ ($\equiv \mathbf{r}_{uvw}$) contained in this plane. The scalar product $\mathbf{N}^*_{hkl} \cdot \mathbf{r}_{uvw}$ is therefore null and the indices of the zone axis row $[u\ v\ w]$ are related to the indices of the planes of the zone by the relation: $h.u + k.v + l.w = 0$.

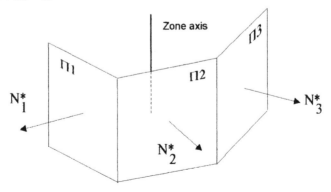

Figure 2.7

Consider two planes $(h_1\, k_1\, l_1)$ and $(h_2\, k_2\, l_2)$. Their zone axis is the row $[u\, v\, w]$ such that $h_1.u + k_1.v + l_1.w = 0$ and $h_2.u + k_2.v + l_2.w = 0$.

We deduce from this the relations:

$$u = k_1.l_2 - l_1.k_2$$
$$v = l_1.h_2 - h_1.l_2$$
$$w = h_1.k_2 - k_1.h_2$$

❏ DIRECT ROWS

Consider the direct row $[u\, v\, w]$ associated with the vector: $\mathbf{r} = u.\mathbf{a} + v.\mathbf{b} + w.\mathbf{c}$. Its modulus is the square root of the scalar product $\mathbf{r}.\mathbf{r}$; the reciprocal of this modulus is equal to the constant distance $\mathbf{D}^*_{uvw}$ between the planes $(u\, v\, w)^*$ of the reciprocal lattice to which the row $[u\, v\, w]$ is normal.

$$r = ||\mathbf{r}|| = \sqrt{(u\mathbf{a} + v\mathbf{b} + w\mathbf{c}).(u\mathbf{a} + v\mathbf{b} + w\mathbf{c})}$$

❏ RECIPROCAL ROWS

Consider the reciprocal row $[h\, k\, l]^*$ associated with the vector:

$$\mathbf{N}^*_{hkl} = h.\mathbf{A}^* + k.\mathbf{B}^* + l.\mathbf{C}^*$$

Its modulus is the square root of the scalar product $\mathbf{N}^*_{hkl} . \mathbf{N}^*_{hkl}$; the reciprocal of this modulus is equal to the constant distance d_{hkl} between the planes of the family $(h\, k\, l)$ of the direct lattice.

❏ ANGLES BETWEEN DIRECT ROWS

Two lattice planes intersect along a row. In a crystal the faces are parallel to the lattice planes and the edges are therefore parallel to the rows. The simplest method for determining the angle between two edges of a crystal is to determine the indices of the rows parallel to the edges in question and then calculate the angle between these edges using the scalar product.

The angle θ between the rows $[u\, v\, w]$ and $[u'\, v'\, w']$ is such that:

$$\cos\theta = \frac{(u\mathbf{a} + v\mathbf{b} + w\mathbf{c}).(u'\mathbf{a} + v'\mathbf{b} + w'\mathbf{c})}{||n_{uvw}||.||n_{n'v'w'}||}$$

❒ ANGLES BETWEEN RECIPROCAL ROWS

Since the reciprocal row $[h\ k\ l]^*$ is orthogonal to the family of lattice planes $(h\ k\ l)$, the angle between two reciprocal rows is the supplementary angle to the dihedral angle between the corresponding planes.

❒ TORSION ANGLE

In descriptions of molecules, the torsion angle is often invoked: in a chain of atoms A, B, C, D, the torsion angle is the dihedral angle between the planes ABC and BCD.

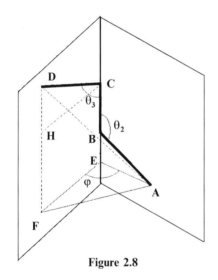

Figure 2.8

We can determine the torsion angle as the angle between the normals to the planes ABC and BCD. These normals are obtained from the vector products $\mathbf{AB} \wedge \mathbf{BC}$ and $\mathbf{CD} \wedge \mathbf{BC}$. We can also use the metric relation in triangle AEF:

$$\cos\varphi = (AE^2 + EF^2 - AF^2)/(2.AE.AF)$$

We also have:

$$AE = l_{12}\sin\theta_2$$
$$EF = l_{34}\sin\theta_3$$
$$AF^2 = AD^2 = DF^2 = l_{14}^2 - DF^2$$
$$DF = EB + BC + DH$$

$$\cos\varphi = \frac{l_{12}^2 + l_{23}^2 + l_{34}^2 - l_{14}^2 - 2l_{12}l_{23}\cos\theta_2 - 2l_{23}l_{34}\cos\theta_3 + 2l_{12}l_{34}\cos\theta_2\cos\theta_3}{2l_{12}l_{34}\sin\theta_2\sin\theta_3}$$

6   THE INTERNATIONAL FRAME

Calculations in the Bravais lattice can often be rather tricky in systems other than tri-orthogonal, and certain calculations are carried out in a direct tri-orthonormal frame $\mathbf{i}$, $\mathbf{j}$, $\mathbf{k}$ called 'the international frame', defined by:

$$\mathbf{i} = \frac{\mathbf{a}}{a}; \mathbf{j} = \frac{\mathbf{a} \wedge \mathbf{C^*}}{a.C^*.\sin\{\mathbf{a},\mathbf{C^*}\}}; \mathbf{k} = \frac{\mathbf{C^*}}{C^*}$$

6.1 Reciprocal Vector in the International Frame

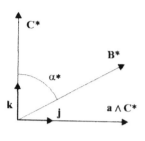

Figure 2.9

Figure 2.9 shows the projection of the international frame on the plane **j**, **k**. Consider the reciprocal row:

$$\mathbf{N}_{hkl}^* = h.\mathbf{A^*} = k.\mathbf{B^*} + l.\mathbf{C^*}$$

The components x, y and z of $\mathbf{N}_{hkl}^*$ in the international frame are such that:

$$x.\mathbf{i} + y.\mathbf{j} + z.\mathbf{k} = h.\mathbf{A^*} + k.\mathbf{B^*} + l.\mathbf{C^*} \qquad (11)$$

Scalar multiplication of (11) by the unit vector **i** gives:

$$x = h.\mathbf{A^*}.\mathbf{i} = h.\mathbf{A^*}.i.\cos\{\mathbf{A^*},\mathbf{a}\} = h.A^*.\cos\{\mathbf{A^*},\mathbf{a}\}$$

$$\cos\{\mathbf{A^*}, \mathbf{a}\} = \frac{\mathbf{A^*}.\mathbf{a}}{||\mathbf{A^*}||.||\mathbf{a}||} = \frac{1}{||\mathbf{A^*}||.||\mathbf{a}||} = \frac{V}{a.b.c.\sin\alpha}$$

$$x = h.A^* \sin\beta^*.\sin\gamma$$

Similarly, y and z are calculated by scalar multiplication of relation (11) by j followed by $\mathbf{C^*}$. We finally obtain:

$$x = h.A^*.\sin\beta.\sin\gamma$$
$$y = -h.A^*.\sin\beta^*.\cos\gamma + k.B^*.\sin\alpha^*$$
$$z = h.A^*.\cos\beta^* + k.B^*.\cos\alpha^* + l.C^*$$

Using the relations between the direct and reciprocal lattices, we can also write:

$$x = h/a$$
$$y = -h/a.\tan \gamma + k/b.\sin \gamma$$
$$z = h.A^*.\cos \beta^* + k.B^*.\cos \alpha^* + l.C^*$$

6.2 Direct Rows in the International Frame

In the direct lattice, consider the row $\mathbf{OD} = u.\mathbf{a} + v.\mathbf{b} + w.\mathbf{c}$

The coordinates of D in the international frame can be calculated by analogy with the previous calculation:

$$x = u.a + v.b.\cos \gamma + w.c.\cos \beta$$
$$y = v.b.\sin \gamma - w.c.\sin \beta.\cos \alpha^*$$
$$z = w.c.\sin \beta.\sin \alpha^*$$

APPLICATION: Calculating the volume of the unit-cell. In the international frame, the components of the base vectors $\mathbf{a}$, $\mathbf{b}$, $\mathbf{c}$ are:

$a, 0, 0$; $b.\cos \gamma, b.\sin \gamma, 0$; $c.\cos \beta, -c.\sin\beta.\cos \alpha^*, c.\sin \beta.\sin \alpha^*$;

Calculation of the scalar product $(\mathbf{a}, \mathbf{b}, \mathbf{c})$ gives:

$$V = a.b.c.\sin \alpha^*.\sin \beta.\sin \gamma$$

EXERCISE: Write the relations of paragraphs 6.1 and 6.2 in matrix form and check that the second matrix is the reciprocal of the transpose of the first.

7 FRACTIONAL COORDINATES

To locate the position of a point P in a cell, we often use fractional coordinates. If the absolute oblique coordinates of a point P in the frame characterised by the base vectors $\mathbf{a}$, $\mathbf{b}$, $\mathbf{c}$ are $x.a$, $y.b$ and $z.c$, then the triplet (x, y, z) is known as the fractional coordinates of P.

It is always possible, by whole lattice translations, to bring the point P to coincide with an identical point inside the original unit-cell. We therefore adopt the following convention for fractional coordinates:

$$0 \leqslant x \leqslant 1, \; 0 \leqslant y \leqslant 1, \; 0 \leqslant z \leqslant 1$$

Summary of notation used

a, **OA**	Vectors in the direct lattice (bold characters)
$\mathbf{C}^*$, $\mathbf{N}^*_{hkl}$	Vectors in the reciprocal lattice (bold and *)
$[u\,v\,w]$	Row in the direct lattice
$(h\,k\,l)$	Plane in the direct lattice
$[h\,k\,l]^*$	Row in the reciprocal lattice
$(u\,v\,w)^*$	Plane in the reciprocal lattice
$\langle h\,k\,l \rangle$	Family of direct rows
$\{h\,k\,l\}$	Family of equivalent planes (form)

Chapter 3

The Stereographic Projection

1 THE STEREOGRAPHIC TRANSFORMATION OF A POINT

DEFINITION: Consider a sphere with centre O, radius R, a diameter NS, a point P on the sphere and the intersection p of SP with the equatorial plane normal to NS. The point p is called the stereographic transformation of point P, and vice versa.

PROPERTIES OF THE TRANSFORMATION:

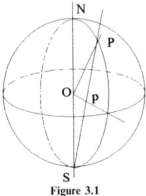

Figure 3.1

♦ It is a positive inversion of the centre S whose power is $\overline{SP}.\overline{Sp} = 2R^2$ denoted:

$$\Im(S, 2R^2).$$

It transforms the sphere into an equatorial plane which is the plane of the projection.

♦ Any circle drawn on the sphere is transformed into a circle (or a straight line) on the equatorial plane.

♦ Angles are conserved during the transformation.

2 THE POLE OF A FACE

We assume the crystal to be at the centre O of the sphere. From this point we take normals to the faces; the points P_i at the intersections of the normals with the sphere are called the **poles** of the faces.

The inversion $\Im(S, 2R^2)$ when applied to the poles P_i gives the points p_i, which are the stereographic transforms of these poles. These points are outside the equatorial circle when the poles are in the hemisphere containing S (southern hemisphere), and are inside when the poles are in the northern hemisphere. The following convention is used so that all the transforms are inside the equatorial circle.

CONVENTION: For poles in the northern hemisphere, we use S as the centre of inversion, and for poles in the southern hemisphere we use N. In order to be able to distinguish the two types of poles easily, we denote those in the northern hemisphere by a **cross** and those in the southern hemisphere by a **circle**.

3 THE STEREOGRAPHIC PROJECTION OF A POLE

The normal to a face has a direction defined by two angles $\{COA\} = \varphi$ (azimuth) and $\{NOP\} = \rho$ (inclination).

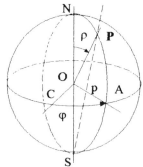
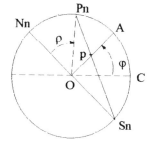

Figure 3.2

On actual crystals, interfacial angles are determined by optical measurements with a two-circle goniometer. One commercial model works as follows:

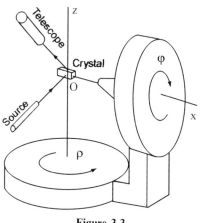

Figure 3.3

The crystal is glued onto a goniometric head mounted on a drum with horizontal axis Ox, graduated to display φ. This drum can turn about a vertical axis Oz. The angle of rotation ρ is measured on a second graduated drum. The viewing system comprises a light source and telescope with optical axes symmetrical about a horizontal plane containing the axis Oz. The light source filament forms an image at infinity which, after being reflected from the face of the crystal, can be observed in the telescope; the corresponding values of φ and ρ are then read off.

CONSTRUCTION OF A POLE: To obtain the transform p (figure 3.2 right), we draw the equatorial circle, then, from the azimuth origin OC, we construct the point A on the circle such that the angle {COA} is equal to φ. We then perform a rotation about OA. After this rotation, S becomes S_n, N becomes N_n and P becomes P_n. The point P_n is such that the angle {N_nOP_n} is equal to ρ.

The point p, which is the intersection of OA and S_nP_n, is the required *stereographic transform*.

4   THE WULFF NET

4.1   Description

In practice, the construction described above can be avoided by using the 'Wulff net'. This net is the stereographic projection of a network of parallels and meridians drawn on the projection sphere viewed along the equator. The resulting network is usually graduated in steps of 2°, forming great circles and small circles orthogonal to the former (figure 3.4). A Wulff net is shown on page 403.

The small circles EFG (figures 3.4 and 3.5) are projections of parallels drawn on the sphere (they are intersections with the sphere of cones of axis CD).

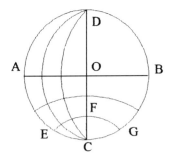

Figure 3.4

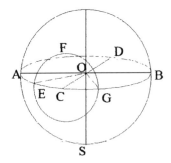

Figure 3.5

The great circles are projections of meridians drawn on the sphere; these are great circles on the diameter CD of the projection sphere (figures 3.4 and 3.6).

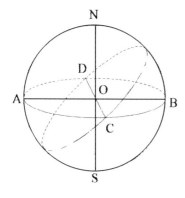

Figure 3.6

SPECIAL CASE: When studying cubic crystals, we have to draw the projections of planes (diagonal mirror planes) with normals characterised by the angles:

$$\rho = \pi/4; \ \varphi = 0, \ \pi/2, \ 3\pi/2.$$

The stereographic projection of the plane defined by $\rho = \pi/4$ and $\varphi = 0$ is the great circle of centre A with radius $R\sqrt{2}$ (see figures 3.6 and 3.7 and paragraph 9).

4.2 Constructing a Stereogram

PRELIMINARY NOTE: Only the angular graduations on the axes AB and CD of the net (figure 4.4) can be used for the constructions.

The stereogram (figure 3.7) is drawn on transparent paper which can be rotated over a Wulff net.

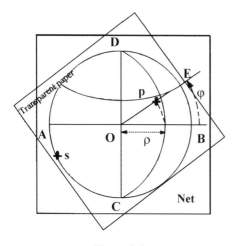

Figure 3.7

We begin by plotting the projection of the axis AB (the origin of the azimuths) on the transparent paper:

A pole of angle $\varphi = 0$ lies on AB at a point p, situated on the great circle of inclination ρ.

If $\rho = 0$ p is at O,

If $\rho = \pi/2$ p is at B.

A pole of angle φ lies on OE and on the great circle perpendicular to OE making an angle $(\pi/2 - \rho)$ with the plane of projection.

The point can be found by bringing the line AB *of the net* to coincide with the line OE *of the tracing* by rotating the latter.

☞ *This rotation must be carried out so as to use an axis of the net (here, AB) for which the angular graduation is correct.*

For any angle ρ, the pole lies at the intersection of OE and the great circle of inclination ρ. If $\rho = 0$ the pole is at O (for any angle φ); if $\rho = \pi/2$ the pole is at E. The poles of inclination $\pi/2$ (such as the pole s in figure 3.7) have their projections located on the circle and are represented with a cross.

4.3   Using the Wulff Net

In practice the Wulff net enables a number of measurements and constructions to be carried out quite simply.

◻   THE ANGLE BETWEEN TWO POLES

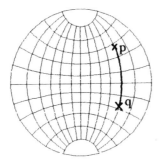

Figure 3.8

The angle between the normals to two faces of a crystal is equal to the angle between the poles p and q of the faces.

By rotating over the net the transparent paper on which the stereogram is drawn, we find the great circle passing through the two poles in question. The angle between the two poles is read off directly on this great circle (figure 3.8).

◻   THE POLE OF A ZONE

We seek the pole (figure 3.9) corresponding to a zone axis. The latter is defined by the great circle (zone circle) which passes through the poles of the planes in the zone. By definition, the zone axis is normal to the plane of the zone ($\delta\rho = \pi/2$).

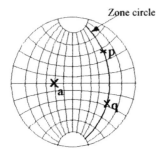

Figure 3.9

We seek the great circle passing through the poles in question (p and q in figure 3.9). This great circle is the *zone circle*. On the axis normal to this zone circle, we move round 90° to obtain the pole *a* which is the zone axis under consideration.

SPECIAL CASE: The centre O of the projection is the zone axis formed by the faces for which the angle ρ is $\pi/2$.

❐ ANGLE BETWEEN TWO ZONE CIRCLES

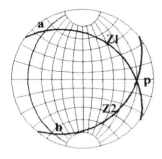

Figure 3.10

We seek the great circle having the point p as pole (or zone axis), p being the intersection of the two circles of the zones Z1 and Z2 under consideration.

The arc ab which is intercepted on this great circle by these two zone circles gives the required angle (figure 3.10).

❐ ROTATION OF φ ABOUT AN AXIS IN THE PLANE OF PROJECTION

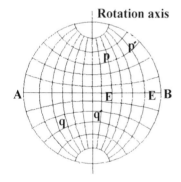

Figure 3.11

The net is rotated to bring the axis in question onto the diameter normal to the small circles. For each rotated pole, we look for the small circle on which it is constrained to move, and we then move through an angle φ on this circle.

To determine φ, we use the intersections E and E' of the great circles, orthogonal to the small circles, with the axis AB (figure 3.11)

$$p \Rightarrow p' \qquad q \Rightarrow q'$$

5   BASIC SPHERICAL TRIGONOMETRY

While spherical trigonometry is not indispensable to geometrical crystallography, it does sometimes simplify calculations.

The most useful relations are given below.

Consider the spherical triangle ABC on the surface of a sphere of unit radius with centre O.

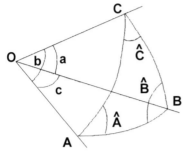

Figure 3.12

The sides of the spherical triangle (see figure 3.12) are the arcs BC, CA and AC with lengths a, b, c respectively.

The angles $\hat{A}$, $\hat{B}$ and $\hat{C}$ of the spherical triangle are respectively equal to the dihedral angles {BAO, CAO}, {ABO, CBO} and {ACO, BCO}.

Consider the point A1 on the great circle AC such that **OA1.OC** $= 0$, and B1 the point such that **OB1.OC** $= 0$.

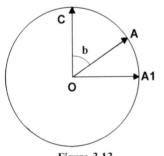

Figure 3.13

We can write:

OA $= \cos b.$ **OC** $+ \sin b.$ **OA1**

OB $= \cos a.$ **OC** $+ \sin a.$ **OB1**

The dihedral angle {OAC, OBC} equal to $\hat{C}$ is also equal to the angle {**OA1, OB1**}

Since $\cos c = $ **OA. OB**, we have:

$$\cos c = \cos a.\cos b + \sin a.\sin b.\cos \hat{C} \qquad (1)$$

Circular permutation gives:

$$\cos a = \cos b.\cos c + \sin b.\sin c.\cos \hat{A} \qquad (2)$$

$$\cos b = \cos c.\cos a + \sin c.\sin a.\cos \hat{B} \qquad (3)$$

The reciprocal relations are of the form:

$$\cos \hat{A} = -\cos \hat{B}.\cos \hat{C} + \sin \hat{B}.\sin \hat{C}.\cos a \qquad (4)$$

Finally from (1) and (3) we have:

$$\sin^2 a(\sin^2 c.\cos^2 \hat{B} - \sin^2 b.\cos^2 \hat{C}) = \cos^2 b - \cos^2 c + \cos^2 a.(\cos^2 c - \cos^2 b)$$

$$\sin^2 a(\sin^2 c.\cos^2 \hat{B} - \sin^2 b.\cos^2 \hat{C}) = (\cos^2 b - \cos^2 c).(1 - \cos^2 a)$$

$$\sin^2 c.\cos^2 \hat{B} + 1 - \sin^2 c = 1 - \sin^2 b + \sin^2 b\cos^2 \hat{C}$$

from which we deduce:

$$\frac{\sin c}{\sin \hat{C}} = \frac{\sin b}{\sin \hat{B}}$$

which can be generalised as:

$$\boxed{\frac{\sin a}{\sin \hat{A}} = \frac{\sin b}{\sin \hat{B}} = \frac{\sin c}{\sin \hat{C}}} \tag{5}$$

6 CRYSTAL MEASUREMENTS BY GONIOMETER

6.1 Principle of the Method

We wish to measure the unit-cell angles α, β, γ and the length ratios of the axes, and to index the faces of the crystal.

To do this, we use a two-circle goniometer to measure the angles of azimuth and inclination for all the faces of the crystal. To make the calculations simpler, we adjust the goniometer head on which the crystal is mounted so that the latter has a symmetry axis coinciding with the origin of the inclination axis of the goniometer. We then draw the corresponding stereogram; the chosen symmetry axis will be at the centre of the diagram. Although this stereogram does not enable very precise calculation of the angles, it does give very useful approximate values.

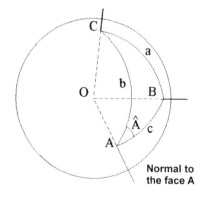

Figure 3.14

On the diagram (figure 3.15) we choose three faces, denoted *arbitrarily* (001), (010) and (100), and a fourth, so-called 'parametric face'. We thus obtain the spherical triangle ABC (figure 3.14).

Since the faces are identified by their normals OA, OB and OC, the lengths of the sides a, b and c of the spherical triangle correspond to the angles between the faces.

The angles $\hat{A}$, $\hat{B}$, $\hat{C}$ between the sides of the spherical triangle are the supplementary angles to those between the edges of the faces.

This is because the zone circles AC, AB, BC are planes which contain the normals to the faces, and the edges between the faces are therefore normal to the zone planes.

6.2   Determination of α, β and γ and the Length Ratios

In general we take as the parametric face one which cuts the three initial faces, indexed (111). The stereogram then has the appearance shown in figure 3.15. The notation is clear: for example, (110) is the face at the intersection of the zones (100)–(010) and (001)–(111).

In what follows, the axes are denoted Ox, Oy and Oz, the parametric face is (111) and the lengths of the sides intersected on the axes by the parametric face are a, b, c.

❏   UNIT-CELL ANGLES

The angles between the sides of a spherical triangle are supplementary to those between the zone edges.

Since γ is equal to the angle $\{Ox, Oy\}$, we can also write that γ is the supplementary angle to that between the zones (001)–(010) and (001)–(100). Similarly:

α is supplementary to the angle between the zones (001)–(100) and (010)–(100).
β is supplementary to the angle between the zones (100)–(010) and (010)–(001).

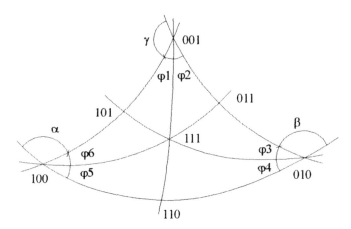

Figure 3.15

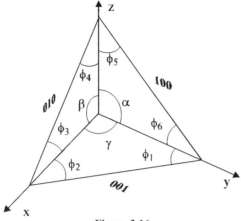

Figure 3.16

◻ BASE VECTOR RATIOS

We can immediately determine the ratios of the base vectors from the elements of the spherical triangle (see figures 3.15 and 3.16).

$$\frac{a}{b} = \frac{\sin \varphi_1}{\sin \varphi_2}, \; \frac{c}{b} = \frac{\sin \varphi_6}{\sin \varphi_5}, \; \frac{c}{a} = \frac{\sin \varphi_3}{\sin \varphi_4}$$

6.3 Indexing the Faces

To index the faces, we choose three axes and a parametric face. Once these are chosen, we can index all the poles of the other faces. Suppose we wish to index the pole (hkl). We pass two zones through this pole: $(h_1k_1l_1)–(h_2k_2l_2)$ and $(h_3k_3l_3)–(h_4k_4l_4)$. The equation of a zone plane is of the form: $h.u + k.v + l.w = 0$, the row $[u \; v \; w]$ being the zone axis. We calculate the indices of the row which is the axis of the first zone:

$$u = k_1l_2 - l_1k_2$$
$$v = l_1h_2 - h_1l_2$$
$$w = h_1k_2 - k_1h_2$$

and from the equation $h.u + k.v + l.w = 0$, we deduce an initial relation between the indices h, k and l.

We repeat the operation with a second zone and deduce a second relation between the indices, and then, for one of the indices, we choose an arbitrary value (there are not always enough faces to be able to pass three zones through a pole) from which we deduce the remaining two.

This method enables all the poles in a stereogram to be indexed. The calculations can also be made by noting that if three planes are in a zone, the determinant Δ of their indices is zero.

$$\Delta = \begin{vmatrix} h_1 & k_1 & l_1 \\ h_2 & k_2 & l_2 \\ h_3 & k_3 & l_3 \end{vmatrix}$$

EXAMPLE: Consider a crystal having a threefold axis which has been placed at the centre of the diagram (figure 3.17). The poles of the faces (100), (010) and (001) are respectively A, B and C. The parametric face (111) has its pole at the origin of the diagram. The faces (011), (101) and (120) have already been identified and we require the indices of the face (hkl).

Two zones are seen to pass through this face: Z1, which also passes through the poles of the faces (101) and (011), and Z2, which passes through (001) and (120).

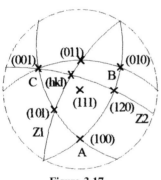

Figure 3.17

The axis of zone Z1 is thus the row $[\bar{1}\,\bar{1}\,1]$ and the indices h, k and l are such that:

$$-h - k + l = 0$$

Similarly, the axis of zone Z2 is the row $[\bar{2}\,\bar{1}\,0]$, hence

$$-2h + k = 0$$

Putting $h = 1$, we obtain the required indices:

$$(hkl) = (123)$$

NOTE: The choice of reference faces and of the parametric face is arbitrary. If this choice is to coincide with the simplest unit-cell of the crystal, the symmetries which appear on the stereogram must be used, remembering that low index faces belong simultaneously to several zones.

Geometric crystallography alone cannot provide a definitive answer to the problem of determining the unit-cell; only the ratios of the lengths of the sides are available by optical methods. The use of X-ray crystallography techniques is necessary to obtain absolute values of the parameters and to confirm the accuracy of choice of unit-cell axes.

7 EXAMPLE OF MEASUREMENTS

The interfacial angles of a crystal of gypsum ($CaSO_4.2H_2O$), measured with a two-circle goniometer, are as follows:

Table 3.1

Faces	d	p	q	h	e	m	n	s
φ	0°	90°	270°	180°	0°	34.58°	325.41°	145.41°
ρ	90°	90°	90°	90°	8.96°	90°	90°	90°

Faces	t	f	g	i	j
φ	214.58	331.61	28.39	151.61	208.39
ρ	90°	41°	41°	139°	139°

7.1 Plotting the Stereographic Projection

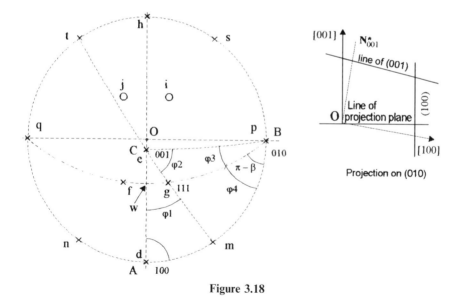

Figure 3.18

7.2 Analysis of this Stereographic Projection

SYMMETRY ELEMENTS: The plane containing the faces d, e and h is a plane of symmetry bringing into relation m and n, q and p, s and t, f and g and so on. The direction OB corresponds to a twofold axis relating m and s, g and i, n and t.

O is a centre of symmetry: the crystal belongs to the class 2/m (monoclinic).

The counterpart of face e cannot be measured as it is the face by which the crystal is attached.

We therefore choose d = (100), p = (010), e = (001), g = (111).

INDEXING THE FACES: Face m belongs to the zone containing e and g; from this we deduce that for the face m: $h = k$. m also belongs to the zone whose axis is [100], hence $l = 0$. m is a (110) face. Using symmetry we can index all the other faces: s = ($\bar{1}$10), i = ($\bar{1}$1$\bar{1}$), h = ($\bar{1}$00) and so on.

UNIT-CELL PARAMETERS: In the spherical triangle ABC the angle $\hat{B}$, equal to $\pi - \beta$, is $\pi/2 - \rho_e$; hence $\beta = \pi/2 + \rho_e = 98°58'$; $\alpha = \gamma = \pi/2$.

The angle φ_1 is equal to ρ_m which is 34°35'. $\varphi_2 = \pi/2 - \varphi_1 = 55°25'$.

$$a/b = \sin \varphi_1 / \sin \varphi_2 = 0.6893$$

Using the Wulff net, we find that the circle of the zone (010)–(111) corresponds to an inclination close to 37°30'. Hence $\varphi_3 \approx 28°30'$; $\varphi_4 \approx 52°30'$ and $c/a° \approx 0.60$. A rigorous calculation is more complex. The following method can be used:

Consider a hypothetical face w (010) and construct WDG, a spherical triangle formed from the poles of w, of g and by [001].

The angle $\hat{W}$ is equal to $\pi/2$.

$$\cos \hat{W} = -\cos \hat{G} \cdot \cos \hat{D} +$$
$$\sin \hat{G} \cdot \sin \hat{D} \cdot \cos \hat{g} = 0$$

Hence: $\cot \hat{G} = \tan \hat{D} \cdot \cos \hat{g}$

$$\hat{D} = \varphi_g \ \hat{g} = \rho_g \Rightarrow \hat{G} = 67.8°$$
$$\frac{\sin \hat{w}}{\sin \hat{G}} = \frac{\sin \hat{g}}{\sin \hat{W}} = \sin \hat{g}$$
$$\hat{w} = 37°24'$$

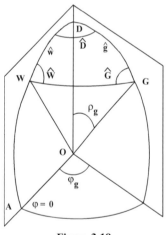

Figure 3.19

The exact value of φ_3 is therefore 28°26'.

8 STEREOGRAPHIC PROJECTIONS OF CUBIC CRYSTALS

Cubic crystals contain oblique symmetry elements, and when stereographic projections are constructed and interpreted, certain features become evident. Take for example a crystal containing the forms {100} (cube), {111} (octahedron) and {110} (rhombic dodecahedron). Figure 3.20 shows how the projection of face (011) is constructed, together with plane D which is both the plane of the zone axis [011] and an oblique plane of symmetry.

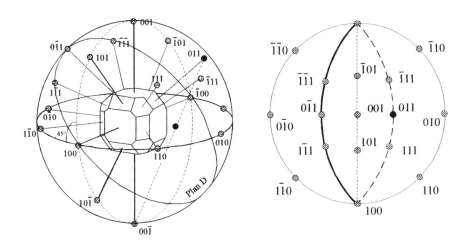

Figure 3.20. Stereographic projection of the poles of the northern hemisphere and plane D (full line: northern hemisphere; dashed line: southern hemisphere).

The following three projections may be used for all cubic crystals. In this system, the position of the poles is independent of the unit-cell parameter, and it is therefore possible to construct the stereographic projections *a priori*.

In figure 3.21, a fourfold axis has been positioned normal to the plane of projection (standard projection). For the sake of clarity, only selected poles of the northern hemisphere are shown. The reader could, as an exercise, finish the projection and calculate the interfacial angles φ and ρ.

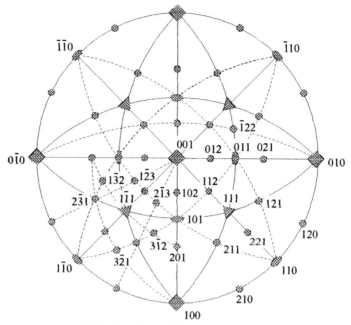

Figure 3.21. Standard cubic stereographic projection

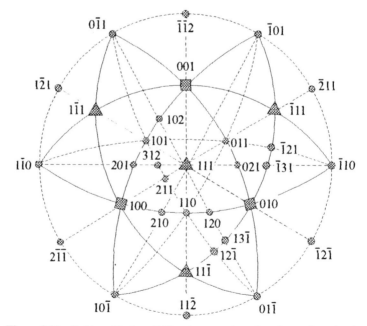

Figure 3.22. Cubic with threefold axis normal to the plane of projection

In figure 22, the projection has been chosen to show the threefold axis. We shall see later that for trigonal crystals, the general arrangement of the poles is identical, but that their positions then become dependent on the angle α of the unit-cell.

The last projection (figure 3.23) is less often used; it corresponds to a crystal with a twofold axis normal to the plane of the projection. Using the properties of the lattice we can show that the poles of the faces (001), (111) and (110) are contained in the plane of the projection.

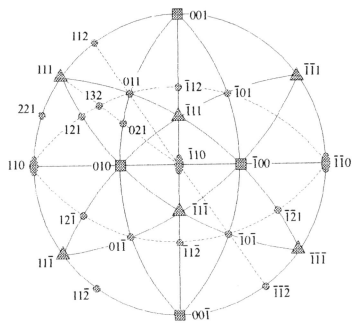

Figure 3.23. Twofold cubic axis normal to the plane of the projection

☐ CHARACTERISTIC ANGLES

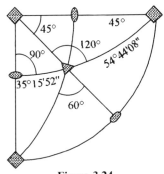

Figure 3.24

Figure 3.24 shows the various angles between the symmetry axes of the cubic system.

These angles can easily be calculated from the dot product of rows parallel to the axes.

Thus the angle between the threefold axes [1 1 1] and [$\bar{1}$ $\bar{1}$ 1] (rows of modulus $a\sqrt{3}$) is:

$$\theta = \text{arc } \cos\frac{(\mathbf{a} + \mathbf{b} + \mathbf{c})(-\mathbf{a} - \mathbf{b} + \mathbf{c})}{a\sqrt{3}.a\sqrt{3}}$$

$$= 109°28'16''.$$

Chapter 4
Symmetry Operations in Crystal Lattices

1 DEFINITION OF A SYMMETRY OPERATION

The fundamental postulate of geometric crystallography is that the crystal lattice remains unchanged during certain 'displacements' in space, i.e. it is transformed into itself with no deformation. These displacements are called tiling operations or symmetry operations. The displacements which have the effect of bringing the lattice back to coincide with itself are, if we limit ourselves to symmetry of orientation:

—*translation*
—*rotation*
—*inversion*
—*improper rotation (rotation combined with inversion)*.

If we include symmetry operations of position, we must add:

—*rotation combined with translation.*

1.1 Translation

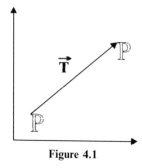

Figure 4.1

In this symmetry operation, there is no fixed point (except for the null translation). Thus in a crystal lattice, translations only constitute symmetry operations if the lattice is infinite.

The vector **T** of the translation must be a vector equivalent to a linear combination of the base vectors of the lattice so as to leave the latter unchanged after the operation. In this symmetry operation, the initial and final objects are strictly superimposable.

An operation which leaves the initial object unchanged will be denoted 'E'; this is the identity operation.

1.2 Rotation

Rotational symmetry operations leave a set of points (the rotation axis) unchanged. Figure 4.2 left shows a rotation in two-dimensional space (the plane of the rotation). In this case there is only one point unchanged: the centre of rotation.

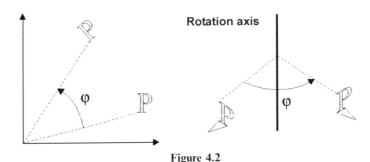

Figure 4.2

Rotations are defined by the axis of rotation **u** and the angle of rotation φ. They are usually denoted: $R(\mathbf{u}, \varphi)$.

If $\varphi = 2\pi/n$ (n being an integer), we refer to an n-fold rotation axis (or an axis of order n) and we denote this C_n. After n operations we have the initial situation again: $(C_n)^n = C_n^n = E$

For $n = 2$ we have a twofold axis (denoted C_2), threefold for $n = 3$ and so on.

In a rotation (figure 4.2 right) the initial and final objects are strictly superimposable after a series of infinitesimal rotations.

1.3 Inversion

Inversion[1] I is a symmetry operation which transforms a vector into its inverse, leaving only one point in space unchanged (this point is the centre of symmetry).

$$\mathbf{I}(\mathbf{u}) = \mathbf{I}.\mathbf{u} = -\mathbf{u}$$

---

[1]Do not confuse the inversion referred to here with the geometrical transformation of the same name used in stereographic projections.

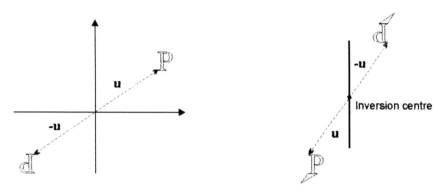

Figure 4.3

Note that after an inversion, it is not possible by a continuous transformation of space (and hence with no change of orientation of that space) to have the initial and final objects perfectly superimposed. (See figure 4.3 left where the arrow on the initial object is pointing forward while that of the final object is pointing the other way).

The final object is the mirror image of the initial object (like right and left hands). Such objects are known as 'enantiomorphs'.

1.4   Products of Symmetry Operations

In the study of crystallography, combinations of symmetry operations are required, and a symmetry product is the symmetry operation resulting from the successive application of two elementary symmetry operations. In general, the final result depends on the order in which the two operations are carried out; the product is then non-commutative.

1.5   Review of Some Products

❐   ROTATION-INVERSION PRODUCT

An inversion I is followed by rotation through an angle φ about a rotation axis **u** which contains the centre of inversion.

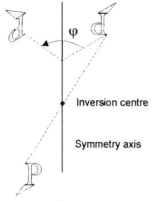

Figure 4.4

This product of rotation and inversion is denoted $\overline{R}$

$$I.R(\mathbf{u},\varphi) = R(\mathbf{u},\varphi).I = \overline{R}(\mathbf{u},\varphi)$$

In this particular case, performing the operations in the reverse order gives the same final result.

Inversion is in fact commutative with all rotations. Again, the initial and final objects are *enantiomorphs*.

If $\varphi = 2.\pi/n$ (where n is an integer), the product operation is denoted $\overline{C}_n$ or I_n.

After performing the operation n times, we obtain the initial element $(\overline{C}_n^{\,n} = I_n^n = E)$.

☐   THE MIRROR PLANE: THE PRODUCT OF INVERSION AND A TWOFOLD AXIS

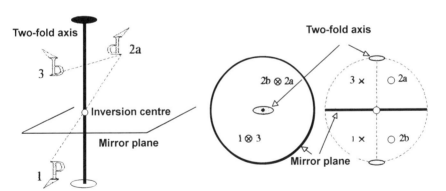

Figure 4.5

The product of a twofold symmetry axis (C_2) and an inversion whose centre is located on the axis, denoted $\overline{C}_2 = I.R(\mathbf{u}, \pi)$, is a plane of symmetry or **mirror** plane (see figure 4.5 left), also denoted σ^2 where:

$$\sigma = I.C_2 = I.R(\mathbf{u}, \pi)$$

---

[2]Horizontal mirror planes are denoted σh and vertical mirror planes σv.

This mirror plane is perpendicular to the axis and contains the centre of inversion. Figures 4.5 right are stereographic representations (with the twofold axis normal to the plane of the figure or in the plane) of this product, which is commutative;

$$\text{The result of the sequence:} \quad 1 \xrightarrow{\text{I}} 2a \xrightarrow{R(\mathbf{u},\, \pi)} 3$$

$$\text{is identical to that of:} \quad 1 \xrightarrow{R(\mathbf{u},\, \pi)} 2b \xrightarrow{\text{I}} 3$$

The relation $\sigma = \text{I.C}_2$ shows that the presence of the third symmetry operation is implied by the presence of the other two.

❐ PRODUCT OF C_n AND A MIRROR PLANE PERPENDICULAR TO THE AXIS

This product is denoted $S(\mathbf{u}, \varphi)$, $\mathbf{u}$ being the vector of the C_n axis and φ the angle of rotation. This operation is sometimes called 'roto-reflection', while that of rotation and inversion is called 'roto-inversion'.

$$S(\mathbf{u}, \varphi) = \sigma . R(\mathbf{u}, \varphi) = R(\mathbf{u}, \varphi). \sigma$$

$$S_n = \sigma . C_n = C_n . \sigma \text{ and } \sigma = \text{I.R}(\mathbf{u}, \pi)$$

$$S(\mathbf{u}, \varphi) = \text{I.R}(\mathbf{u}, \pi). R(\mathbf{u}, \varphi) = \overline{R}((\mathbf{u}, \varphi + \pi)$$

Thus a 'roto-reflection' is a 'roto-inversion' through an angle $\sigma + \pi$.

In descriptions of symmetry properties, one or other of the two systems may be used. Generally, physicists use the 'roto-reflections' of the Schönflies system whereas crystallographers tend to use the 'roto-inversions' of the Hermann–Mauguin system.

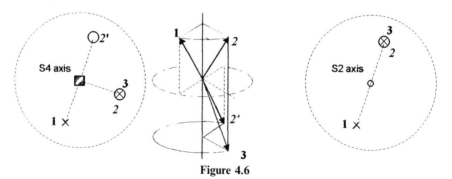

Figure 4.6

As an example, figure 4.6 shows the stereographic projections of the S_4 and S_2 axes.

For the S_4 axis, a rotation of $\pi/2$ ($1 \Rightarrow 2$) followed by the operation of a mirror plane normal to the axis ($2 \Rightarrow 3$) is seen to be equivalent to inversion ($1 \Rightarrow 2'$) followed by rotation through $3\pi/2$ ($2' \Rightarrow 3$). We thus have $S_4^1 = I_4^3$ and by permutation of the values of the rotation angles we can show that $S_4^3 = S_4^{-1} = I_4^1$. Since S_4^2 and I_4^2 are equivalent to a twofold axis, for $n = 4$ roto-inversions correspond to roto-reflections.

— We can similarly show the correspondence between S_6 and I_3^2 and between S_3 and I_6^{-1}.
— An S_6 axis is equivalent to a C_3 axis normal to a mirror plane. (The S_{2n} axes where n is odd are equivalent to a C_n axis normal to the mirror plane).
— The S_2 axis is equivalent to a pure inversion I (figure 4.6 right).
— An S_1 axis is equivalent to a mirror plane.

❏ PRODUCT OF TWO INTERSECTING TWOFOLD AXES

Consider two twofold axes C_2 and C_2', intersecting at O and defining a plane Π. These two axes are at an angle φ to each other (figure 4.7).

Their product is a rotation through 2φ about an axis **u** normal at O to the plane Π. The direction of rotation is that in which we encounter the first axis (written to the right) in the product with the second (written to the left).

If the rotation angle φ is equal to π/n where n is an integer, the rotation axis is a C_n axis. We can then write: $C_n = C_2'.C_2$.

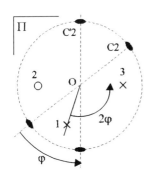

Figure 4.7

Furthermore, $C_2.C_2 = C_2^2 = E$, so on multiplying on the right by C_2 the two terms of the relation above, we have:

$$C_n.C_2 = C_2'.C_2.C_2 = C_2'.$$

Similarly: $C_2'.C_n = C_2$

The inverse element of the product is thus:

$$C_2.C_2' = C_2.C_n.C_2 = C_n^{-1}$$

$$\text{where: } C_n^{-1}.C_n^1 = C_n^1.C_n^{-1} = E$$

Note that (2) is below the plane of the figure.

Thus the product of two intersecting twofold axes at an angle of $\pi/4$ is a C_4 axis.

❑ PRODUCT OF TWO INTERSECTING MIRROR PLANES

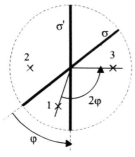

Consider two mirror planes σ and σ' whose planes intersect along a line u. These two planes are at a dihedral angle of φ. Their product is a rotation by angle 2φ about the axis $\mathbf{u}$. The direction of this rotation is that in which we encounter the first axis in the product with the second.

If $\varphi = \pi/n$, the axis is C_n: $\sigma'.\sigma = C_n$ and we also have: $C_n.\sigma = \sigma'$ and $\sigma'.C_n = \sigma$.

Note that (2) is above the plane of the figure.

Figure 4.8

❑ PRODUCT OF A C_2 AXIS AND A C_n AXIS PERPENDICULAR TO C_2

We assume that $\varphi = \pi/n$, n being an *integer*. The results above show that the product is a twofold axis perpendicular to the C_n axis at an angle of $\pm \varphi/2$ with the initial twofold axis, the sign depending on the order of the factors in the product. The same analysis can be performed for the product of a mirror plane by a C_n axis contained in the mirror plane. The product is a mirror plane, also containing the axis, and at a dihedral angle of $\pm\varphi/2$ to the initial mirror plane.

☞ | *If there exists* **one** C_2 *axis normal to a* C_1 *axis, then there exist* **n** *axes. Similarly, if there exists* **one mirror plane** *containing a* C_n *axis, then there exist* **n** *axes.*

❑ PRODUCT OF TWO ROTATIONS ABOUT INTERSECTING AXES

Consider the two rotations $R(\mathbf{OA}, 2\alpha)$ and $R(\mathbf{OB}, 2\beta)$ whose axes intersect at O. We let $\{AOB\} = \Psi$. Consider the sphere of centre O and A and B the traces of the rotation axes (rotation poles) on this sphere.

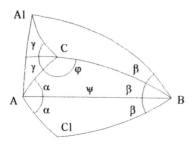

The product of the two rotations is a rotation about an axis OC through an angle 2γ. We draw on the sphere the great circles AC at an angle $+\alpha$ to AB and BC an angle $-\beta$ to AB. Similarly we draw the great circles AC_1 and BC_1 at angles $-\alpha$ and $+\beta$ to AB. The rotation $R(OA, 2\alpha)$ brings C to C_1 and the rotation $R(OB, 2\beta)$ then brings C_1 to C. C is unchanged in the operation and is therefore the rotation axis product.

Figure 4.9

On applying the product of rotations to the vector **OA** we obtain the vector **OA$_1$**, producing a rotation angle of 2γ. In the spherical triangle ABC, the angles φ and γ are supplementary. From trigonometric relations we have:

$$\cos \gamma = -\cos \varphi = \cos \alpha . \cos \beta - \sin \alpha . \sin \beta . \cos \Psi$$

This product is not commutative; to achieve the reverse operation, the trace of the rotation axis will be C$_1$ and the angle of rotation produced is again $+2\gamma$.

NOTE: If both axes are twofold ($2\alpha = 2\beta = \pi$), we find the rotation angle product is equal to 2Ψ.

1.6 Proper and Improper Rotations

A pure rotation can be replaced by a *continuous* transformation of space in which spatial orientation is unchanged. An object and its image after the operation are strictly superimposable. Such a rotation is called a **proper rotation**. On the other hand, operations such as inversion, which change the orientation of space, and for which object and image are not superimposable, are called **improper rotations**.

1.7 Product of a Rotation and a Translation

We denote by (R, T) the product of a rotation R(**u**, φ) and a translation of vector **T**. In this operation, the vector **X** is associated with its image **Y** such that:

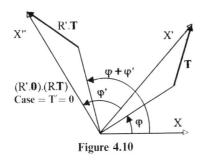

Figure 4.10

$$Y = (R, T).X$$

NOTE: If we apply in turn (R, T) then (R', T') to **X** we obtain:

$$X' = (R, T).X$$
$$X'' = (R', T').X' = (R', T').(R, T).X$$
$$(R', T').(R, T) = (R'.R, R'.T + T') \text{ where:}$$
$$R'.R = R(u, \varphi + \varphi')$$

❐ EQUIVALENT OPERATIONS

If we perform a translation of the origin of the initial frame, characterised by a vector **S**, this translation modifies the operator (R, T) and, in the new frame, we have: $Y' = (R', T').X'$.

The vector **S** is chosen such that (R', T') is equivalent to an operator $(R'', 0)$ containing no translation operation:

$$(R', T') = (R'', 0) = R(u', \varphi')$$

In the change of frame we have: $X = X' + S$.

If E is the identity rotation operation, this relation can be written in the form:

$$X = X' + S + (E, S).X' \Rightarrow X' = X - S = (E, -S).X$$
$$Y' = (E, -S).Y = (R', T').X'$$
$$Y' = (E - S).Y = (E, -S).(R, T).X = (E, -S).(R, T).(E, S).X'$$
$$(E, -S).(R, T).(E, S) = (R', T')$$

Taking the product of the three symmetry operations, we have:

$$\boxed{(R', T') = (R, R.S - S + T)}$$

We look for the vectors **S** which, if possible, enable elimination of the translation part $T' = R.S - S + T$ from the product operation (R', T'); *for if* **T'** *is null, the product of the rotation $R(u, \varphi)$ and the translation **T** is equal to a pure rotation $R(u', \varphi)$.*

We can decompose the vectors **S** and **T** into a component parallel to the rotation axis **u** and a perpendicular component:

$$S = S_\| + S_\perp \quad T = T_\| + T_\perp$$

This decomposition of the translation vectors gives rise to the following four possibilities for composing a rotation, which can be proper or improper, with parallel or perpendicular translation.

❐ THE PRODUCTS OF A PROPER ROTATION AND A TRANSLATION

◆ *S parallel to **u**.*

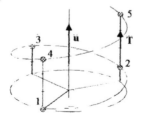

Figure 4.11

In this case, the effect produced by the rotation upon the translation vector is the invariant $R.S_{//} \equiv S_{//}$. It is then impossible to eliminate **T'** by changing the origin. Successive images lie on a helix of axis **u** and pitch **T**.

Figure 11 corresponds to an axis $R(u, \varphi = 2\pi/3)$,**T** (C_3 helical) applied four times in succession.

☞ | *The product of a rotation R(**u**, φ) and a translation **T** parallel to the rotation axis **u** is a 'screw axis' with axis **u** and an angle equal to φ.*

In a crystal, a screw axis can only be a symmetry element if the values of φ and **T** are compatible with the lattice tiling operations (In figure 4.11, if 1 is a node, 4 must *also* be a node).

◆ *S* perpendicular to *u*.

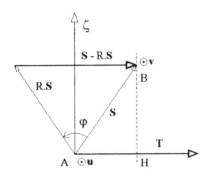

Figure 4.12

Figure 4.12 is drawn in the plane perpendicular to the symmetry axis **u** at A containing the vector **T**.

If **S** is such that $\mathbf{S}_\perp - R.\mathbf{S}_\perp = \mathbf{T}$, then:

$$R(\mathbf{u}, \varphi), \mathbf{T} = R(\mathbf{v}, \varphi), \mathbf{0}$$

By taking B, the trace of **v** on the plane of the figure, on the bisector of **T**, with the angle $(BA\zeta) = \varphi/2$, we transform the product $R(\mathbf{u}, \varphi), \mathbf{T}$ into a pure rotation through an angle φ, but with the vector **v**//**u** as axis.

In this product, B is an invariant point and $BH = \dfrac{\mathbf{T}}{2\tan(\varphi/2)}$.

☞ | *The product of a rotation R(**u**, φ) and a translation **T**, normal to the rotation axis **u**, is a rotation through an angle φ about an axis **v** lying on the bisector of the vector **T**.*

❐ PRODUCT OF AN IMPROPER ROTATION AND A TRANSLATION

◆ *S is parallel to **u**.*

$$\mathrm{I}.R(\mathbf{u}, \varphi) = \overline{R}(\mathbf{u}, \varphi). \quad \Rightarrow \quad \overline{R}.\mathbf{S}_\parallel = -\mathbf{S}_\parallel$$

$$\mathbf{T}' = R.\mathbf{S}_\parallel - \mathbf{S}_\parallel + \mathbf{T} = -2\mathbf{S}_\parallel + \mathbf{T}_\parallel$$

It is thus always possible to eliminate **T**' by changing the origin.

☞ | *There are no improper screw axes.*

◆ *S is perpendicular to **u**.*

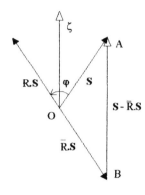

Figure 4.13

Figure 13 is drawn in the plane perpendicular to the symmetry axis containing the vector **T**. In the general case $\mathbf{S} - \overline{R}.\mathbf{S} \neq \mathbf{0}$ can be eliminated.

On the other hand, in the special case of a mirror plane [axis $\overline{C}_2 = \sigma = \overline{R}(\mathbf{u}, \pi)$], we have:

$$\mathbf{S} = \overline{R}.\mathbf{S}$$

It is therefore impossible, in the product of a translation and a mirror plane, to eliminate the translation by changing the origin.

☞ | *The product of a mirror plane and a translation is a **glide plane**.*

In a crystal lattice, only values compatible with the lattice tiling operations are allowed for the glide vector **T**.

2 THE MATRIX REPRESENTATION OF SYMMETRY OPERATIONS

2.1 Rotation Matrices

Consider a vector **OP** and its image after a rotation operation **OP′**. This rotation can be represented by a matrix which enables the coordinates of P′ to be calculated in terms of those of P. In a *cartesian coordinate frame*, rotation matrices[3] are particularly simple. If **OQ**, the projection of **OP** on xOy, is at an angle θ to Ox, its image **OQ′** after rotation through an angle φ about the axis Oz will be at an angle $\theta + \varphi$ to Ox. The coordinates of the point Q are $x = R\cos\theta$ and $y = R\sin\theta$, while those of Q′ are $x' = R\cos(\theta + \varphi)$ and $y' = R\sin(\theta + \varphi)$. In cartesian coordinates, the rotation matrix about the axis Oz is then written:

$$R(\varphi) = \begin{pmatrix} \cos\varphi & -\sin\varphi & 0 \\ \sin\varphi & \cos\varphi & 0 \\ 0 & 0 & 1 \end{pmatrix}$$

The determinant of this orthogonal matrix is equal to $+1$ and its **trace** (the sum of the terms on the main diagonal) is equal to: $\mathrm{Tr} = 1 + \cos 2\varphi$.

---

[3]Do not confuse rotation matrices (new coordinates of a point in the frame after rotation) with frame-change matrices related to a rotation of the axes (new axes which are functions of the old axes).

☞ | *In linear algebra, it can be shown that the determinant and the trace of a matrix are invariant during a change of coordinate frame.*

The determinant Δ of the matrix representing a proper rotation is always equal to $+1$.

2.2 Inversion Matrices

In a cartesian frame, the inversion matrix is written:

$$I = \begin{pmatrix} -1 & 0 & 0 \\ 0 & -1 & 0 \\ 0 & 0 & -1 \end{pmatrix}$$

Its determinant is equal to -1. Since an improper rotation is the product of a proper rotation ($\Delta = 1$) and an inversion ($\Delta = -1$), it can be represented by a matrix whose determinant is also equal to -1.

2.3 Affine Transformations

In general, a geometrical symmetry operation can be represented by an affine transformation of the type:

$$\begin{pmatrix} x_1' \\ x_2' \\ x_3' \end{pmatrix} = \begin{pmatrix} r_{11} & r_{12} & r_{13} \\ r_{21} & r_{22} & r_{23} \\ r_{31} & r_{32} & r_{33} \end{pmatrix} \cdot \begin{pmatrix} x_1 \\ x_2 \\ x_3 \end{pmatrix} + \begin{pmatrix} t_1 \\ t_2 \\ t_3 \end{pmatrix}$$

which can also be written:

$$\mathbf{x}' = \mathbf{R}.\mathbf{x} + \mathbf{t}$$

The elements r_{ij} of the matrix $\mathbf{R}$ represent a proper or an improper rotation and t_i a translation.

2.4 Homogeneous Matrices

To represent symmetry operations, homogeneous matrices can also be used; these are 4×4 matrices which enable the new coordinates to be calculated in terms of the old ones according to the relation:

$$(x'_1 \ x'_2 \ x'_3 \ 1) = (x_1 \ x_2 \ x_3 \ 1). \begin{pmatrix} r_{11} & r_{12} & r_{13} & 0 \\ r_{21} & r_{22} & r_{23} & 0 \\ r_{31} & r_{32} & r_{33} & 0 \\ t_1 & t_2 & t_3 & 1 \end{pmatrix}$$

The elements r_{ij} represent a proper or an improper rotation and t_i a translation.

3 THE SYMMETRY AXES POSSIBLE IN A CRYSTAL LATTICE

The postulate of the invariance of crystal lattices implies that during a rotation through an angle φ characterised by a matrix $(R\varphi)$, any lattice vector (which will have integer coordinates) will be transformed into another lattice vector which also has integer coordinates. Thus, all the elements of the matrix $(R\varphi)$, *expressed in the frame of the base vectors*, are integers and hence the trace of $(R\varphi)$ is also an integer. The trace of a rotation matrix, which is unchanged in any change of frame, is equal to:

$$\text{Tr}(R(\varphi)) = \pm(1 + 2.\cos\varphi)$$

The + sign corresponds to proper rotations and the − sign to improper rotations. The values of φ compatible with the nature of the crystal lattice must satisfy the relation:

$$1 + 2.\cos\varphi = m \text{ (integer)}$$

which possesses only five solutions of the form $\varphi = -2\pi/n$, where $n=1, 2, 3, 4$ and 6:

Table 4.1

$m = 3$	$\cos\varphi = +1$	$\varphi = 0, 2\pi\ldots$	Identity
$m = 2$	$\cos\varphi = +1/2$	$\varphi = \pm 2\pi/6$	C_6
$m = 1$	$\cos\varphi = 0$	$\varphi = \pm 2\pi/4$	C_4
$m = 0$	$\cos\varphi = -1/2$	$\varphi = \pm 2\pi/3$	C_3
$m = -1$	$\cos\varphi = -1$	$\varphi = 2\pi/2$	C_2

☞ | *The only possible symmetry axes in a crystal lattice are thus, apart from identity, 2, 3, 4 and 6-fold axes.*

The following equivalent proof can also be used:

Consider a lattice vector $\mathbf{T}$ normal to the axis of rotation of the angle φ. If O is a lattice node, the ends of the 4 vectors $\mathbf{T}$, $-\mathbf{T}$, $\mathbf{T}' = R(\mathbf{u}, \varphi).\mathbf{T}$ and $\mathbf{T}'' = R(\mathbf{u}, -\varphi). - \mathbf{T}$ are also nodes.

The vector $\mathbf{T}' - \mathbf{T}''$ is thus a lattice vector parallel to $\mathbf{T}$ such that:

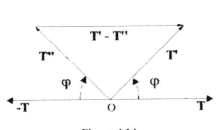

Figure 4.14

$$\mathbf{T}' - \mathbf{T}'' = m\mathbf{T} \ (m \text{ an integer}).$$

By projection onto an axis parallel to $\mathbf{T}$ we obtain: $2.\cos\varphi = m$.

It is possible to tile a plane with parallelograms, rectangles, squares and regular hexagons. On the other hand, tiling is impossible with regular pentagons, or regular polygons with more than six sides.

4 SYMMETRY OPERATIONS AND SYMMETRY ELEMENTS

Symmetry operations are transformations of space in which an object is transformed into a corresponding object strictly superimposable on the original or its mirror image. If we consider only orientation symmetries, the operations comprise pure rotations, reflections, inversion, roto-reflections and roto-inversions. In crystals, the possible values for the order of a rotation are 1, 2, 3, 4 and 6. In all of these symmetry operations, there exist fixed points (unchanged in the operation).

The set of fixed points (points, lines or planes) in a symmetry operation is called a symmetry element.

These symmetry elements are used in the graphic representation of the operation with which they are associated. The most effective way of representing symmetry operation in a crystal is to draw the stereographic projection of its symmetry elements; those involved in crystallography are:

—the *centre of symmetry* associated with an inversion;
—the *mirror plane* associated with reflections;
—the proper (C_n) or improper (S_n) *rotation axes*, which may be interpreted as the combination of several simpler symmetry elements (mirror plane, inversion, axis with a lower index).

Examples of symmetry elements

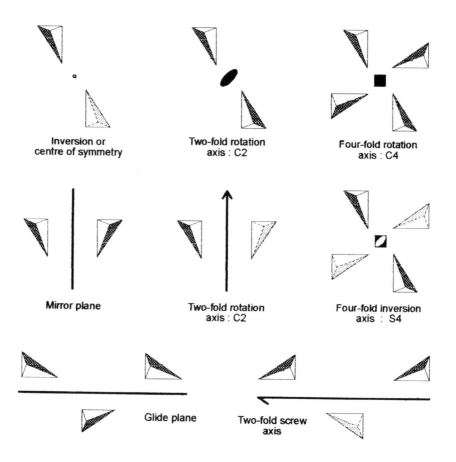

Inversion or
centre of symmetry

Two-fold rotation
axis : C2

Four-fold rotation
axis : C4

Mirror plane

Two-fold rotation
axis : C2

Four-fold inversion
axis : S4

Glide plane

Two-fold screw
axis

Chapter 5

Crystallographic Point Groups

Numerous classification schemes for the crystallographic point groups have been developed since the work of Bravais, and here we have chosen the method of Burckhardt[1]. This elegant method, which only uses elementary notions of group theory, is easily handled by non-specialists. The review below is for the benefit of readers unfamiliar with group theory; the examples given all relate to crystal symmetry operations.

1 GROUP STRUCTURE

1.1 Definitions

A set G of elements X, Y, Z . . . is a group if:

◆ it is subject to an *internal associative composition law* which relates to the ordered couple (X, Y) of elements in G, another element from G, called the product, and denoted X.Y (e.g.: *The product of two rotations*).
This product may be non-commutative (X.Y $\neq$ Y.X). If the product is commutative, the group is called an *abelian group*.

◆ G contains a *neutral* or *identity element* E such that:

$$\forall X \in G, \ E.X = X = X.E$$

◆ Any element G can be associated with another element in the set, called the *inverse* of G:

$$\forall X \in G, \ \exists X^{-1} \in G \ \text{with: } X.X^{-1} = E = X^{-1}.X$$

(e.g.: X = *Rotation* $\Re(\mathbf{u}, \varphi)$; X^{-1} = *Rotation* $\Re(\mathbf{u}, -\varphi)$.)
The *order* g of a group G is equal to the number of elements it contains.

---

[1]J.J. BURCKHARDT *Die Bewegungsgruppen der Kristallographie*, Basel (1947).

❏ EXAMPLES OF GROUPS:

◆ Consider a vector **r**. The two operations E and I form the group {E,I}:

E: identity (**r** = E.**r**)   I: inversion (−**r** = I.**r**).

◆ Group {E, C_2}

E: identity C_2: rotation $\mathfrak{R}(\mathbf{u},\pi)$ about an axis in $\mathbb{R}_3$.

◆ A group formed from an element and its powers {A, $A^2 = $ A.A, $A^3 \ldots A^n = $ E} is called a cyclic group. The rotations $2\pi/n$ (*n* an integer) about an axis are C_n cyclic groups. For example, the group C_3 contains the elements:

$$C_3 = \{C_3^1 = \mathfrak{R}(\mathbf{u}, 2\pi/3), C_3^2 = \mathfrak{R}(\mathbf{u}, 4\pi/3), C_3^3 = \mathfrak{R}(\mathbf{u}, 2\pi) = E\}$$

◆ The groups formed from A and B and containing the elements {A, $A^2 \ldots A^n = B^2 = $ E, B, B.A $= A^{n-1}$.B $\ldots$ } are dihedral groups $\mathcal{D}_n$.

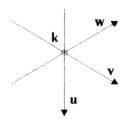

Figure 5.1

Thus the 6-element group:

$$\{C_3^1 = \mathfrak{R}(\mathbf{k}, 2\pi/3), C_3^2, C_3^3 = C_2^2 = E,$$

$$C_2 = \mathfrak{R}(\mathbf{u}, \pi), C_3.C_2 = \mathfrak{R}(\mathbf{v}, \pi),$$

$$C_3^2.C_2 = C_2.C_3 = \mathfrak{R}(\mathbf{w}, \pi)\}$$

is the dihedral group $\mathcal{D}_3$ (formed from a C_3 and an orthogonal C_2.)

NOTE: groups are abstract objects, but we can associate *representations* with them. For example, for a group composed of symmetry operations, the matrices associated with each element in the group form a representation which is related to the choice of origin and frame used.

☞ *It is important to make the distinction between a* **symmetry element** *and the* **associated operator(s)** *which are element(s) of the group of symmetry operators. For example, in a group containing a fourfold axis (a symmetry element) there are the four operators* C_4^1, C_4^2, C_4^3, $C_4^4 = $ E *(elements in the group).*

1.2 Subgroups and Co-sets

◆ Suppose G is a group of order g. A subset H of G is a **subgroup** of G if H is itself a group subject to the same rule of composition defining the group G. EXAMPLE: C_3 is a subgroup of $\mathcal{D}_3$

◆ Let G be a group of finite order g and H a subgroup of G of order h distinct from G; there exists at least one element A of G, not contained in H.

$$\forall X \in G, \ A.X \in G, \ A.X \notin H$$

All the A.X (when X describes H) are external to this subgroup and form a set of h distinct elements, denoted AH and called a **co-set**.

G can be decomposed into a union of disjoint subsets, defined from H:

$$G = E.H + A.H + B.H + \ \cdots$$
$$A \notin H \qquad B \notin H \qquad B \notin A.H \ \ldots$$

The order h of a subgroup H is a divisor of g, the order of the group $(i = g/h)$. i is the *index* of the subgroup H with respect to the group G. EXAMPLE: H = C_3 is a subgroup of index 3 of G = $\mathcal{D}_3$. $(\mathcal{D}_3 = E.C_3 + C_2.C_3)$

1.3 The Orthogonal Group O(3)

We consider the set of rotations leaving one point unchanged. The determinants of the matrices associated with these matrices are equal to $+1$. We also consider the inversion operator which transforms the vector **r** into $-\mathbf{r}$ and which is commutative with the rotations. The set of rotations and rotation-inversions makes up the orthogonal group O(3).

1.4 Direct Product of Two Subgroups of a Group

Suppose G is a group and H and K are two subgroups of G. G is said to be the direct product of these two subgroups if:

◆ All elements g of G appear as the product of an element $h \in H$ and an element $k \in K : g = h.k$;
◆ This decomposition is unique for a given element g of G;
◆ The elements of H and K are commutative.

The usual notation for the direct product is as follows:

$$G = H \otimes K, \ (H \subset G, \ K \subset G)$$

2   PROPER AND IMPROPER POINT GROUPS

These are the subgroups of the group O(3), and there are an infinite number. Here, we shall list the crystallographic point groups, i.e. those compatible with the symmetry of a crystal lattice.

We distinguish two types of group:

◆ proper groups which only contain rotations (determinants of the matrices are equal to $+1$);
◆ improper groups which contain rotations (determinants of the matrices are equal to $+1$) and rotation-inversions or roto-inversions (determinants of the matrices are equal to -1).

❒   THEOREM OF IMPROPER GROUPS

In an improper group G_i, the proper operators constitute a proper subgroup of index 2.

Consider an improper group G_i. In G_i there exist proper symmetry operations (at least the identity operation E). Consider the proper operators in G_i: E, R, R_j ... R_n; they constitute a proper subgroup: $G_p = \{E, R, R_j \ldots R_n\}$ the determinants Δ of whose matrices are equal to $\neq 1$.

Consider $\bar{R}$, an improper element, $(\Delta = -1)$ and $\bar{R}^{-1}$ its inverse: $\bar{R}.\bar{R}^{-1} = E$; $(\bar{R})^2$ is a proper element (since $\Delta = +1$) hence $(\bar{R})^2 = R_j$

$$\bar{R}^{-1}.\bar{R}.\bar{R} = \bar{R}^{-1}.R_j \rightarrow \bar{R} = \bar{R}^{-1}.R_j$$

$$\bar{R}.R_j^{-1} = \bar{R}^{-1}.R_j.R_j^{-1} \rightarrow \bar{R}^{-1} = \bar{R}.R_j^{-1}$$

We now form the co-set associated with the subgroup G_p:

$$\bar{R}.G_p = \{\bar{R}, \bar{R}.R_1, \ldots, \bar{R}.R_j, \ldots, \bar{R}.R_n\}$$

Consider $\bar{R}$, any improper element:

$$\bar{R}.\bar{R}' \text{ is a proper element } (\Delta = -1. -1 = +1) \Rightarrow \bar{R}.\bar{R}' = R_k$$

$$\bar{R}^{-1}.\bar{R}.\bar{R}' = \bar{R}^{-1}.R_k \qquad \text{but} \qquad \bar{R}^{-1} = \bar{R}.R_j^{-1}$$

$$\bar{R}' = \bar{R}.R_j^{-1}.R_k = \bar{R}.R_m$$

$\bar{R}'$ belongs to the associated co-set $\bar{R}. G_p$ and all the improper elements of G_i belong to the co-set associated with G_p.

□ TYPES OF IMPROPER GROUPS

◆ **Improper groups containing an inversion**
The decomposition of G_i into co-sets can be written $G_i = G_p + I.G_p$
The group $\{e, I\}$, like G_p, is a subgroup of G_i. G_i appears as the direct product $G_i = \{E, I\} \otimes G_p$. This relation leads to an obvious method for constructing improper groups containing the inversion.

◆ **Improper groups not containing the inversion**
Consider $\bar{R}$ ($\neq I$) as the element chosen for the second co-set in the decomposition of G_i, hence $G_i = G_p + \bar{R}.G_p$ ($I \notin G_i$). Now, $R = I.R$ ($\bar{R} = I.R$), where R is a proper operation not contained in G_p; it is not contained in G_i either, since its inverse R^{-1} would also be contained together with $\bar{R}.R^{-1} = I.R.R^{-1} = I$, which would be contrary to the initial hypothesis.
Consider the set: $G' = G_p + R.G_p$; we show that it constitutes a group of the same type (It is said to be *isomorphous* with G_i).
Let P, Q ... be the elements of G_p. By hypothesis G_i is a group, hence:

$$(\bar{R}.P).Q, \; Q.(\bar{R}.P) \in G_i \text{ and } \in \bar{R}.G_p$$

$$(\bar{R}.P)(\bar{R}.Q) \in G_i \text{ and } \in G_p$$

Furthermore, $\forall P, Q \in G_p$ we have:

$$\bar{R}.P.Q = I.(R.P.Q) \text{ and } Q.\bar{R}.P = I.(Q.R.P) \in I.R.G_p \Rightarrow R.P.Q, \; Q.R.P \in R.G_p$$

Therefore:

$$(\bar{R}.P).(\bar{R}.Q) = (I.R.P).(I.R.Q) = (R.P).(R.Q) \in G_p$$

And finally:

$$(\bar{R}.P)^{-1} \in \bar{R}.G_p \Rightarrow (R.P)^{-1} \in R.G_p$$

G' thus forms a group which contains only proper rotations.
Construction of the improper groups G_i which do not contain the inversion could be done by starting with proper groups G' which admit an invariant proper subgroup G_p of index 2, replacing the element R, the factor of the second co-set of the decomposition, by the product of R and inversion.

3   ENUMERATION OF POINT GROUPS

3.1   Method of Enumeration

To enumerate the crystallographic point groups, it is only necessary to determine all the proper groups and follow the above reasoning to deduce the improper groups.

The proper groups correspond to rotation axes. Since the rotation axes compatible with lattice symmetry are the 1, 2, 3, 4 and 6-fold axes, the crystallographic point groups must contain the cyclic groups C_1, C_2, C_3, C_4 and C_6. The purpose of enumeration is to find which associations of rotation axes C_n, C_m . . . are compatible with a group structure.

NOTATION: We consider a sphere of centre O, the meeting point of the rotation axes of the group under study. A ray coming from O crosses the sphere at P. We denote by $|P>$ the vector **OP**. By analogy with the stereographic projection, P is called the *pole* of the ray.

3.2   Finding Proper Groups of Order n

❐   CONJUGATE POLES

Let ${}^1C_{n1}$ be an axis of order n whose operators belong to G_p, and $|{}^1P_1>$ its pole. The group of rotations about this axis is:

$$^1C_{n1} = \{E, \ {}^1R_1, \ {}^1R_1^2, \ . . ., \ {}^1R_1^{n_1-1}\}$$

This is a cyclic subgroup of index $j_1 = n/n_1$.

If all the symmetry operations are applied to the point M, we obtain $(n_1 - 1)$ points different from M, hence $(n_1 - 1)$ different identity operations correspond to the pole $|{}^1P_1 >$.(With a four-fold axis are associated C_4^1, C_4^2, C_4^3, $C_4^4 = E$).

We decompose G_p into co-sets associated with the group ${}^1C_{n1}$:

$$G_p = \{{}^1C_{n1} + {}^1R_2.{}^1C_{n1} + \ . . . \ + {}^1R_j^1.{}^1C_{n1}\} \quad {}^1R_i \in G_p, \ {}^1R_i \notin {}^1C_{n1}$$

The operator ${}^1R_i$ is a rotation about an axis with a pole different from $|{}^1P_1>$ which brings this pole into a position $|{}^1P_i>$, equivalent to $|{}^1P_1>$. The pole $|{}^1P_i>$ is thus also the pole of an axis of order n_1.

Using the $(j_1 - 1)$ operators ${}^1R_i$, we can construct $(j_1 - 1)$ distinct poles equivalent to $|{}^1P_1>$.

Let us suppose

$$^1R_i.|{}^1P_1> = {}^1R_j.|{}^1P_1>$$

Then

$$|{}^1P_1> = E.|{}^1P_1> = ({}^1R_i)^{-1}.{}^1R_i.|{}^1P_1> = ({}^1R_i)^{-1}.{}^1R_j.|{}^1P_1>$$

Hence

$$({}^1R_i)^{-1}.{}^1R_j \in {}^1C_{n1} \text{ (Since the product leaves } |{}^1P_1\rangle \text{ unchanged)}$$

As the group is cyclic, we have:

$$({}^1R_i)^{-1}.{}^1R_j = ({}^1R_1)^p \text{ and } {}^1R_i.({}^1R_i)^{-1}.{}^1R_j = {}^1R_i.({}^1R_1)^p$$

$$^1R_j = {}^1R_i.({}^1R_1)^p \in {}^1R_i.{}^1C_{n1}$$

This is contrary to the hypothesis, therefore:

$$^1R_i.|{}^1P_1\rangle \neq {}^1R_j.|{}^1P_1\rangle$$

The set of j poles $|{}^1P_i\rangle$ forms a 'conjugate' system of poles defining axes of the same order.

For example, we shall see that in group 432 there exist six poles which correspond to the three tetragonal axes.

❐ PARTITION INTO CONJUGATE SYSTEMS

Let ${}^1C_{n2}$ be an axis of order n_2 whose operators belong to G_p and whose pole $|{}^2P_1\rangle$ does not belong to the conjugate system of $|{}^1P_i\rangle$. This axis defines a cyclic subgroup of G_p of order n_2, ${}^2C_1$ and with index $j_2 = n/n_2$.

From ${}^2C_1$ it is possible to define a new system of conjugate poles formed from the set of j_2 $|{}^2P_j\rangle$ poles. The two systems have no common pole: for if ${}^1R_i.|{}^1P_1\rangle = {}^2R_j|{}^2P_1\rangle$ we would then have:

$$({}^2R_j)^{-1}.({}^1R_i).|{}^1P_1\rangle = |{}^2P_1\rangle$$

$|{}^2P_1\rangle$ would be one of the poles $|{}^1P_i\rangle$, which is contrary to the hypothesis.

Using this method, all the poles of axes can be partitioned into *h conjugate systems*.

❐ ENUMERATION OF PROPER ROTATIONS OF G_p

For the system of poles $|{}^1P_i\rangle$, there exist $(n_1 - 1)$ different neutral-element operators per pole. There are $j_1 = n/n_1$ poles, i.e. $j_1/2 = n/2n_1$ rotation axes and therefore $n.(n_1 - 1)/2n_1$ different identity operators of order n_1. The group C_1 is trivial and so summation over the h sets of poles gives:

$$\sum_{r=1}^{h} \frac{1}{2} \frac{n}{n_r}(n_r - 1) = n - 1 \tag{1}$$

$$n \geq n_r \geq 2 \tag{2}$$

Dividing both sides of (1) by n/2 gives:

$$\sum_{r=1}^{h} \frac{1}{n_r} = h - 2 + \frac{2}{n} \qquad (3)$$

The conditions imposed by (2) are very restrictive:

◆

$$n_r \geq 2 \Rightarrow \frac{1}{n_r} \leq \frac{1}{2} \Rightarrow \sum_{r=1}^{h} \frac{1}{n_r} \leq \frac{h}{2}$$

Hence:

$$h - 2 + \frac{2}{h} \leq \frac{h}{2} \qquad \text{since } h \text{ is an integer we have: } h \leq 3$$

◆

$$n \geq n_r \Rightarrow \frac{1}{n_r} \geq \frac{1}{n} \Rightarrow \sum_{r=1}^{h} \frac{1}{n_r} \geq \frac{h}{n}$$

Hence:

$$h - 2 + \frac{2}{n} \geq \frac{h}{n} \qquad \text{since } h \text{ is an integer we have: } h \geq 2$$

The only possible values for the number h of conjugate systems are 2 and 3.

For the case h = 2

Equation (3) gives:

$$\frac{1}{n_1} + \frac{1}{n_2} = \frac{2}{n} \quad \text{which has the unique solution:}$$

$$n_1 = n_2 = n$$

There are two systems of conjugate poles each formed from a single pole, which corresponds to an *n*-fold axis (the axis crosses the sphere at two points). This cyclic group can have, *a priori*, any order whatever, but to enumerate the crystallographic groups we need only the groups C_1, C_2, C_3, C_4, C_6.

We can draw up a table of these five groups, showing the two crystallographic notations in common use:

Table 5.1. Proper Groups

Cyclic group	Schönflies	Hermann–Maugin
C_1	C_1	1
C_2	C_2	2
C_3	C_3	3
$C_4 \neq C_2 \otimes C_2$	C_4	4
$C_6 \equiv C_3 \otimes C_2$	C_6	6

☞ *It is suggested that the reader should construct the stereographic projection of the symmetry elements of groups being studied and check the results against the illustrations in the next chapter or the atlas.*

For the case h = 3

Equation (3) is written:

$$\frac{1}{n_1} + \frac{1}{n_2} + \frac{1}{n_3} = 1 + \frac{2}{n}$$

◆ Let n_1 be the smallest of the n_i:

$$\frac{1}{n_1} + \frac{1}{n_2} + \frac{1}{n_3} < \frac{3}{n_1} \Rightarrow 1 + \frac{2}{n} < \frac{3}{n_1}$$

The left hand side is strictly greater than 1 and the right hand side is greater than 1 if n_1 is less than 3, hence: $\boldsymbol{n_1 = 2}$.

◆ Let n_2 be the smallest of the n_2, n_3:

$$\frac{1}{n_2} + \frac{1}{n_3} \leqslant \frac{2}{n_2} \Rightarrow \frac{1}{2} + \frac{2}{n} \leqslant \frac{2}{n_2}$$

but n_2 is strictly less than 4: $\boldsymbol{n_2 = 2}$ **or** $\boldsymbol{3}$

◆ $\boldsymbol{h = 3, n_1 = 2, n_2 = 2}$
Equation (3) gives $n_3 = n/2$. The indices of the subgroups will be:

$$j_1 = n/2 \quad j_2 = n/2 \quad j_3 = 2$$

The pole diagram thus contains:

— 2 poles diametrically opposed to each other which are the poles of an axis of order $n_3 = n/2$;
— 2 conjugate pole systems each containing $n/2$ twofold axis poles.

These twofold axes are equivalent if j is odd ($\mathcal{D}_3$) but form two different classes if j is even ($\mathcal{D}_2, \mathcal{D}_4, \mathcal{D}_6$).

These groups are thus dihedral groups $\mathcal{D}_n$. In this way we obtain four additional crystallographic proper groups:

Table 5.2. Dihedral Groups

Dihedral group	Schönflies	Hermann–Maugin
$\mathcal{D}_2$	D_2	222
$\mathcal{D}_3$	D_3	32
$\mathcal{D}_4$	D_4	422
$\mathcal{D}_6$	D_6	622

EXAMPLES:

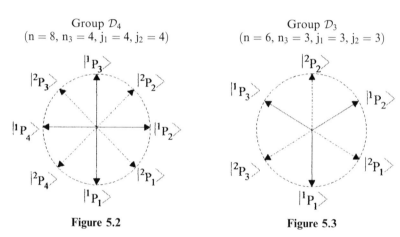

Group $\mathcal{D}_4$
($n = 8$, $n_3 = 4$, $j_1 = 4$, $j_2 = 4$)

Group $\mathcal{D}_3$
($n = 6$, $n_3 = 3$, $j_1 = 3$, $j_2 = 3$)

Figure 5.2   Figure 5.3

◆ $h = 3$, $n_1 = 2$, $n_2 = 3$

Equation (3) gives

$$\frac{1}{n_3} = \frac{1}{6} + \frac{2}{n} \Rightarrow n_3 = \frac{6n}{12+n} \Rightarrow n_3 < 6$$

n_3 can take the values 2, 3, 4 or 5.

◆ $n_3 = 2$

$$n_1 = 2 \quad n_2 = 3 \quad n_3 = 2$$

This case corresponding to two subgroups of index 2 has already been met, and need not be studied further here.

◆ $n_3 = 3$

$$n_1 = 2 \quad n_2 = 3 \quad n_3 = 3 \quad n = 12$$

The subgroup indices are:

$$j_1 = 6 \quad j_2 = 4 \quad j_3 = 4$$

Thus in this group we find six poles of twofold axes, four poles of threefold axes and four poles of threefold axes.

The group, of order 12, contains three twofold and four threefold axes.

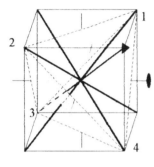

Figure 5.4

The four poles of threefold axes lie symmetrically on the sphere, since a rotation of $2\pi/3$ about one of the axes (e.g. No.1) must make the other three axes coincide:

$$(4) \rightarrow (2), (2) \rightarrow (3), (3) \rightarrow (4)$$

The poles of the twofold axes must also bring the threefold axes into correspondence two by two. The four threefold axes are oriented along the diagonals of a cube or along the normals to the faces of a regular tetrahedron.

The twofold axes are normal to the faces of the cube. The angle between two threefold axes is 109°28′ and the group is that of the *tetrahedron* (denoted T or 23) which in group theory is associated with an abstract group denoted $\mathcal{A}_4$.

◆ $n_3 = 4$

$$n_1 = 2 \quad n_2 = 3 \quad n_3 = 4 \quad n = 24.$$

The subgroup indices are:

$$j_1 = 12 \quad j_2 = 8 \quad j_3 = 6$$

In this group are found twelve twofold axis poles, four threefold axis poles and six fourfold axis poles.

The group, of order 24, contains a total of three fourfold axes, four threefold axes and six twofold axes. Symmetry considerations require the fourfold axes to be rows in a cube, of the type [100], the threefold axes to be rows of the type [111] and the twofold axes to be rows of the type [110]. This is the *octahedron* group (denoted O or 432) with which the abstract group Π_4 is associated.

◆ $n_3 = 5$

$$n_1 = 2 \quad n_2 = 3 \quad n_3 = 5 \quad n = 60$$

This group (the *icosahedron* group) has fivefold axes. It must therefore be excluded from the list as it cannot be a crystallographic group.

❒ SUMMARY OF PROPER GROUPS FOUND

In enumerating the crystallographic groups, we have found five cyclic groups, four dihedral groups and two special groups making a total of:

eleven crystallographic proper groups.

3.3 Finding Improper Groups G_p

❒ IMPROPER GROUPS CONTAINING INVERSION

According to the theorem of improper groups we have:

$$G_i = G_p + I.G_p = \{E, I\} \otimes G_p \text{ i.e.: } G_i = G_p \otimes C_2$$

From the eleven proper groups, we can construct eleven improper groups containing the inversion.

The product of the inversion and a C_{2n} axis is a mirror plane normal to the axis of symmetry. For the groups of classes C_2, C_4, C_6, D_2, D_4 and D_6, we obtain a mirror plane σ_h normal to the principal axis. For the classes T and O we obtain three σ_h type mirror planes normal to the twofold axes of class T or to the fourfold axes of class O.

Table 5.3. Improper groups containing the inversion

	Proper groups			Improper groups	
Schönflies	Hermann–Maugin	Abstract group	Abstract group	Hermann–Maugin	Schönflies
C_1	1	C_1	$C_1 \otimes C_2$	$\bar{1}$	C_i
C_2	2	C_2	$C_2 \otimes C_2$	$2/m$	C_{2h}
C_3	3	C_3	$C_3 \otimes C_2$	$\bar{3}$	$S_6 = C_{3i}$
C_4	4	C_4	$C_4 \otimes C_2$	$4/m$	C_{4h}
C_6	6	C_6	$C_6 \otimes C_2$	$6/m$	C_{6h}
D_2	222	$\mathcal{D}_2$	$\mathcal{D}_2 \otimes C_2$	mmm	D_{2h}
D_3	32	$\mathcal{D}_3$	$\mathcal{D}_3 \otimes C_2$	$\bar{3}$m	D_{3d}
D_4	422	$\mathcal{D}_4$	$\mathcal{D}_4 \otimes C_2$	$4/mmm$	D_{4h}
D_6	622	$\mathcal{D}_6$	$\mathcal{D}_6 \otimes C_2$	$6/mmm$	D_{6h}
T	23	$\mathcal{A}_4$	$\mathcal{A}_4 \otimes C_2$	m3	T_h
O	432	Π_4	$\Pi_4 \otimes C_2$	m3m	O_h

For the groups from classes D_2, D_3, D_4, D_6 and O we also obtain σ_v mirror planes normal to the twofold axes.

❒ IMPROPER GROUPS NOT CONTAINING THE INVERSION

We showed in paragraph 2 that: $G_i = G_p + R.G_p$ ($I \notin G_i$) and that, to construct the improper groups G_i which do not contain the inversion, we can start with proper groups G′ admitting an invariant subgroup G_p of index 2, replacing the element R, a factor in the second co-set of the decomposition, by its product with inversion.

We can thus decompose the group $G = C_6$ into $G = C_3 + C_6.C_3$ and the corresponding improper group is: $G_i = C_3 + \bar{C}_6.C_3$

The groups C_1, C_2 and T have no subgroup of index 2, whereas D_4 and D_6 can both be decomposed into different subgroups. We thereby obtain $(11 - 3 + 2) = 10$ new improper groups which do not contain the inversion.

The group C_2 gives the group S_1 (mirror plane), C_4 gives S_4 and C_6 yields C_{3h} (a C_3 axis plus a normal mirror plane σ_h). From the groups D_2, D_3, D_4 and D_6 we obtain the groups C_{2v}, C_{3v}, C_{4v} and C_{6v} (comprising a number n of σ_v mirror planes containing the principal axis). The second decompositions D_4 and D_6 give respectively the groups D_{2d} (principal axis S_4, two twofold axes and two twofold σ_v axes at 45°) and D_{3h} (principal axis S_3, three twofold axes and three σ_v containing the twofold axes). Finally, the group O gives the group T_d (three S_4 axes, four threefold axes and six diagonal mirror planes at 45° to the S_4).

Table 5.4. Improper groups not containing the inversion

	Group G'$_p$		Improper group G$_i$		
Schönflies	G_p + R.G_p	G_p + R̄.G_p	Schönflies	Hermann–Maugin
C_2	$C_2 = C_1 + C_2.C_1$	$C_1 + \bar{C}_2.C_1$	S_1, C_s	$\bar{2} = m$
C_4	$C_4 = C_2 + C_4.C_2$	$C_2 + \bar{C}_4.C_2$	S_4	$\bar{4}$
C_6	$C_6 = C_3 + C_6.C_3$	$C_3 + \bar{C}_6.C_3$	S_4, C_{3h}	$\bar{6}$
D_2	$\mathcal{D}_2 = C_2 + C_2.C_2$	$C_2 + \bar{C}_2.C_2$	C_{2v}	mm2
D_3	$\mathcal{D}_3 = C_3 + C_2.C_3$	$C_3 + \bar{C}_2.C_3$	C_{3v}	3m
D_4	$\mathcal{D}_4 = C_4 + C_2.C_4$	$C_4 + \bar{C}_2.C_4$	C_{4v}	4mm
D_4	$\mathcal{D}_4 = \mathcal{D}_2 + C_4.\mathcal{D}_2$	$\mathcal{D}_2 + \bar{C}_4.\mathcal{D}_2$	D_{2d}	$\bar{4}2m$
D_6	$\mathcal{D}_6 = C_6 + C_2.C_6$	$C_6 + \bar{C}_2.C_6$	C_{6v}	6mm
D_6	$\mathcal{D}_6 = \mathcal{D}_3 + C_6.\mathcal{D}_3$	$\mathcal{D}_3 + \bar{C}_6.\mathcal{D}_3$	D_{3h}	$\bar{6}2m$
O	$\mathit{\Pi}_2 = \mathcal{A}_4 + C_4.\mathcal{A}_4$	$\mathcal{A}_4 + \bar{C}_2.\mathcal{A}_4$	T_d	$\bar{4}3m$

3.4 Summary of Point Groups

The following crystallographic point groups have been found:

— 11 proper groups
— 11 improper groups with inversion
— 10 improper groups without inversion

making a total of **32 point groups**.

To fully understand the notions involved in point groups, it is an invaluable exercise to construct them according to the directions given. Constructing stereographic projections of proper groups is simple. For improper groups, where the products of symmetry elements are involved, the composition laws given in chapter 4 should be used.

The study of cubic groups is rather tricky, and it can help to make models with cardboard and glue. After appendix D the reader will find instructions for several models.

These include a pentagonal dodecahedron. The attentive reader will find the following symmetry elements: 1 centre of inversion, 15 mirror planes, 15 twofold axes, 10 threefold axes and 6 fivefold axes! However, since the indices of the faces are irrational, this polyhedron cannot represent a crystal.

Apart from these 32 groups which are compatible with operations of orientation symmetry in crystals, there does exist an infinite number of non-crystallographic point groups. In appendix A is a brief account of the way they are listed. They are extremely relevant to molecular physics, since the absence of constraints imposed by the lattice allows the existence in molecules of rotation axes of order 5, 7, 8 etc.

The symmetry class of a crystal, which is related to the nature of its lattice, should not be confused with any symmetry in the objects making up its repeating unit. Thus, in benzene, which crystallises in the class mmm (group Pbca), the repeating unit is the molecule of benzene with the symmetry 6/mmm. Similarly, crystals of fullerene are cubic close packed (class m3m) whereas the molecule (C_{60}) has the symmetry $\bar{5}3\frac{2}{m}$ (an icosahedral group having 6 A_5, 10 A_3. 15 A_2 axes, 15 mirror planes and a centre of symmetry).

Chapter 6

Classes, Systems and Crystal Lattices

1 CRYSTAL CLASSES AND CRYSTAL SYSTEMS[1]

Each of the 32 point groups forms a crystal class.

All the operations in a point group G_p to which a crystal structure belongs transform the crystal into an entity which must be superimposable on the initial structure through lattice translations.

Consider now a point group H of the lattice. Some operations of H, operations which bring all *the lattice nodes into coincidence with lattice nodes*, may transform the structure in such a way that it is impossible to return it to its initial position by a simple translation.

In general the point group G_p to which the structure belongs is only a subgroup of H; the orientation symmetry of the lattice is greater than that of the structure.

1.1 Enumerating Lattice Point Groups

Since the crystallographic point groups are all known, the quickest way of enumerating the point groups H of a lattice is by considering them as point groups having special properties.

❑ THE GROUPS H CONTAIN THE INVERSION

If the vector $\mathbf{T}$ is a lattice translation, then so is the vector $-\mathbf{T}$; all the groups H must necessarily contain the inversion.

---

[1]Syngonies is also used.

❏ IF H CONTAINS A C_n $(n>2)$, THEN IT ALSO CONTAINS THE GROUP C_{nv}

Consider a lattice vector **a** in a plane normal to the C_n axis and the vector **b** $= C_n(\mathbf{a})$.

If $n = 4$, the mirror plane σ_v defined by the C_4 axis and **a** transforms the vectors **a** and **b** into **a** and $-\mathbf{b}$, leaving unchanged the set of lattice vectors normal to the C_4 axis. Since the vectors parallel to an axis are unaffected by this rotation, the set of lattice translations in the group is invariant in C_{4v}.

If $n = 3$ or $n = 6$, the mirror plane σ'' defined by the C_3 axis (for the case $n = 3$, or contained in the C_6 axis for the case $n = 6$) and normal to **a**, leaves the set of lattice translations unchanged. The set of lattice translations of the groups is thus invariant in C_{3v} and C_{6v}.

Among the thirty two point groups, eleven contain inversion and there are seven which also satisfy the above condition. These are:

$\bar{1}$ (C_1); $\frac{2}{m}$ (C_{2h});   mmm (D_{2h}); $\bar{3}$m (D_{3d});

$\frac{6}{m}$ mm (D_{6h}); $\frac{4}{m}$ mm (D_{4h});   m3m (O_h).

Each of these seven groups is associated with a **crystal system**.

Each of the seven systems possesses a particular set of dimensions corresponding to the symmetry of the lattice.

Each lattice is also characterized by one or more particular directions: those of the lattice symmetry elements.

Table 6.1. The seven crystal systems

System	Lattice group H	Lattice characteristics	Lattice geometry
Triclinic	$\bar{1}$ (C_i)	1 centre	$a \neq b \neq c$ $\alpha \neq \beta \neq \gamma \neq \pi/2$
Monoclinic	$\frac{2}{m}$ (C_{2h})	1 twofold direction (axis or mirror plane normal to this direction)	$a \neq b \neq c$ $\alpha = \gamma = \pi/2$ $\beta > \pi/2$
Orthorhombic	mmm (D_{2h})	3 twofold directions	$a \neq b \neq c$ $\alpha = \beta = \gamma = \pi/2$
Trigonal	$\bar{3}$m (D_{3d})	1 threefold direction	$a = b = c$ $\alpha = \beta = \gamma \neq \pi/2$
Tetragonal	$\frac{4}{m}$ mm (D_{4h})	1 fourfold direction	$a = b \neq c$ $\alpha = \beta = \gamma = \pi/2$
Hexagonal	$\frac{6}{m}$ mm (D_{6h})	1 sixfold direction	$a = b \neq c$ $\alpha = \beta = \pi/2$ $\gamma = 2\pi/3$
Cubic	m3m (O_h)	4 fourfold directions	$a = b = c$ $\alpha = \beta = \gamma = \pi/2$

We can establish a **hierarchy** between these systems:

The system S1, characterised by the group H1, is said to be inferior to the system S2 characterised by the group H2, if: H1 $\subset$ H2.

Consider a class of which the group G_p is a subgroup of H1, and such that there exists no system S2 inferior to S1 for which G_p is also a subgroup of the group H2 defining S2.

The class in question is said to belong to the system S1. From this definition of the hierarchy, we can assign the 32 crystal symmetry classes to the seven systems as shown in table 6.2:

Table 6.2. Classification of point groups into systems

Triclinic	1, $\bar{1}$
Monoclinic	2, m, **2/m**
Orthorhombic	222, mm2, **mmm**
Trigonal	3, $\bar{3}$, 32, 3m $\bar{3}$m
Tetragonal	4, $\bar{4}$, 4/m, 4mm, 422, $\bar{4}$ 2m, **4/mmm**
Hexagonal	6, $\bar{6}$, 6/m, 6mm, 622, $\bar{6}$ 2m, **6/mmm**
Cubic	23, m3, 432, $\bar{4}$ 3m, **m3m**

International notation (Hermann-Mauguin)

Triclinic	C_1, $\mathbf{C_i}$
Monoclinic	C_2, C_s, $\mathbf{C_{2h}}$
Orthorhombic	D_2, C_{2v}, $\mathbf{D_{2h}}$
Trigonal	C_3, C_{3i}, D_3, C_{3v}, $\mathbf{D_{3d}}$
Tetragonal	C_4, S_4, C_{4h}, C_{4v}, D_4, D_{2d}, $\mathbf{D_{4h}}$
Hexagonal	C_6, C_{3h}, C_{6h}, C_{6v}, D_6, D_{3h}, $\mathbf{D_{6h}}$
Cubic	T, T_h, O, T_d, $\mathbf{O_h}$

Schönflies notation (lattice groups in bold)

1.2   The Conventions of International Nomenclature

The symbols used to name the classes are as follows:

$$1, 2, 3, 4, 6, \bar{1}, m, \bar{3}, \bar{4}, \bar{6}, 2/m, 4/m \text{ and } 6/m.$$

The symmetry axes are oriented along the axes of the coordinate frame of the system being studied. In the case of mirror planes, the direction of the normal to the plane is taken. In systems with a symmetry axis higher than twofold, (principal axis), the direction of the vector **c** is that of the highest-order symmetry axis in the group. The classes in the trigonal system do not, however, follow this rule. Here, use is made of the 'Miller scheme' which gives prominence to the threefold axis. In this scheme, the stereographic projections of the trigonal classes are drawn using the *direction of the threefold axis normal to the plane of projection*.

For historical reasons, in the monoclinic system, the direction of the vector **b** is taken along the direction of the twofold axis and the vectors **a** and **c** are then chosen to have $\beta > \pi/2$.

In international nomenclature (Hermann–Mauguin), the name of the group contains three symbols. They are written in an order indicated in the table below which also specifies the directions of the symmetry operators.

Note: For centro-symmetric groups, the symbol $\frac{2}{m}$, sometimes written 2/m for typographical reasons, is often replaced by the symbol m. For example, $\frac{2}{m}\frac{2}{m}\frac{2}{m}$ is written mmm.

If, in naming a class, we wish to specify the choice of convention for the orientation of the unit-cell axes, the axes of order 1 may be added to the name.

– The notations 121 and 1m1 correspond respectively to the classes 2 and m if the axis **b** is chosen parallel to the twofold axis direction.
– 112 and 11m correspond to the same classes when axis **c** is chosen parallel to the twofold direction.

For the tetragonal and hexagonal systems, the second and third symbols enable the distinction to be made between two twofold classes and two mirror planes σ_v.

Table 6.3. The order of symbols and orientations

Systems	1st symbol	2nd symbol	3rd symbol
Triclinic	1 or $\bar{1}$		
Monoclinic	**b**		
Orthorhombic	**a**	**b**	**c**
Tetragonal	**c**	**a, b**	**a + b, a − b**
Hexagonal and			
Trigonal (P cell)	**c**	**a, b**	**2a + b** . . .
Cubic	**a** . . .	**a + b + c** . . .	**a ± b** . . .
	2- or 4-fold axes	3-fold axes	oblique 2-fold axes

1.3 Holohedral and Merohedral Classes

The seven classes having the same point group as the lattice of their system are called **holohedral** classes (they are shown in bold in table 6.2). The other classes, with symmetry lower than that of the lattice, are called the **merohedral** classes. If the merohedral class is a subgroup of a holohedral class of order 2, it is a hemihedral class; subgroups of order 4 and 8 are named **tetartohedral** and **ogdohedral**. This nomenclature, used mainly by mineralogists, enables another classification to be defined for point groups, where classes are gathered together according to the nature of their generators. The classes may thus be regrouped into centred hemihedral, (4/m, 6/m . . .), pyramidal hemihedral (mm2, 4mm, 6mm, 3m), enantiomorphic hemihedral (1, 2, 222, 32, 422, 622, 432) and so on.

This re-classification brings out the similarities in physical properties of crystals belonging to the classes under study. For example, crystals in the enantiomorphic hemihedral class can be optically active. (see page 214).

For space groups other than cubic, the classification proposed in table 4, based on the nature of the group generators and their associations, also shows analogies which may exist between different groups.

Table 6.4. Point symmetry classes

$n = 1, 2, 3,$ $4, 6$	Triclinic	Monoclinic	Orthorhombic	Trigonal	Tetragonal	Hexagonal
n	1	2		3	4	6
$\bar{n}$	$\bar{1}$	m		$\bar{3}$	$\bar{4}$	$\bar{6}$
$\frac{n}{m}$		$\frac{2}{m}$			$\frac{4}{m}$	$\frac{6}{m}$
$n2$			222	32	422	622
nm			$mm2$	$3m$	$4\,mm$	$6\,mm$
$\bar{n}\,m$				$\bar{3}\,m$	$\bar{4}2m$	$\bar{6}2m$
$\frac{n}{m}mm$			mmm		$\frac{4}{m}mm$	$\frac{6}{m}mm$

1.4   Stereographic Projections of the 32 Classes

In presenting a crystal class, we generally use the stereographic projection of all its symmetry elements. From this projection it is easy to determine all the directions which are equivalent to a given direction.

> ☞ *In a given class, the set of planes* (or faces) *equivalent to a family of planes (or to the face) of indices (h k l) is called a form and is written {h k l}. Similarly the set of rows equivalent to a row of indices [u v w] is written ⟨u v w⟩.*

It should be remembered that the projection of a symmetry axis has only one or two points, these being the projections of the intersections of the axis with the projection sphere. The dotted lines joining the two points are simply there for clarity. The circles and full lines correspond to mirror plane projections. The following diagrams show how symmetry elements and their stereographic projections are related for two simple systems.

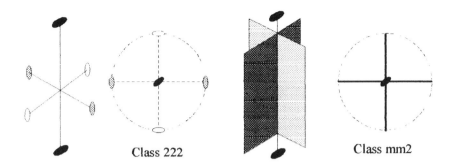

The international symbols used on projections to show axes of symmetry are as follows:

The stereographic projections of the symmetry elements of each of the 32 classes are shown on pages 78 and 79. Also shown on each projection are the poles of the most general form in the group {*h k l*}. Thus the multiplicity of the class (the number of equivalent general directions) can readily be determined. In appendix A will be found the detailed stereographic projections and representations of the forms possible in each of the classes.

2 THE LAUE CLASSES

Obtaining experimental evidence for the presence or absence of a centre of symmetry in a crystal is often tricky. In particular, in the classical diffraction methods used in X-ray crystallography, a centre of symmetry is systematically introduced into the diffraction pattern, even when the crystal being studied is *non centrosymmetric* (by Friedel's law). We consequently have to re-group symmetry classes which differ only by the *presence or absence of inversion*.

The result of classification following these criteria is the **Laue classes**. The 32 point groups are distributed among these 11 classes as shown in table 5. In this table, the group at the top of each column listing the classes is the *centrosymmetric group*; it is this which defines the point symmetry of the Laue class being considered.

Non-cubic point groups

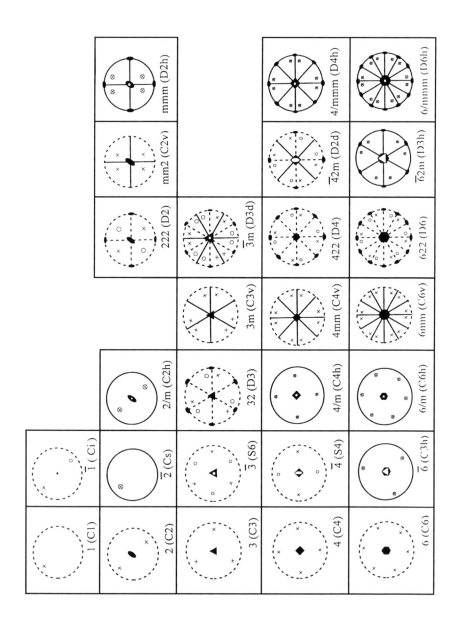

Cubic point groups

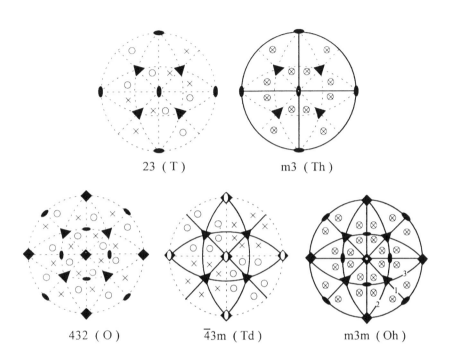

23 (T)   m3 (Th)

432 (O) $\overline{4}$3m (Td)   m3m (Oh)

Symmetry elements of the cube

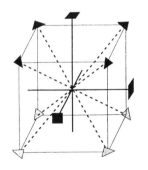

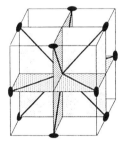

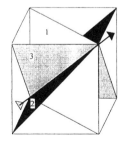

4- and 3-fold symmetry axes

2-fold axes and mirror planes
normal to the 4-fold axes

Diagonal mirrors intersecting
along [111]

(The numbers on the planes are
those of the projection of class
m3m)

Table 6.5. The 11 Laue classes

$\bar{1}$, 1	$\bar{3}$, 3	4/m, 4, $\bar{4}$
2/m, m, 2	$\bar{3}$m, 32, 3m	4/mmm, 422, 4mm, $\bar{4}$2m
mmm, 222, mm2	6/m, 6, $\bar{6}$	m3, 23
	6/mmm, 622, $\bar{6}$2m, $\bar{6}$2m	m3m, $\bar{4}$3m, 432

3   THE BRAVAIS LATTICES

If the base vectors are chosen in accordance with the lattice symmetries, a
simple unit-cell, i.e. one containing a single node, is not necessarily obtained. (A
parallelepiped in a lattice has eight corners, and each corner is shared among
eight unit-cells).

For simplicity, we shall illustrate this in a two-dimensional rectangular
lattice having a mirror plane parallel to the direction Ox.

Let **T1** be a single lattice translation, i.e. where the vector $\frac{1}{2}$**T1** is not a lattice
translation. **T2**, the mirror image of **T1**, is a lattice translation. **T1** + **T2** and
T1 − **T2** are two orthogonal vectors defining a rectangular unit-cell. If
T1 + **T2** and **T1** − **T2** are two single translations (figure 6.1 right), the lattice can
be described either by a single diamond cell or a multiple rectangular cell (a
centred cell).

If $\frac{1}{2}$(**T1** + **T2**) and $\frac{1}{2}$(**T1** − **T2**) are two single translations (figure 6.1 left), we
obtain a simple rectangular unit-cell. Only by considering a multiple unit-cell
can we bring out the entire lattice symmetry.

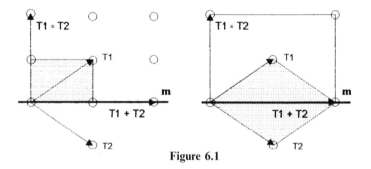

Figure 6.1

For each system, we find we have to consider not only the primitive lattice
constructed entirely from whole lattice translations, but also lattices containing
semi-translations which conserve the symmetry of the system. To specify the

nature of the lattice obtained, we associate a letter characteristic of the lattice **type** with the basic name of the system.

In addition to the primitive lattice (type **P**), we have to examine the face-centred lattices comprising type **A**: (100) faces; type **B**: (010) faces; type **C**: (001) faces; type $\mathbf{F}^2$: all faces centred; and type $\mathbf{I}^3$: body centred. In addition to whole lattice translations, we include in type **C** the translation $\mathbf{T} = \frac{1}{2}(\mathbf{a} + \mathbf{b})$, in type **I** the translation $\mathbf{T} = \frac{1}{2}(\mathbf{a} + \mathbf{b} + \mathbf{c})$, and in type **F**, the translations $\mathbf{T}_1 = \frac{1}{2}(\mathbf{a} + \mathbf{b})$, $\mathbf{T}_2 = \frac{1}{2}(\mathbf{b} + \mathbf{c})$ and $\mathbf{T}_3 = \frac{1}{2}(\mathbf{c} + \mathbf{a})$.

These various types are not all necessarily present in each system; with a suitable choice of base vectors, it is sometimes possible to obtain a unit-cell of lower multiplicity which conserves the lattice symmetry. The **14 lattice types** were first identified by Bravais around 1850.

THE TRICLINIC SYSTEM

Multiple unit-cells that can be constructed in this system possess no more symmetry than the initial cell.

Only type **P** need be considered.

THE MONOCLINIC SYSTEM

Two types are possible: **P** and **C**.

The transformation $\mathbf{a}_1 = -\mathbf{c}$, $\mathbf{c}_1 = \mathbf{a}$ changes type A into type C.
The transformation $\mathbf{a}_2 = \mathbf{a} + \mathbf{c}$, $\mathbf{c}_2 = \mathbf{c}$ changes type I into type C.
The transformation $\mathbf{a}_3 = \mathbf{a}$, $\mathbf{c}_3 = \frac{1}{2}(\mathbf{a} + \mathbf{c})$ changes type F into type C.
Type B is equivalent to type P.

THE ORTHORHOMBIC SYSTEM

There exist four possible types: **P, C, I, F**.
Types A and B are equivalent to type C after permutation of the base vectors.

---

[2]Initial letter of the German word Flächenzentrierte.
[3]Initial letter of Innenzentrierte.

THE TRIGONAL SYSTEM (Rhombohedral unit-cell)

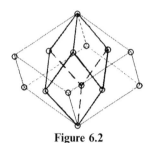

In this system, a single type is possible: the primitive type, denoted R (for rhombohedron).

Type C (face centred) lattices are incompatible with threefold symmetry.

Types F and I are equivalent to type R. Figure 6.2 shows the transformation of an F type cell into an R type cell.

Figure 6.2

THE TETRAGONAL SYSTEM

Two types are possible: **P** and **I**.

Types A and B are incompatible with tetragonal symmetry.

Type C is equivalent to type P for the transformation $a_1 = \frac{1}{2}(a+b)$, $c_1 = c$. This same transformation converts type F into type I.

THE HEXAGONAL SYSTEM

Only one type is possible: primitive, type **P**.

Types A, B, C, I and F are incompatible with the sixfold symmetry of the lattice. On the other hand, the P type hexagonal unit-cell is compatible with trigonal symmetry elements.

THE CUBIC SYSTEM

Three types are possible: **P**, **F** and **I**.

Types A, B and C are incompatible with the lattice symmetry.

Table 6.6. The 14 Bravais lattice types

Triclinic	P	Tetragonal	P, I
Monoclinic	P, C	Hexagonal	P
Orthorhombic	P, C, I, F	Cubic	P, F, I
Trigonal	R		

For F and I type lattices, the primitive unit-cell is rhombohedral (cf. § 5.6)

The 14 Bravais lattice types are given in table 6.6 and illustrated on page 83, where the main characteristics of each system can be seen.

The Bravais lattices

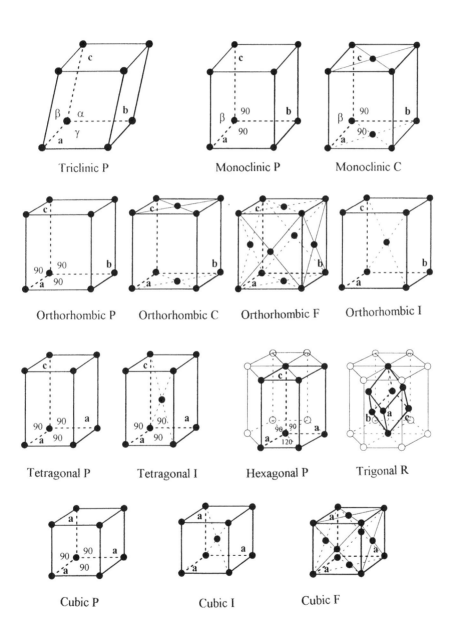

Triclinic P   Monoclinic P   Monoclinic C

Orthorhombic P   Orthorhombic C   Orthorhombic F   Orthorhombic I

Tetragonal P   Tetragonal I   Hexagonal P   Trigonal R

Cubic P   Cubic I   Cubic F

4 RECIPROCAL BRAVAIS LATTICES

Reciprocal lattices have the same symmetry as the lattices from which they are derived. In the case of direct lattices, the symmetry is not always evident in the unit-cell, and the same is true for reciprocal lattices. We shall begin with type C lattices, where the result is easy to visualise.

❐ RECIPROCAL LATTICE OF A TYPE C LATTICE

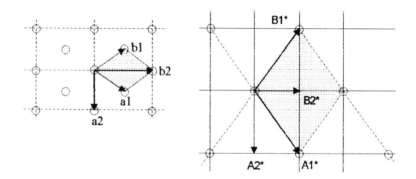

Figure 6.3

Let a_2, b_2 and c_2 be the base vectors of the face-centred unit-cell and a_1, b_1 and c_1 the base vectors of the primitive unit-cell.

$$a_2 = a_1 - b_1; \qquad b_2 = a_1 + b_1; \qquad c_2 = c_1$$

On constructing the reciprocal lattices ($A_1^* \perp b_1$, $c_1 \cdots$), we obtain:

$$A_2^* = \frac{1}{2}(A_1^* - B_1^*) \qquad B_2^* = \frac{1}{2}(A_1^* + B_1^*) \qquad C_2^* = C_1^*$$

The lattice constructed using the vectors A_2^*, B_2^* and C_2^* has the correct symmetry, but certain nodes are missing. Only the nodes such that $h + k = 2n$ (where n is an integer) are present. There is a systematic absence of node if $h + k$ is odd.

◻ ANALYSIS

We have shown that the base vectors of the reciprocal lattice are contravariant with those of the direct lattice. If $(\mathbf{A})$ and $(\mathbf{A}^*)$ denote the transformation matrices for the base vectors, we have:

$$(\mathbf{A}^*) = (\mathbf{A}^T)^{-1}$$

For any C type lattice, we can write:

$$\begin{pmatrix} \mathbf{a}_2 \\ \mathbf{b}_2 \\ \mathbf{c}_2 \end{pmatrix} = \begin{pmatrix} 1 & \bar{1} & 0 \\ 1 & 1 & 0 \\ 0 & 0 & 1 \end{pmatrix} \cdot \begin{pmatrix} \mathbf{a}_1 \\ \mathbf{b}_1 \\ \mathbf{c}_1 \end{pmatrix}$$

After inverting the transpose of matrix $(\mathbf{A})$, we have:

$$\begin{pmatrix} \mathbf{A}_2^* \\ \mathbf{B}_2^* \\ \mathbf{C}_2^* \end{pmatrix} = \begin{pmatrix} 1/2 & \overline{1/2} & 0 \\ 1/2 & 1/2 & 0 \\ 0 & 0 & 1 \end{pmatrix} \cdot \begin{pmatrix} \mathbf{A}_1^* \\ \mathbf{B}_1^* \\ \mathbf{C}_1^* \end{pmatrix}$$

We can therefore write:

$$\mathbf{A}_2^* = 1/2(\mathbf{A}_1^* - \mathbf{B}_1^*) \qquad \mathbf{B}_2^* = 1/2(\mathbf{A}_1^* + \mathbf{B}_1^*) \qquad \mathbf{C}_2^* = \mathbf{C}_1^*$$

For the primitive unit-cell, the nodes of the reciprocal lattice are such that:

$$\mathbf{R}_1^* = h.\mathbf{A}_1^* + k.\mathbf{B}_1^* + l.\mathbf{C}_1^*$$

Consider the vector: $\mathbf{R}_2^* = h'.\mathbf{A}_2^* + k'.\mathbf{B}_2^* + l'.\mathbf{C}_2^*$

We can also write this: $\mathbf{R}_2^* = \dfrac{1}{2}(h' + k').\mathbf{A}_1^* + \dfrac{1}{2}(k' - h').\mathbf{B}_1^* + l'.\mathbf{C}_1^*$

In this second coordinate frame, the points defined by $\mathbf{R}_2^*$ are *nodes* if the coefficients and $\mathbf{A}_1^*$, $\mathbf{B}_1^*$ and $\mathbf{C}_1^*$ are *integers*, i.e. if $h' + k'$ is even.

◻ RECIPROCAL LATTICES OF F AND I LATTICES

Within the F type unit-cell we can define a primitive unit-cell (figure 6.4) characterised by the base vectors:

$$\mathbf{a}_1 = \frac{1}{2}(\mathbf{b}_2 + \mathbf{c}_2) \qquad \mathbf{b}_1 = \frac{1}{2}(\mathbf{a}_2 + \mathbf{c}_2) \qquad \mathbf{c}_1 = \frac{1}{2}(\mathbf{a}_2 + \mathbf{b}_2)$$

$$\mathbf{a}_2 = -\mathbf{a}_1 + \mathbf{b}_1 + \mathbf{c}_1 \qquad \mathbf{b}_2 = \mathbf{a}_1 - \mathbf{b}_1 + \mathbf{c}_1 \qquad \mathbf{c}_2 = \mathbf{a}_1 + \mathbf{b}_1 - \mathbf{c}_1$$

The matrix form of these relations is:

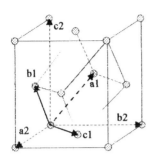

$$\begin{pmatrix} \mathbf{a}_2 \\ \mathbf{b}_2 \\ \mathbf{c}_2 \end{pmatrix} = \begin{pmatrix} \bar{1} & 1 & 1 \\ 1 & \bar{1} & 1 \\ 1 & 1 & \bar{1} \end{pmatrix} \cdot \begin{pmatrix} \mathbf{a}_1 \\ \mathbf{b}_1 \\ \mathbf{c}_1 \end{pmatrix}$$

For the reciprocal lattices we can therefore write:

$$\begin{pmatrix} \mathbf{A}_2^* \\ \mathbf{B}_2^* \\ \mathbf{C}_2^* \end{pmatrix} = \begin{pmatrix} 0 & 1/2 & 1/2 \\ 1/2 & 0 & 1/2 \\ 1/2 & 1/2 & 0 \end{pmatrix} \cdot \begin{pmatrix} \mathbf{A}_1^* \\ \mathbf{B}_1^* \\ \mathbf{C}_1^* \end{pmatrix}$$

Figure 6.4

We write: $\mathbf{R}_2^* = h'.\mathbf{A}_2^* + k'.\mathbf{B}_2^* + l'.\mathbf{C}_2^*$

i.e.: $\mathbf{R}_2^* = \frac{1}{2}(k' + l').\mathbf{A}_1^* + \frac{1}{2}(h' + l').\mathbf{B}_1^* + \frac{1}{2}(h' + k').\mathbf{C}_1^*$

In the multiple unit-cell, the points defined by $\mathbf{R}_2^*$ are nodes if the coefficients of $\mathbf{A}_1^*$, $\mathbf{B}_1^*$ and $\mathbf{C}_1^*$ are integers, i.e. if h', k' and l' all have the same parity. We can see by construction that a lattice whose nodes all satisfy this condition

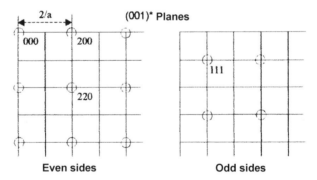

Figure 6.5

is a type I lattice with parameter $A^* = 2/a$.

A similar analysis can be made of a direct type I lattice; it is easier in fact to note that the reciprocal of the reciprocal is identical to the original. The reciprocal lattice of an I lattice is thus an F type lattice (only reciprocal nodes such as $h + k + l = 2.n$ exist in the lattice).

The reciprocal lattice of an F lattice is an I lattice and vice-versa.

5 GEOMETRICAL RELATIONSHIPS IN THE LATTICES

5.1 The Triclinic System

The triclinic lattice has the most general unit-cell possible:

$$a \neq b \neq c, \alpha \neq \beta \neq \gamma \neq \pi/2$$

The lattice plane spacing of a family $(h\ k\ l)$, d_{hkl}, is equal to the reciprocal of the modulus of the reciprocal vector N^*_{hkl}. In the general case we can thus write:

$$\frac{1}{d^2_{hkl}} = h^2.A^{*2} + k^2.B^{*2} + l^2.C^{*2} + 2.h.k.A^*.B^* + 2.k.l.B^*.C^* + 2.l.h.C^*.A^*$$

In all the other systems, this general formula for the interplanar spacing of lattice planes can be simplified. For the triclinic system, the reciprocal values used should be obtained from the following general relations, established in chapter 2 on lattices:

$$\cos \alpha^* = \frac{\cos \gamma . \cos \beta - \cos \alpha}{\sin \gamma . \sin \beta}, \cos \beta^* = \frac{\cos \alpha . \cos \gamma - \cos \beta}{\sin \alpha . \sin \gamma}, \cos \gamma^* = \frac{\cos \alpha . \cos \beta - \cos \gamma}{\sin \alpha . \sin \beta}$$

$$A^* = \frac{1}{a . \sin \beta . \sin \gamma^*}, \quad B^* = \frac{1}{b . \sin \alpha^* . \sin \gamma}, \quad C^* = \frac{1}{c . \sin \alpha . \sin \beta^*}$$

$$V = a.b.c. \sin \alpha . \sin \beta . \sin \gamma^* = a.b.c. \sin \alpha^* . \sin \beta . \sin \gamma = a.b.c. \sin \alpha . \sin \beta^* . \sin \gamma$$

5.2 The Monoclinic System

Monoclinic lattices have the following unit-cell characteristics:

$$a \neq b \neq c, \alpha = \gamma = \pi/2, \beta > \pi/2$$

For this system we therefore have:

$$A^* = \frac{1}{a . \sin \beta}, \quad C^* = \frac{1}{c . \sin \beta}, \quad B^* = \frac{1}{b}$$

$$\beta^* = \pi - \beta, \cos \beta^* = -\cos \beta, \alpha^* = \gamma^* = \frac{\pi}{2}$$

From this can be deduced the following expression for the lattice interplanar spacing in terms of the direct lattice parameters:

$$d_{hkl} = \frac{\sin \beta}{\sqrt{\dfrac{h^2}{a^2} + \dfrac{l^2}{c^2} + \dfrac{k^2 . \sin^2 \beta}{b^2} - \dfrac{2.h.l.\cos\beta}{a.c}}}$$

The volume of the unit-cell is: $V = a.b.c. \sin \beta$

5.3 The Orthorhombic System

For orthorhombic lattices, the unit-cell is defined by:

$$a \neq b \neq c, \ \alpha = \beta = \gamma = \pi/2$$

$$A^* = 1/a, \ B^* = 1/b, \ C^* = 1/c, \ \alpha^* = \beta^* = \gamma^* = \pi/2$$

Here, calculation of the lattice spacing is simple, giving:

$$d_{hkl} = \frac{1}{\sqrt{\dfrac{h^2}{a^2} + \dfrac{k^2}{b^2} + \dfrac{l^2}{c^2}}}$$

The volume of the orthorhombic unit-cell is: $V = a.b.c.$

5.4 Hexagonal and Rhombohedral Lattices

The hexagonal lattice P (unit-cell $a = b \neq c, \ \alpha = \beta = \pi/2.\gamma = 2\pi/3$) is compatible with all the trigonal and hexagonal groups. On the other hand, the rhombohedral unit-cell R ($a = b = c, \ \alpha = \beta = \gamma \neq \pi/2$) is only compatible with the five trigonal groups. As calculations are generally trickier to carry out for a rhombohedral cell R than for a hexagonal cell P, rhombohedral structures are often represented with a multiple hexagonal unit-cell.

❏   RELATIONS BETWEEN THE R AND P LATTICES

It is possible to construct a multiple hexagonal cell P which contains a single rhombohedral cell R.

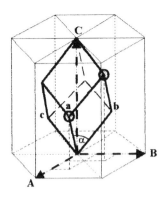 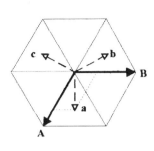

Figure 6.6

Inside the P lattice (figure 6.6 left), in the prism defined by the vectors **A**, **B** and **C**, we add two nodes with fractional coordinates (2/3, 1/3, 1/3) and (1/3, 2/3, 2/3). From these nodes we can define the cell R characterised by the base vectors **a**, **b** and **c**. Figure 6.6 right is a projection of the base vectors of the two lattices onto a plane normal to the threefold axis.

The matrices for passing from one coordinate system to the other are thus as follows:

Hexagonal ⇒ Rhombohedral | *Rhombohedral ⇒ Hexagonal*

$$\begin{pmatrix} \mathbf{a} \\ \mathbf{b} \\ \mathbf{c} \end{pmatrix} = \frac{1}{3} \begin{pmatrix} 2 & 1 & 1 \\ \bar{1} & 1 & 1 \\ \bar{1} & \bar{2} & 1 \end{pmatrix} \cdot \begin{pmatrix} \mathbf{A} \\ \mathbf{B} \\ \mathbf{C} \end{pmatrix} \qquad \begin{pmatrix} \mathbf{A} \\ \mathbf{B} \\ \mathbf{C} \end{pmatrix} = \begin{pmatrix} 1 & \bar{1} & 0 \\ 0 & 1 & \bar{1} \\ 1 & 1 & 1 \end{pmatrix} \cdot \begin{pmatrix} \mathbf{a} \\ \mathbf{b} \\ \mathbf{c} \end{pmatrix}$$

From these matrices, the relations between the cell parameters can be deduced:

$$a = \frac{1}{3} \sqrt{3.A^2 + C^2} \qquad\qquad A = 2.a.\sin\frac{\alpha}{2}$$

$$\sin\frac{\alpha}{2} = \frac{3.A}{2\sqrt{3.A^2 + C^2}} \qquad\qquad C = a.\sqrt{3 + 6\cos\alpha}$$

All that is necessary to express the Miller indices of a family of planes or of a direct row in the two frames, is to make use of the covariance of the lattice plane indices and the contravariance of the row indices.

❑ GEOMETRICAL RELATIONS

TRIGONAL LATTICE:

Two parameters determine the cell geometry:

$$a = b = c, \ \alpha = \beta = \gamma \neq \pi/2$$

For the trigonal lattice, the parameters of the reciprocal lattice are:

$$A^* = \frac{1}{a. \sin \alpha. \sin \alpha^*}$$

$$\cos \alpha^* = \frac{\cos^2 \alpha - \cos \alpha}{\sin^2 \alpha}$$

The expression for d_{hkl} is thus:

$$d_{hkl} = \frac{a.(1 - 3. \cos^2 \alpha + 2. \cos^3 \alpha)}{\sqrt{(h^2 + k^2 + l^2). \sin^2 \alpha + 2.(h.k + k.l + l.h).(\cos^2 \alpha - \cos \alpha)}}$$

The volume of the rhombohedral cell is: $V = a^3.(1 - 3. \cos^2 \alpha + 2. \cos^3 \alpha)^{1/2}$.

HEXAGONAL LATTICE

The geometry of the unit-cell again depends on two parameters:

$$a = b \neq c, \ \alpha = \beta = \pi/2, \ \gamma = 2\pi/3$$

$$A^* = B^* = \frac{2}{a\sqrt{3}}, \quad C^* = \frac{1}{c}, \quad \alpha^* = \beta^* = \frac{\pi}{2}, \quad \gamma^* = \frac{\pi}{3}$$

The expression for d_{hkl} is much simpler than for the rhombohedral cell:

$$d_{hkd} = \frac{a}{\sqrt{\frac{4}{3}(h^2 + k^2 + hk) + l^2.(a/c)^2}}$$

The volume of the hexagonal cell is: $V = \frac{\sqrt{3}}{2}.a^2.c$.

❑ BRAVAIS–MILLER INDICES

R and P lattices present an additional problem: in all the other lattices, it is easy to determine the indices of equivalent faces by considering the symmetry operations of the class. (In lattices where the principal axis is oriented along Oz, a mirror plane (001) changes l into $-l$, a twofold axis [100] changes k into $-k$ and l into $-l$, in the cubic classes the threefold axis [111] leads to circular permutation of the 3 indices h, k and l, and so on.)

In a hexagonal unit-cell, a system with four indices is required to denote the Miller indices of the faces.

In the plane (100) we take a fourth axis with coordinates (figure 6.7):

$$\mathbf{d} = -(\mathbf{a} + \mathbf{b})$$

One face is then redundantly denoted ($h\ k\ j\ l$). To establish the relation between the four indices, consider a plane with the classical notation ($h\ k\ 0$).

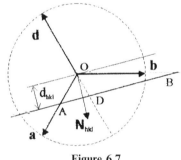

Figure 6.7

This plane intersects the axes of the (001) plane at A, B and D; d_{hkl}, which is invariant in the change of coordinate frame, is equal to the projection of the vector **OA** or the vector **OD** onto the unit vector normal to the plane.

$$d_{hkl} \cdot \| \mathbf{N}^*_{hkl} \| = 1$$

And considering $\mathbf{a}.\mathbf{A}^* = 1$, $\mathbf{a}.\mathbf{B}^* = 0$ . . . we have:

$$d_{hkl} = \frac{(h\mathbf{A}^* + k\mathbf{B}^* + l\mathbf{C}^*)\,\mathbf{a}}{N^*_{hkl}}\frac{}{h} = \frac{(h\mathbf{A}^* + k\mathbf{B}^* + l\mathbf{C}^*)\,\mathbf{d}}{N^*_{hkl}}\frac{}{j} = \frac{(h\mathbf{A}^* + k\mathbf{B}^* + l\mathbf{C}^*)}{N^*_{hkl}}\frac{-(\mathbf{a}+\mathbf{b})}{j}$$

$$j = -(h + k)$$

NOTE: The calculation is proposed as an exercise; in fact the relation is obvious given the covariant character of the Miller indices of the faces.

With this choice of indices, equivalent faces are deduced from each other by a circular permutation over the *first three indices*.

NOTATION CONVENTION: Since the index j is a linear combination of h and k, it is omitted from the notation of the face and replaced by a full stop. A $(11\bar{2}1)$ face will be denoted (11.1).

EXAMPLE: The table below indicates the three types of notation possible in the hexagonal system for a (100) face. In appendix A will be found the notations of all the special hexagonal forms in $(hk.l)$ notation.

$(h\,k\,l)$	(100)	(010)	$(\bar{1}10)$	$(\bar{1}00)$	$(0\bar{1}0)$	$(1\bar{1}0)$
$(h\,k\,j\,l)$	$(10\bar{1}0)$	$(01\bar{1}0)$	$(\bar{1}110)$	$(\bar{1}010)$	$(0\bar{1}10)$	$(1\bar{1}00)$
$(h\,k\,.\,l)$	(10.0)	(01.0)	$(\bar{1}1.0)$	$(\bar{1}0.0)$	$(0\bar{1}.0)$	$(1\bar{1}.0)$

5.5 The Tetragonal System

For tetragonal lattices the unit-cell is defined by:

$$a = b \neq c,\ \alpha = \beta = \gamma = \pi/2$$

$$A^* = B^* = 1/a,\ C^* = 1/c,\ \alpha^* = \beta^* = \gamma^* = \pi/2$$

From which we deduce: $d_{hkl} = \dfrac{a}{\sqrt{h^2 + k^2 + l^2.(a/c)^2}}$ and $V = a^2.c$.

5.6 The Cubic System

❏ GEOMETRICAL RELATIONSHIPS

In cubic lattices the geometry only depends on one parameter:

$$a = b = c,\ \alpha = \beta = \gamma = \pi/2$$

$$A^* = B^* = C^* = 1/a,\ \alpha^* = \beta^* = \gamma^* = \pi/2$$

We obtain: $d_{hkl} = \dfrac{a}{\sqrt{h^2 + k^2 + l^2}}$ and $V = a^3$

In the cubic system the reciprocal lattice is homothetic with the direct lattice. Now, the reciprocal row $[h\,k\,l]^*$ is normal to the direct planes $(h\,k\,l)$, and so:
   In the direct lattice any direct row $[u\,v\,w]$ is normal to the planes of the direct lattice $(u\,v\,w)$. In the reciprocal lattice any reciprocal row $[h\,k\,l]^*$ is normal to the reciprocal planes $(h\,k\,l)^*$.

In the other systems this property does not in general hold. Only rows parallel to the symmetry axes are normal to the lattice planes (having low indices and hence high node densities).

❏ THE FACE CENTRED CUBIC LATTICE

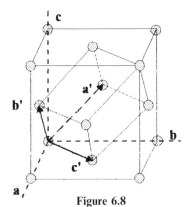

Figure 6.8

The unit-cell of a face centred cubic lattice (figure 6.8) is a rhombohedron with:

$$a_R = a.\frac{\sqrt{2}}{2}, \ \alpha = 60°$$

(The rhombohedral angle is the angle between the cubic rows [110] and [101].) The base vectors of the rhombohedral cell are obtained by joining a corner of the cube to the face centres.

$$\mathbf{a}' = \frac{1}{2}(\mathbf{b} + \mathbf{c})$$

$$\mathbf{b}' = \frac{1}{2}(\mathbf{a} + \mathbf{c})$$

$$\mathbf{c}' = \frac{1}{2}(\mathbf{a} + \mathbf{b})$$

❏ THE BODY CENTRED CUBE

The unit-cell of a body centred cubic lattice (figure 6.9 left) is a rhombohedron with:

$$a_R = a\frac{\sqrt{3}}{2}, \ \alpha = 109°28'$$

(The rhombohedral angle is the angle between the threefold axes of the cube).

The base vectors of the rhombohedral cell are obtained by joining a corner of the cube to the centres of adjacent cubes.

$$\mathbf{a}' = \frac{1}{2}(-\mathbf{a} + \mathbf{b} + \mathbf{c}) \qquad \mathbf{b}' = \frac{1}{2}(\mathbf{a} - \mathbf{b} + \mathbf{c}) \qquad \mathbf{c}' = \frac{1}{2}(\mathbf{a} + \mathbf{b} - \mathbf{c})$$

The unit-cell is obtained by completing the rhombohedron (figure 6.9 right).

NOTE: The presence of species at $(0, 0, 0)$ and $(1/2, 1/2, 1/2)$ in cell in a given structure does not imply that its lattice is type I. The two species must be *identical* and have the *same environment*.

Thus the CsCl lattice is primitive. A lattice is type I if, to any object with coordinates (x, y, z), there corresponds an identical object with coordinates $(x+\frac{1}{2}, y+\frac{1}{2}, z+\frac{1}{2})$, i.e. if the vector $\mathbf{T} = \frac{1}{2}(\mathbf{a}+\mathbf{b}+\mathbf{c})$ is a lattice translation.

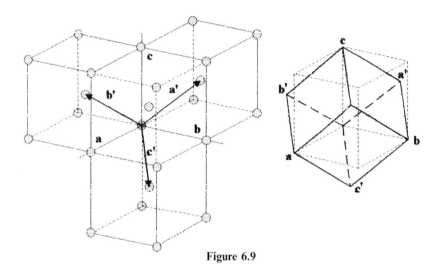

Figure 6.9

6   POINT GROUP RELATIONSHIPS

Relationships can be established between one class and those which are derived from it by the loss of one or more symmetry elements. The derived classes are subgroups of the initial class. This relationship is obvious when it is between a holohedral class and the merihedral classes of a system. Loss of a symmetry element may also involve a change of crystal system in the resulting group. The relationships between the holohedral classes of the various systems are as follows:

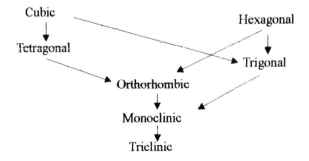

EXAMPLES OF RELATIONSHIPS BETWEEN GROUPS:

m3m	$\Rightarrow$	$\bar{3}$m	(4 A_3 → single A_3)
m3m	$\Rightarrow$	4/mmm	(loss of threefold axes)
4/mmm	$\Rightarrow$	mmm	(A_4 → A_2)
mmm	$\Rightarrow$	mm2	(loss of centre of symmetry)
mm2	$\Rightarrow$	m	(loss of twofold axis)
6/mmm	$\Rightarrow$	6mm	(loss of centre of symmetry)
6/mmm	$\Rightarrow$	mmm	(A_6 → A_2)

The table of relationships is reproduced on page 357.

Chapter 7

Space Groups

Point groups enable us to describe the macroscopic symmetry of crystals. To study microscopic symmetry we have to use the Schönflies–Fédorov postulate and the resulting description of symmetry through space groups. The 230 space groups were first enumerated by Fédorov in 1895, and later independently by Schönflies.

1 THE SPACE GROUP OF A CRYSTAL

The most general type of symmetry operation in a lattice, where we pass from a point to another equivalent point, can be described as the product of a proper or improper point symmetry operation R and a translation $\mathbf{t}$. This general operation is denoted $(R, \mathbf{t})$. We recall[1] that the action of this operation on a vector $\mathbf{X}$ is:

$$\mathbf{Y} = (R, \mathbf{t}).\mathbf{X} = R.\mathbf{X} + \mathbf{t}$$

A pure translation is denoted $(E, \mathbf{t})$ and a pure rotation by $(R, 0)$

The space group of a crystal is the set $G_E = \{(R, \mathbf{t})\}$ of symmetry operations which transform any point in the crystal to an equivalent point.

1.1 Group Properties

◆ There exists a law of internal composition: $(R', \mathbf{t}').(R, \mathbf{t})$

$$(R', \mathbf{t}').(R, \mathbf{t}).\mathbf{X} = (R', \mathbf{t}').(R.\mathbf{X} + \mathbf{t}) = R'(R.\mathbf{X} + \mathbf{t}) + \mathbf{t}' = R'.R.\mathbf{X} + R'.\mathbf{t} + \mathbf{t}'$$

$$(R', \mathbf{t}').(R, \mathbf{t}) = (R'R, R'.\mathbf{t} + \mathbf{t}')$$

---

[1]Cf. § 1.7 of chapter 4 on lattice symmetry elements.

◆ The law is associative.
◆ There exists a neutral element $(E, \mathbf{0})$.
◆ There exists an inverse for all elements:
Suppose that $(R', t') = (R, t)^{-1}$ hence: $(R', t').(R, t) = (E, 0) = (R'.R, R'.t + t')$
The inverse exists if:

$$R'.R = E \text{ or } R' = R^{-1} \text{ and if: } R'.t + t' = 0 \text{ or } t' = -R'.t = -R^{-1}.t$$
$$(R, t)^{-1} = (R^{-1}, -R^{-1}.t)$$

The set $G_E = \{(R, t)\}$ is an infinite non-commutative group.

1.2   Associated Point Group

If $G_E = \{(R, t)\}$ is a space group, the associated point group is the subgroup of G_E:

$$G_P = \{(R, \mathbf{0})\}$$

1.3   Crystal Space Groups

The periodicity of the crystal lattice imposes restrictions on the symmetry operations allowed in the space groups.

❏   ROTATION RESTRICTIONS

If the space group is that of a crystal, the associated point group is a crystal point group which therefore contains only axes of order 1, 2, 3, 4 or 6.

☞ | *The only rotations possible in crystal space groups are those of order 1, 2, 3, 4 or 6.*

❏   TRANSLATION RESTRICTIONS

Consider (R, t), an operation of G_E.

$$(R, t)^2 = (R^2, R.t + t)$$
$$(R, t)^3 = (R^3, R^2.t + R.t + t)$$
$$(R, t)^m = (R^m, R^{m-1}.t + R^{m-2}.t + \ldots + R.t + E.t) = (R^m, [R].t)$$

$$\boxed{[R] = R^{m-1} + R^{m-2} + \ldots + R + E}$$

If the axis is n-fold, then $(R, t)^n = (E, [R].t)$ must be a lattice translation $\mathbf{T}$.

☞ | *The translations* **t** *in crystal space group operations* (**R**, **t**) *must satisfy the condition:* [R].t=T.

T is necessarily a lattice translation, whereas **t** is not necessarily one.

2   SPACE GROUP SYMMETRY ELEMENTS

We must add to the orientation symmetry operations for point groups, those operations which are the product of orientation and translation. We have already studied the general laws of composition of rotation–translation products, with the following results:

Consider a symmetry operation characterised in a given coordinate system by (**R**, **t**) and a new coordinate system defined by a translation vector **S**, with the origin of the original coordinate system. In the new system, the operation is characterised by:

$$(R', t') = (R, R.S - S + t)$$

If the translation part of the operations can be eliminated by suitable choice of the vector **S**, we return to an orientation symmetry operation or else we induce a new symmetry operation.

◆ The product of a proper rotation of an axis **u** and a translation **t** parallel to **u** is a *screw axis*.
◆ The product of a proper rotation of an axis **u** and a translation **t** perpendicular to **u** is a proper rotation through the same angle about an axis lying on the bisector of **t**.
◆ The product of an improper rotation of an axis **u** and a translation **t** parallel to **u** is an improper rotation whose centre of inversion is translated along the axis of the vector **t**/2.
◆ The product of a mirror plane and a translation **t** parallel to the mirror plane is a *glide plane*.

When enumerating the crystal space groups, we must also take into account the constraints set by the relation [R].t = T, imposed by the lattice periodicity.

3   SCREW AXES IN THE CRYSTAL SPACE GROUPS

❑ ALLOWED TRANSLATIONS
For an *n*-fold axis, we must have: $[R].t = T$ where $[R] = \Sigma_{p=1}^{n} R^p$, $R^n = E$
Consider an *n*-fold symmetry axis parallel to O*z*, characterised by a lattice translation of vector **c**. [R] is the sum of matrices R^P

$$R^p = \begin{pmatrix} \cos\frac{2\pi p}{n} & -\sin\frac{2\pi p}{n} & 0 \\ \sin\frac{2\pi p}{n} & \cos\frac{2\pi p}{n} & 0 \\ 0 & 0 & 1 \end{pmatrix}$$

We write $S = \sum_{p=1}^{n} e^{2j\pi p/n} = e^{2j\pi/n} + e^{4j\pi/n} + \ldots + e^{2j\pi n/n}$

Now: $S.e^{2j\pi/n} = e^{4j\pi/n} + e^{6j\pi/n} + \ldots + e^{2j\pi n/n} + e^{2j\pi/n} = S$

Hence: $S.e^{2j\pi/n} = S$, $S.(e^{2j\pi/n} - 1) = 0 \Rightarrow S = 0$ (if $n \neq 1$)

A similar calculation with $S^{\dagger}$, the complex conjugate of S, leads to $S^{\dagger} = 0$. In the matrix representing $[R]$, the sums of the n cosines and n sines are null:

$$[R] = \begin{pmatrix} 0 & 0 & 0 \\ 0 & 0 & 0 \\ 0 & 0 & n \end{pmatrix}$$

The translation $\mathbf{t}$ is a vector parallel to the rotation axis and $\mathbf{T}$ is a lattice translation equal to $m.\mathbf{c}$ (where m is an integer). The relation $[R].\mathbf{t} = \mathbf{T}$ becomes:

$$[R].\mathbf{t} = \begin{pmatrix} 0 & 0 & 0 \\ 0 & 0 & 0 \\ 0 & 0 & n \end{pmatrix} . \begin{pmatrix} 0 \\ 0 \\ t \end{pmatrix} = \begin{pmatrix} 0 \\ 0 \\ m.c \end{pmatrix} \Rightarrow nt = mc$$

The possible values of the vectors $\mathbf{t}$ are such that:

$$\boxed{\mathbf{t} = \frac{m}{n}\mathbf{c}}$$

We can set $m < n$ since, by hypothesis, two whole lattice translations cannot be distinguished.

We can now enumerate the screw axes compatible with the symmetry properties of the lattice. According to international conventions, the principal axes of the group being studied are shown perpendicular to the plane of projection, which is by convention the (001) plane. For two-fold axes, axes parallel to the plane of projection, are also shown. The graphic symbols used to represent the rotation axes in these projections are those in the International Tables.

❐ TWO-FOLD AXES
$n = 2 \Rightarrow m = 0, 1.$
$m = 0 \Rightarrow$ 'normal' two-fold axis   Symbol **2**
$m = 1 \Rightarrow$ Two-fold 'screw' axis   Symbol **2₁**

Representation of axes parallel to (001).

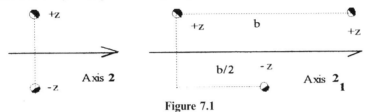

Figure 7.1

The axes are in the plane (001) and thus of height 0. Since the axis 2_1 is oriented along Oy, the translation **t** is **b**/2.

Representation of axes perpendicular to (001)

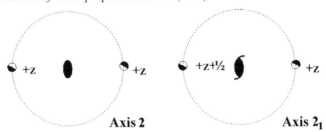

Figure 7.2

The axes are $\perp$ to the (001) plane. Since the 2_1 axis is oriented along Oz, the translation **t** is **c**/2.

❐ THREE-FOLD AXES
$n = 3 \quad \Rightarrow \quad m = 0, 1, 2.$
$m = 0 \quad \Rightarrow$ 'Normal' three-fold axes   Symbol **3**
$m = 1, 2 \Rightarrow$ Three-fold 'screw' axes   Symbols **3₁, 3₂**

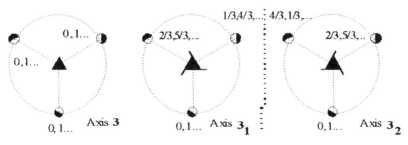

Figure 7.3

For the 3_1 axis, successive heights of equivalent positions are: 0, 1/3, 2/3, 1, $4/3 \equiv 1/3$... (Two positions which only differ by whole lattice translations are indistinguishable.) For the 3_2 axis the heights are: 0, 2/3, $4/3 \equiv 1/3$, $2 \equiv 1$...

The 3_2 axis is the image of a 3_1 axis in a mirror parallel with the rotation axes. A 3_1 axis corresponds to a clockwise rotation and a 3_2 axis to an anticlockwise rotation. The two axes are said to be *enantiomorphs*.

◻ FOUR-FOLD AXES

$n = 4$ $\Rightarrow m = 0, 1, 2, 3$.

$m = 0$ $\Rightarrow$ 'Normal' four-fold axes   Symbol **4**

$m = 1, 2$ $\Rightarrow$ Four-fold 'screw' axes   Symbols 4_1, 4_2, 4_3.

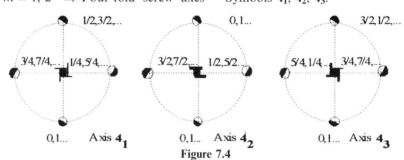

Figure 7.4

The 4_1 axis corresponds to a clockwise rotation and the 4_3 to an anticlockwise rotation. These two axes are enantiomorphs.

◻ SIX-FOLD AXES

$n = 6$ $\Rightarrow m = 0, 1, 2, 3, 4, 5$

$m = 0$ $\Rightarrow$ 'Normal' six-fold axes   Symbol **6**

$m = 1 \ldots 5 \Rightarrow$ six-fold 'screw' axes   Symbols $6_1 \ldots 6_5$.

The graphic symbols for these axes are as follows:

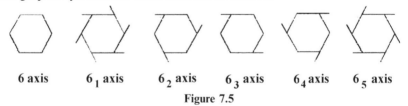

6 axis 6_1 **axis** 6_2 **axis** 6_3 **axis** 6_4 **axis** 6_5 **axis**

Figure 7.5

Axes 6_1 and 6_5 are enantiomorphic, as are axes 6_2 and 6_4.

4   GLIDE PLANES

◻ ALLOWED TRANSLATIONS

In the case of mirror planes, a translation *parallel to the plane of the mirror* causes a glide plane to appear.

Let a_1, a_2 and a_3 be the lattice vectors which define the given symmetry plane. The condition for the existence of the translational symmetry element is written:

$$[R].t = T = n_1 a_1 + n_2 a_2 + n_3 a_3$$

For a mirror plane we have: $[R] = \sigma + E$. Suppose the mirror plane is (001); the matrix representation of $[R]$ is then:

$$[R] = \begin{pmatrix} 1 & 0 & 0 \\ 0 & 1 & 0 \\ 0 & 0 & \bar{1} \end{pmatrix} + \begin{pmatrix} 1 & 0 & 0 \\ 0 & 1 & 0 \\ 0 & 0 & 1 \end{pmatrix} = \begin{pmatrix} 2 & 0 & 0 \\ 0 & 2 & 0 \\ 0 & 0 & 0 \end{pmatrix}$$

Now, $t = \alpha.a_1 + \beta.a_2$ (vectors parallel to the (001) plane).

$$[R].t = \begin{pmatrix} 2 & 0 & 0 \\ 0 & 2 & 0 \\ 0 & 0 & 0 \end{pmatrix}.\begin{pmatrix} \alpha.a_1 \\ \beta.a_2 \\ 0 \end{pmatrix} = \begin{pmatrix} 2\alpha.a_1 \\ 2\beta.a_2 \\ 0 \end{pmatrix} = \begin{pmatrix} n_1.a_1 \\ n_2.a_2 \\ 0 \end{pmatrix} \Rightarrow 2\alpha = n_1, \ 2\beta = n_2$$

We can impose the conditions $0 \leqslant (n_1, n_2) < 1$ since whole lattice translations are indistinguishable. The translations possible for mirror glide planes parallel to (001) are thus:

$$\boxed{t = 0, \ t = \frac{a_1}{2}, \ t = \frac{a_2}{2}, \ t = \frac{a_1 + a_2}{2}}$$

This reasoning must be repeated for all mirror plane orientations allowed by point symmetry. There must also be taken into account non-integral lattice translations allowed in certain Bravais types. In particular, for orthorhombic lattices F, tetragonal I and cubic F and I, the so-called 'diamond' translations must be considered; these are equal to:

$$t_F^1 = \frac{1}{4}(a+b), \ t_F^2 = \frac{1}{4}(c+b), \ t_F^3 = \frac{1}{4}(a+c) \text{ and } t_I = \frac{1}{4}(a+b+c)$$

The various types of mirror planes possible are shown in the following table:

Conventions for showing glide planes

Notation	Type	Representation ⊥ to plane of projection	Representation // to plane of projection
m	Normal mirror plane	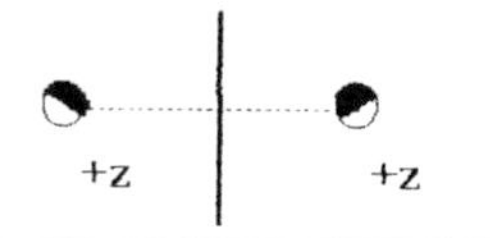 +z +z	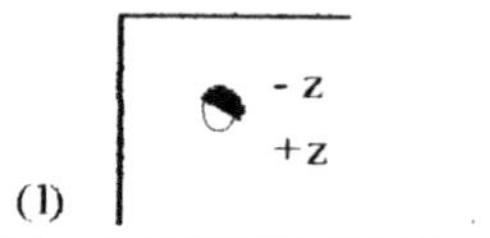 - z / + z (1)
a , b	Glide plane //	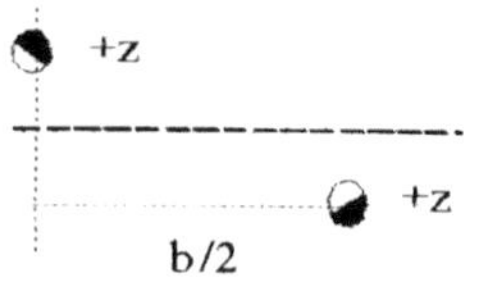 +z +z b/2	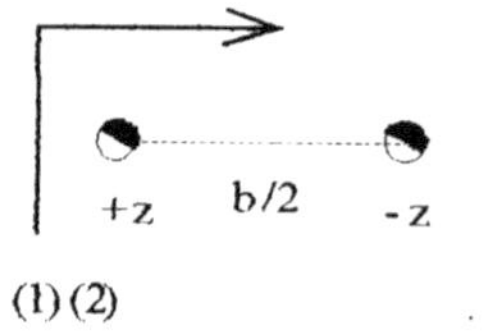 +z b/2 - z (1)(2)
c	Glide plane ⊥	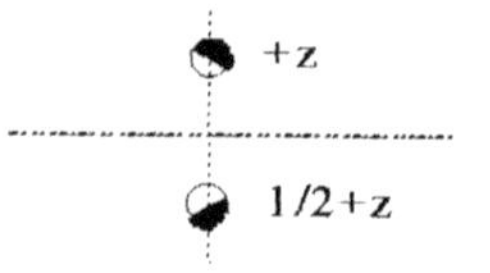 +z 1/2+z	
n	Diagonal glide plane	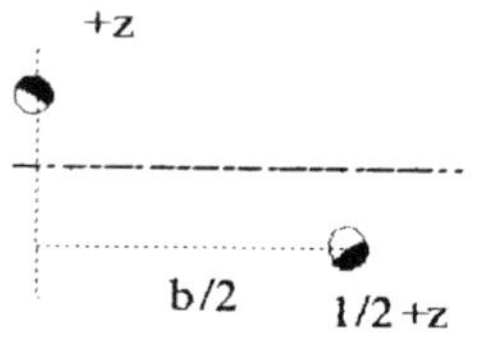 +z b/2 1/2 +z	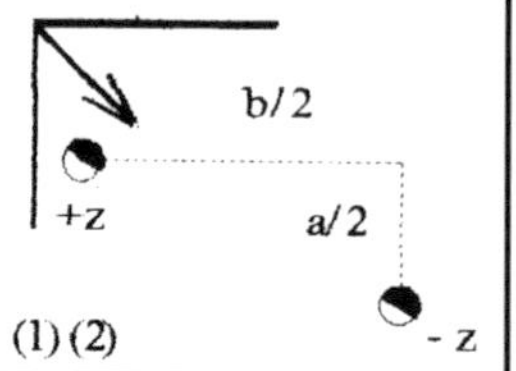 b/2 +z a/2 - z (1)(2)
d	Diamond glide plane	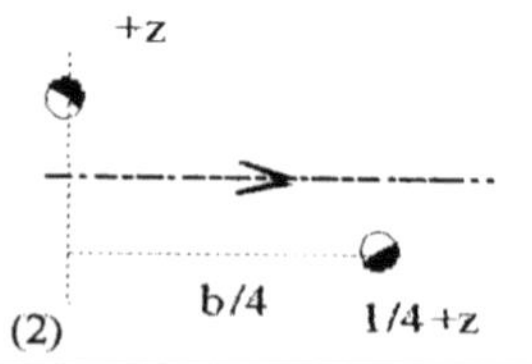 +z b/4 1/4 +z (2)	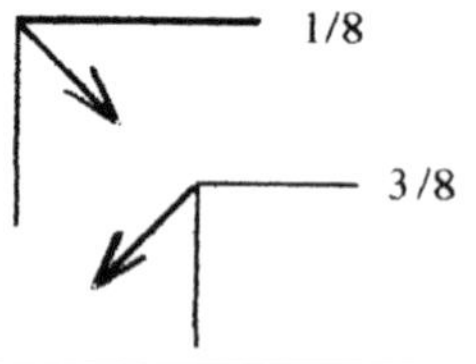 1/8 3/8

Figure 7.6

(1) It is assumed that the plane of the mirror lies at a height 0.

(2) The arrow indicates the direction of translation.

The mirror planes **a** and **b** are shown by dashed lines, **c** by dotted lines, **n** by dash-dot lines and **d** by dash-dot lines with arrow. Heights of mirror planes parallel to (001) are only included on the projections if they are other than zero.

In the latest edition of the International Tables, further specific symbols are found for oblique mirror planes and cubic groups.

5 SPACE GROUP NOTATION

We shall identify space groups using the international (Hermann-Mauguin) symbols, whose meaning is much clearer than those of Schönflies, which consists of adding a number of arbitrary size to the name of the point group from which the given space group is derived (e.g. (C_{2v}^{20}, D_{6h}^3, D_{2d}^5...)etc.).

◆ The name of the space group is prefixed with a capital letter (lower case letters are reserved for the 17 plane groups) indicating the type of lattice:

$$P, A, B, C, I, F, R.$$

◆ In the name of the point group, the symbols 2, 3, 4, 6 and m may be replaced by symbols corresponding to symmetry operations of translation existing in the given space group.

Symbol in the crystal class	Symbols in the space group
2	2, 2_1
3	3, 3_1, 3_2
4	4, 4_1, 4_2, 4_3
6	6, 6_1, 6_2, 6_3, 6_4, 6_5
m	m, a, b, c, n, d

Conversely, the symbol for the crystal class is derived from that of the space group by eliminating references to the translational parts of symmetry elements. The letter characterising the lattice is eliminated, screw axes are replaced by rotation axes and glide planes are replaced by mirror planes (m).

EXAMPLES:

Group **Pnma** or D_{2h}^{16}

    The class is orthorhombic mmm; the lattice is primitive.
    The mirror plane (100) normal to Ox is type $n : t = \frac{1}{2}$ (**b**+**c**).
    The mirror plane (010) normal to Oy is an ordinary mirror plane.
    The mirror plane (001) normal to Oz is type $a : t = \frac{1}{2}$ **a**.

Group **I$\frac{4_1}{a}$md** or D_{4h}^{19}

    The class is tetragonal **4/mmm**; the lattice is body-centred.
    The four-fold (quadratic) axis is type $4_1 : t = \frac{1}{4}$ **c**.
    The mirror plane (001) is a glide plane: $t = \frac{1}{2}$ **a**.

The mirror planes (100) and (010) are ordinary mirror planes.
The diagonal mirror planes (110) are type $d : t = \frac{1}{4} (a+b+c)$.

6   CONSTRUCTION OF SPACE GROUPS

To construct space groups, we combine the symmetry operations of point groups with the infinite set of lattice translations. We can do this in two steps:

◆ For each of the 32 classes, we consider the effect of translations related to the lattice types, for each lattice type of the system to which the class belongs. We thus obtain the 73 *symmorphic groups* which have been generated solely from simple symmetry elements.[2]
◆ We repeat the process, systematically replacing each simple symmetry element in the group by all the derived elements of translational symmetry, again taking account of possible lattice types.

For example, to obtain the groups derived from the class mmm, we must consider the types P, A, B, C, I and F. Then, for each type, we can replace the first mirror plane m by m, b, c or n, the second by m, a, c or n and the third by m, a, b or n. (no type a mirror plane normal to Ox can be found). If the lattice is F mode, mirror planes d must also be included.

☞ | *The groups so obtained are not all distinct since the combination of different symmetry elements can yield the same space group.*

As an example, we shall determine which space groups are derived from the monoclinic class '2' by deducing the group symmetry elements from equivalent positions of an object in the lattice.

Space groups derived from class 2

For this class, we must consider, *a priori*, the groups P2, P2$_1$, C2 and C2$_1$.

The projections are made onto the (001) plane and the direction of the two-fold axis is [010]. The position of the origin is arbitrary in a lattice, and here we shall take it as lying on the two-fold axis.

If the row [010] is a two-fold axis, all parallel rows are also two-fold axes: two-fold axes pass through $x = 1$, $z = 0$; $x = 0$, $z = 1$; $x = 1$, $z = 1$ etc.

---

[2]These groups may nevertheless contain translational symmetry elements.

Group P2

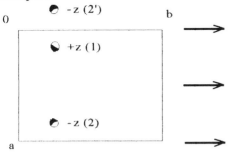

Figure 7.7

We consider atom 1 lying at a height $+z$ and in a general position (it is not lying on a symmetry element).

Its image (2′), given by the two-fold axis, is at a height $-z$. Atom (2), deduced from atom (2′) by a lattice translation **a** is equivalent to atom (1). (1) is transformed into (2) by a two-fold axis passing through $x = 1/2$, $z = 0$.

In this space group there exist two equivalent general positions:

$$(x, y, z) \text{ and } (-x, y, -z) \equiv (1 - x, y, -z) \equiv (1 - x, y, 1 - z)$$

The equivalent atoms (x, y, z) and $(1 - x, y, 1 - z)$ are both contained in the lattice and can be deduced from each other by a two-fold axis passing through $x = 0$, $z = 1/2$.

Group P2₁

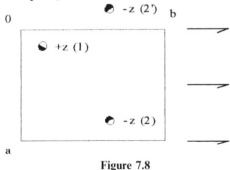

Figure 7.8

Consider an atom (1) lying at a height $+z$ in a *general position*. Its image (2′), given by the 2_1 axis, is at a height $-z$. Atom (2) deduced from (2′) by lattice translation **a** is equivalent to atom (1); we transform (1) into (2) by a 2_1 axis passing through $x = 1/2$, $z = 0$. The coordinates of the two equivalent atoms are: (x, y, z) and $(-x, y, +1/2 - z)$

Group C2

For group C2, the translation mode C equal to $\frac{1}{2} (\mathbf{a} + \mathbf{b})$ must be added to the whole lattice translations. If there exists an atom (1) lying at (x, y, z), then there also exists an equivalent atom (3) lying at $(x + \frac{1}{2}, y + \frac{1}{2}, z)$.

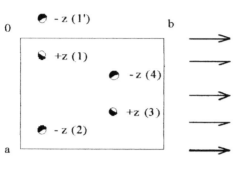

Figure 7.9

The image (1') of (1) given by axis 2 is the height $-z$, Atom (2) deduced from (1') by lattice translation **a** is equivalent to atom (1); similarly (3) may be transformed into (4) by symmetry about the two-fold axis followed by a translation **a**. Atom (4) is similarly deduced from atom (1) by an axis 2_1 passing through $x = 1/4$, $z = 0$.

The coordinates of the four equivalent atoms are:

$$(x, y, z); \ (-x, y-z); \ (x + 1/2, y + 1/2, z); \ (1/2 - x, y + 1/2, -z)$$

Similar operations performed on the group C2_1 give an identical result to that of C2. The symmetry elements are identical (2 and 2_1 axes); only the lattice origin is modified, and C$2_1 \equiv$ C2.

The space groups derived from class 2 are thus: P2, P2_1 and C2.

REMARKS:

◆ To construct the group we do not need to know all of the symmetry elements since some of the latter (called the group generators) imply the presence of the others.

◆ In this example, the only symmetry elements to be sought are the 2 or 2_1 axes (the product is obtained of the two-fold axis of the point group and the translations).

◆ This simplified method of seeking symmetry elements of space groups, based on a study of equivalent positions of atoms in the unit-cell, works correctly in simple cases (a single symmetry element). However, it cannot be used for more complex groups, since the relative positions of the various symmetry elements in the unit-cell are, *a priori*, unknown.

7   POSITION OF SYMMETRY ELEMENTS IN THE UNIT-CELL

◆ The product of a proper rotation through an angle 2φ and a perpendicular translation t is a rotation through the same angle about an axis lying on the bisector of the translation vector at a distance $d = t/2.\tan\varphi$

◆ Equidistant from two two-fold axes, equivalent by a whole lattice translation, there exists another two-fold axis.
◆ Equidistant from two mirror planes, equivalent by a whole lattice translation, there exists another mirror plane.

7.1 Symmorphic Groups with Primitive Cell

We shall take as an example the group $P\frac{4}{m}mm$
The class is tetragonal $4/mmm$; the lattice is primitive.
The symmetry elements are:

Four-fold axis [001] $\Rightarrow$ (E, 000), $(C_4, 000)$, $(C_4^2, 000)$, $(C_4^3, 000)$
Two-fold axes [100] $\Rightarrow$ $(C_{2x}, 000)$, $(C_{2y}, 000)$
Two-fold axes [110] $\Rightarrow$ $(C_{2xy}\ 000)$, $(C_{2\bar{x}y}, 000)$
Mirror plane (001) $\Rightarrow$ $(\sigma_h, 000)$
Mirror plane (100) $\Rightarrow$ $(\sigma x, 000)$, $(\sigma y, 000)$
Mirror planes (110) $\Rightarrow$ $(\sigma_{xy}, 000)\ \sigma_{\bar{x}y}, 000)$
$\bar{4}$-axis [001] $\Rightarrow$ $(S_4, 000)$, (I, 000), $(S_4^3, 000)$

In the corresponding point group, there are 16 equivalent general directions. There must exist 16 equivalent general positions (and 16 symmetry operators). The lattice translations (E, 100), (E, 010), (E, 001) must also be included. To construct the group, we proceed in two steps:

◆ At the origin, place all the symmetry elements in the point group, followed by all the elements deduced from them by whole translations of the lattice.

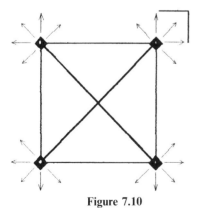

Figure 7.10

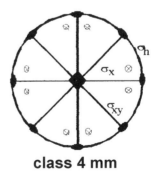

class 4 mm

Figure 7.11

◆ Add the elements resulting from the product of symmetry operations and translations.

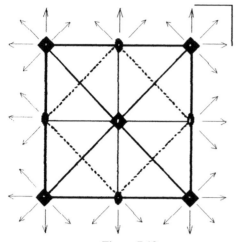

The product of C_4 passing through the origin and a translation **a** is C_4 passing through the centre of the cell.

The product of an inversion I and a translation **t** is an inversion at the centre of the vector **t**.

The product of $(C_4)^2 = C_2$ and the translation **a** is C_2 passing through the centre of vector **a**.

The position of the mirror planes m is obvious.

Figure 7.12

The product of $(\sigma_{xy}, 000)$ and a translation by the vector **a** can be written:

$$(\sigma_{xy}, \mathbf{a}) = \left(\sigma_{xy}, \frac{\mathbf{a}+\mathbf{b}}{2} + \frac{\mathbf{a}-\mathbf{b}}{2}\right) = (\sigma_{xy}, \mathbf{t}_\perp + \mathbf{t}_{//})$$

The normal component of the translation can be eliminated by placing the mirror plane at $x = 1/4$, $y = 1/4$; the parallel component cannot be eliminated, and gives a mirror plane n.

NOTE: Symmetry elements form a group and so the product of two elements in the group is itself an element in the group. Thus to generate a group it is not necessary to use all of its elements—it is sufficient to select a suitable number.

7.2   Symmorphic Groups of a Centred Cell

We place all the symmetry elements of the class at the origin and position the elements which can be deduced from these by whole lattice translations. Next we place all the symmetry elements of the class at the extremity of the translations which are proper to the lattice type. Finally, we take the product of all these elements and the translations related to the lattice type, and then their product with whole translations.

EXAMPLE:

The group **Cmm2** (Class mm2).

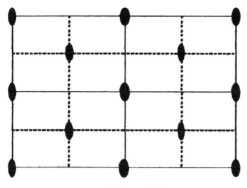

Figure 7.13

In addition to the whole translations we must include the translation $\mathbf{t} = \frac{1}{2}(\mathbf{a} + \mathbf{b})$, whose product with the mirror planes m produces mirror planes of type a at $y = 1/4$ and type b at $x = 1/4$.

7.3 Non-symmorphic Groups

Here the presence of translation means that the symmetry elements are not necessarily concurrent, and so we must begin by determining *the relative positions of the symmetry elements* chosen to generate the group.

Group **P4bm** (Class 4mm).

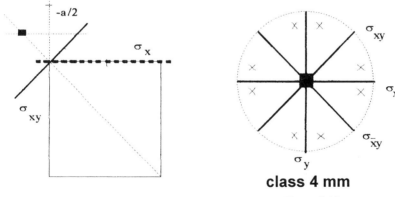

Figure 7.14 **Figure 7.15**

The mirror planes m and b are positioned at the origin and we take their product:

$$(\sigma_{xy}, \mathbf{0}) \cdot \left(\sigma_x, \frac{\mathbf{b}}{2} \right) = \left(\sigma_{xy} \cdot \sigma_x, \sigma_{xy} \left(\frac{\mathbf{b}}{2} \right) \right) = \left(\mathbf{C}_4, -\frac{\mathbf{a}}{2} \right)$$

The product of a C_4 axis and a translation normal to the axis is a C_4 axis passing through

$$x = -1/4, \quad y = -1/4$$

If the new origin is taken on the fourfold axis, the symmetry elements become:

$$(\mathbf{C}_4, 0), \left(\sigma_x, \frac{\mathbf{a} + \mathbf{b}}{2}\right), \left(\sigma_{xy}, \frac{\mathbf{a} + \mathbf{b}}{2}\right)$$

The position of the other mirror planes m, b and n is obtained by taking the product of the 4-fold axis and the mirror planes. The product with mirror planes m gives the two-fold axes.

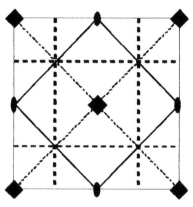

Figure 7.16

Group Ama2 (Class mm2).

We take the translation $\mathbf{t} = 1/2 \ (\mathbf{b} + \mathbf{c})$. The origin is placed at O, the intersection of mirror planes m and a.

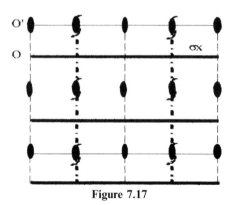

Figure 7.17

The product of mirror plane m (σ_x, 0) and mirror plane a (σ_y, $\mathbf{a}/2$) gives the two-fold axis (C_2, $\mathbf{a}/2$).

The product of mirror plane a (σ_y, $\mathbf{a}/2$) and the translation $\frac{1}{2}(\mathbf{b} + \mathbf{c})$ gives a mirror plane (σ_y, $(\mathbf{a} + \mathbf{b} + \mathbf{c})/2$), and hence a mirror plane n at $y = 1/4$.

The product of the axis 2 (C_2, 0) and the translation $\frac{1}{2}(\mathbf{b} + \mathbf{c})$ gives an axis 2_1 situated at $x = 1/4$, $y = 1/4$.

We can now place the origin at O, on the two-fold axis.

8 GENERAL AND SPECIAL POSITIONS

When we know the nature and position of all the symmetry elements in a space group, we can determine the m equivalent positions of objects placed in the unit-cell (the *general orbit* of the group).

If n is the number of equivalent general directions in the class from which the group comes, the number m of equivalent general positions is n if the lattice type is P or R, 2n if the mode is A, B, C or I and 4n if the mode is F.

EXAMPLES:

Group Fm3m $48 \times 4 = 192$ equivalent general positions.
Group Ama2 $4 \times 2 = 8$ equivalent general positions.
Group P4/mmm $16 \times 1 = 16$ equivalent general positions.

If an object is situated on one or more *non-translational* symmetry elements it is not repeated by those symmetry elements: the object is then in a *special position*. The set of equivalent positions constitutes a *special orbit* with multiplicity m' which is a sub-multiple of the multiplicity of the general orbit. The symmetry of objects placed on special sites must correspond to the symmetry of the sites.

EXAMPLES:

Group Ama2 (m = 8).
Atom on the two-fold axis at $(0, 0, z)$: four equivalent positions:

$(0, 0, z)$; $(1/2, 0, z)$; $(0, 1/2, 1/2+z)$; $(1/2, 1/2, 1/2+z)$

Atom in a mirror plane m at $(\frac{1}{4}, y, z)$: four equivalent positions:

$(1/4, y, z)$; $(-1/4, -y, z)$; $(3/4, 1/2, -y, z)$; $(1/4, 1/2,+y, z)$

If the object is placed on a screw axis or a glide plane, its position remains general owing to the translational part of the symmetry element.

9 CONCLUSIONS

This brief analysis of a few simple examples shows the principle by which space groups are constructed. These space groups are distributed as follows among the different systems:
(Appendix D contains the list of standard names for the 230 groups).

Triclinic system	2	Tetragonal system	68
Monoclinic system	13	Hexagonal system	27
Orthohombic system	59	Cubic system	36
Trigonal system	25		

For certain groups (and especially cubic groups) working out the construction can be long and tricky. Use should be made of the 'International Crystallographic Tables' which contain all the information relating to the 230 space groups.

Classification of structures according to geometrical crystallography

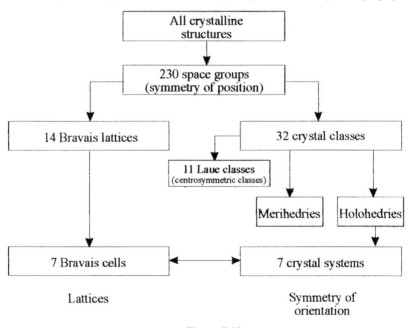

Figure 7.18

Chapter 8

Using the International Tables

The information below is designed to help the reader who may be unfamiliar with the 'International Tables for Crystallography'. A considerable amount of further information is to be found in the explanatory articles in the introduction to volume A of the tables.

References here are to volume A of the third edition of the tables, published in 1983 and revised in 1989 by the 'International Union of Crystallography'[1].

This third edition should be used rather than the second (1952 edition) as it contains numerous improvements:

❏ For monoclinic groups, the projections corresponding to both approved conventions (two-fold axis oriented along **b** or along **c**) are shown. In both cases, three orthogonal projections are drawn with either [100], [010] or [001] normal to the plane of projection.

❏ For orthorhombic groups, the unit-cell axes are usually chosen so as to obtain the standard name as the name of the group. This standard name, which is meant to be as good an indication as possible to the symmetry of the crystal, is not always the most appropriate. Projections are given for the six direct coordinate frames possible, with the corresponding names.

❏ Projections of cubic groups are now drawn, and new symbols specific to cubic groups have been introduced.

❏ Group symmetry operations are listed and the optimum choice of generators to use is specified.

On the following pages, explanations of the various features of the tables will be found facing the reproductions of each of the two pages of the International Tables devoted to the group Pma2, taken here as an example.

---

[1]International Tables for Crystallography edited by Theo Hahn, *Kluwer Academic Publishers*, Dordrecht, Holland (1989).

① Pma2 C_{2v}^{4} mm2   **Orthorhombic**
② N°. 28   Pma2   Patterson symmetry Pmmm

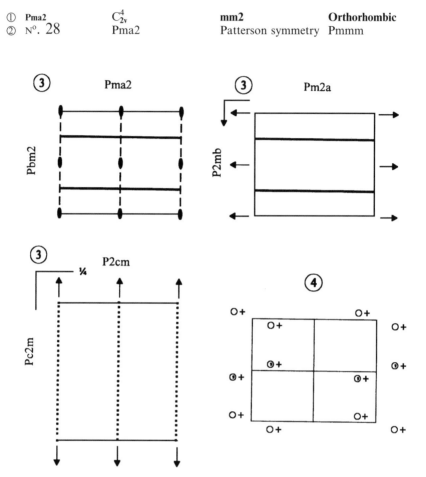

⑤ Origin on 1a2
⑥ Asymmetric unit $0 \leqslant x \leqslant 1/4$; $0 \leqslant y \leqslant 1$; $0 \leqslant z \leqslant 1$
⑦ Symmetry operations
(1) 1   (2) 0, 0, z (3) a x, 0, z (4) m 1/4, y, z

From International Tables for Crystallography, Volume A
Reproduced by permission of the International Union of Crystallography

① The heading comprises:
   The standard name of the group in shortened Hermann–Mauguin notation
   (Pma2).

The Schönflies symbol C_{2v}^4 of the group.
The class or point group (mm2).
The crystal system (Orthorhombic).

② The group number (the initial choice is arbitrary).
The group symbol in full Hermann–Mauguin notation:
e.g. the group **P2₁/c** is denoted $P1\frac{2}{c}1$
The group symmetry of the Patterson function.
(Always centro-symmetric and symmorphic).

③ Diagram(s) of the symmetry elements of the group; the number of diagrams depends on the crystal system.
If it is the *projection* of the axis which is shown on the diagram, then the name of the axis is indexed with a 'p' (a_p is the projection of **a** onto the plane of projection).

④ Illustration of a set of equivalent general positions.
The position of an atom is shown by the symbol ○ to which is affixed the height. In the expression for the height, the letter z is systematically omitted. Thus + and − correspond to $+z$ and $-z$; similarly, $1/2+$ denotes $1/2 + z$.

○   Initial object.

⊙   Object deduced from the initial object by an inversion, a roto-inversion or a mirror reflection (enantiomorph of the initial object).

⊙   Notation used for two superimposed positions in the case of a mirror plane parallel to the plane of projection.

⑤ Position of the origin.
The position of the origin is stated through its symmetry (symmetry elements intersecting at the point in question). In the example, the origin has been chosen as the intersection of the two-fold axis and the mirror plane a.

⑥ Definition of the smallest volume whose repetition by the group symmetry elements will generate the whole crystal.

⑦ List of the symmetry elements of the group. Each element is identified by an ordering number: (1), (2) etc.
The nature of each symmetry element is specified: 1, 2, a etc.
The position of the element in the cell is given: 0, 0, z etc.

① CONTINUED   No. 28   Pma2
② **Generators**   t(1, 0, 0)   t (0, 1, 0);   t(0, 0, 1);   (2);   (3)
   selected (1);

③ **Positions**

Multiplicity	Coordinates	Reflection
Wyckoff letter		conditions
Site symmetry		

		General:
4 d 1	(1) x, y, z (2) $\bar{x}, \bar{y}, z$ (3) $x+1/2, \bar{y}, z$ (4) $\bar{x}+1/2, y, z$	$h01 : h=2n$
		$h00 : h=2n$
		Special : as
		above plus
2 c m..	$1/4, y, z$ $3/4, \bar{y}, z$	no extra
		condition
2 b ..2	$0, 1/2, z$ $1/2, 1/2, z$	$hkl : h=2n$
2 a ..2	$0, 0, z$ $1/2, 0, z$	$hkl : h=2n$

④ **Symmetry of special projections**

Along [001] p2 mg   Along [100] p1m1   Along [010] p11m
$\mathbf{a'=a}$ $\mathbf{b'=b}$ $\mathbf{a'=b}$ $\mathbf{b'=c}$ $\mathbf{a'=c}$ $\mathbf{b'=1/2\ a}$
Origin at 0, 0, z Origin at x, 0, 0   Origin at 0, y, 0

⑤ **Maximal non-isomorphic subgroups**

I   [2]P112(P2) 1;2
           ○   [2]P1a1(Pc) 1;3
                               [2]Pm11(P- 1;4
                               m)
IIa   none
IIb   [2]Pba2 (**b'=2b**); [2]Pmn2₁ (**c'=2c**); [2]Pca2₁ (**c'=2c**); [2]Pcn2 (**c'=2c**)(Pnc2);
           ☐   [2]Ama2 (**b'=2b, c'=2c**); [2]Aba2 (**b'=2b, c'=2c**)

⑥ **Maximal isomorphic subgroups of lowest index**
IIc   [3]Pma2 (**a'=3a**); [2]Pma2 (**b'=2b**); [2]Pma2 (**c'=2c**);

⑦ **Minimal non-isomorphic supergroups**
I   [2]Pccm; [2]Pmma; [2]Pmna; [2]Pbcm
II   [2]Ama2; [2]Bma2 (Abm2); [2]Cmm2; [2]Ima2; [2]Pmm2 (**2a'=a**)

① Simplified heading.

② A minimal set of generators is specified. The symmetry operations are
   noted by their ordering number in the list of operations in the group, and
   the translations are indicated by the vector components.

③ List of general and special positions.
   The following information is given for each set of positions:

The multiplicity (in order of decreasing multiplicity).
The Wyckoff symbol for the site (see note 1).
The local symmetry of the site.
The coordinates of the equivalent positions preceded by the number of the symmetry operation which generates the positions (see note 2).
The existence conditions for diffraction spots (see note 3).

④ For each group, three orthographic projections along the symmetry axes are given. Each projection is accompanied by the direction of the projection, the name of the corresponding space group, its axes and origin.

⑤ ⑥ Maximum-order subgroups.
   I: Translations are the same as for the initial group.
   II: The crystal class is identical to that of the initial group.
      a: The same unit-cell (centred groups); **b, c**: larger unit-cell.
      c (⑥): Subgroups with the same standard symbol as the initial group.
      ○ [2] Order of the subgroup; **P1a1**: complete symbol for the subgroup;
      □ (**b'=2b**): base of the lattice; (For IIa, **b** and **c**);
      ○ (**Pc**): conventional symbol for the subgroup;
      ○ **1;3**: list of symmetry operations.

⑦ Minimal order supergroups.
   These tables are the inverse of those for the subgroups. The notation is identical to that used for the subgroups.

Additional remarks:

1—Following the work of Schönflies and Fédorov, Wyckoff determined the coordinates of the general and special equivalent positions for all 230 groups. His classification appeared in the first 'International Tables' and has been kept ever since. The notation is still used in crystallography and solid state physics as it enables a simple characterisation of the sites in a crystal.

2—The coordinates of the equivalent positions are fractional coordinates. To display as clearly as possible the nature of the elements in a group, the coordinates are given by points which do not all belong to the initial unit-cell. Coordinates of equivalent points in the unit-cell are obtained by adding whole lattice translations.

3—The general existence conditions for diffraction spots of indices h, k, l depend on the translations and translation elements of the group.

The notation '$h0l : h=2n$' means that reflections of indices $h0l$ are only allowed if h is even. If the atoms are in a special position, new systematic absences may occur; these will be related to the symmetry of the lattice and not to the contents of the repeating unit. These supplementary conditions are listed for each of the special positions.

Lattice-related absences

Type of cell	Reflection conditions
Primitive P	None
Face centred C	$h+k=2n$
Face centred A	$k+l=2n$
Face centred B	$h+l=2n$
Body centred I	$h+k+l=2n$
Face centred F	h, k, l all even or odd

Examples of absences related to translation symmetry elements:

Type of element	Reflection conditions
2_1 axis along [001]	$00l : l=2n$
4_1 axis along [001]	$00l : l=4n$
2_1 axis along [100]	$h00 : h=2n$
Mirror plane a (001)	$hk0 : h=2n$
Mirror plane a (010)	$h0l : h=2n$
Mirror plane n (001)	$hk0 : h+k=2n$

Proof of absences engendered by symmetry
elements will be given in chapter 10.

In volumes B and C of the numerical tables are to be found wavelengths, atomic diffusion factors, coefficients of absorption etc. together with copious information on the experimental methods of X-ray diffraction.

Part 2
X-RAY CRYSTALLOGRAPHY

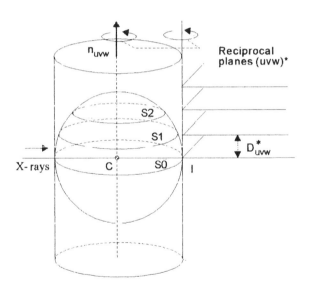

Chapter 9

X-rays

1 PRODUCTION OF X-RAYS

X-rays were discovered by Röntgen in 1895, and their wave nature was shown in 1913 when the first diffraction experiments suggested by von Laue were carried out. Later, Barkla demonstrated that X-rays are transverse waves, thereby establishing that they are electromagnetic.

X-rays range in wavelength from 0.1 Å (the lower limit of γ-rays) to 100 Å (the upper limit in the far ultra-violet); this corresponds in energy to the range 0.1–100 keV. The energy (in electron-volts) of an X-ray photon of wavelength γ (in Å) is:

$$E = \frac{12400}{\lambda}$$

$(E = hv = hc/\gamma$ and 1 eV $= 1.6 \times 10^{-19}$ joules$)$

X-ray crystallography uses X-rays of wavelength between 0.5 and 2.5 Å.

❒ THE ORIGIN OF X-RAYS

X-rays are produced when electrons which have been accelerated by an electric field strike a target (the anode). The efficiency η is low as can be seen from the following empirical formula:

$$\eta = \frac{\text{photon energy}}{\text{electron energy}} = 1.1 \times 10^{-9}.\text{Z.V}$$

where Z is the atomic number of the target metal and V the accelerating potential of the electrons (in volts). For a tungsten anode at a potential of 100 kV, the efficiency is of the order of 0.8%.

☐ SEALED X-RAY TUBES

A typical X-ray tube configuration is illustrated in figure 9.1.

The anode block is made of steel. It is equipped with four windows made of thin sheets of beryllium, which is transparent to X-rays (figure 9.1). The metal disc which is the target is brazed to a water-cooled copper block. The tube terminates in a glass neck at the bottom of which are fixed the electrical contacts. The tube is under high vacuum. A tungsten filament, heated by a variable current (to set the temperature which controls the level of emission and hence the tube current), is made negative with respect to the anode.

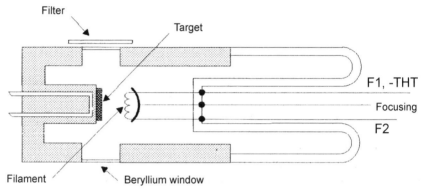

Figure 9.1

For safety reasons the anode is kept at earth potential. A focusing cup concentrates the electron beam onto a small rectangular zone on the target. On leaving the tube, the X-ray beam is given a defined geometry by the use of collimators.

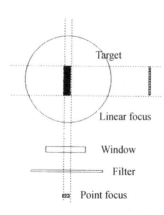

Figure 9.2

The X-ray source is 'seen' over an angle of incidence of about 6°, and both spot and line foci are available. (figure 9.2).

The electrical power dissipated in a normal X-ray tube is around 1.5–2 kW. Since nearly all the power is converted to heat, the anode must be cooled. To improve the cooling, the anode can be rotated; the dissipating capacity of rotating anodes is of the order of 20 kW. However, the high cost of the apparatus (pump, rotating seals, power supply) limits its use to applications where high energy beams are necessary.

After emerging from the beryllium windows, the beam can be passed through filters.

◻ POWER SUPPLIES

An X-ray tube is self-rectifying, but for reasons of stability it is supplied with DC, adjustable over the range 30 kV–100 kV. The tube current can be set from a few mA to 60 mA.

The consumption depends on the current flowing through the X-ray tube and modern equipment uses current and voltage feedback control. Highly insulated transformers and wiring are required to handle the very high voltages. Modern switching power supplies enable the construction of compact and reliable equipment.

In a few specialised centres (Orsay (Lure), Grenoble (ESRF), Hamburg, Daresbury, Brookhaven, Stanford), synchrotron radiation is used to produce a very intense beam of X-rays. Synchrotron radiation is generated through the motion of electrons moving close to the speed of light in a storage ring. The radiation is emitted tangential to the trajectory in a white spectrum. The beam is 10^4–10^5 times as intense as that of a conventional generator.

2   THE ANODE SPECTRUM

Figure 9.3 shows the spectrum emitted by a tungsten anode at an anode-cathode potential of 100 kV.

This emission is made up of a continuous spectrum on which is superimposed a line spectrum (K_α, K_β, L_α etc.)

The lines occur in series (K, L, M etc.) and close study shows that they have quite a complex structure. The intensity of these lines is very much higher than that of the continuous spectrum (by a factor of > 100 for the K_α line of a copper anode). The continuous spectrum is characterised by a sudden discontinuity on the low wavelength side.

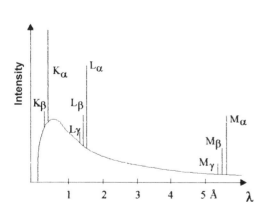

Figure 9.3

2.1   Continuous Radiation

This radiation is known as Bremstrahlung[1]. It is the result of the emission of an electromagnetic wave by electrons in the incident beam which undergo a sudden deceleration on interacting with electrons in the target. The discontinuity at low wavelengths is due to the total transfer of the energy of the incident electron to the emitted X-ray photon:

$$W = e.V = h\nu_{max} = \frac{h.c}{\gamma_{min}} \Rightarrow \gamma_{min}\text{Å} = \frac{h.c}{e.V} = \frac{12394}{V_{(volt)}}$$

As the efficiency depends on the atomic number of the anode metal, to obtain 'white' radiation, a target of high atomic number at high voltage must be used.

2.2   Characteristic Radiation

The line spectrum is *characteristic of the target metal*, and is the result of electronic transitions between atomic energy levels in the target. The photons from the continuous radiation have enough energy to cause the inner electrons of an atom to ionise. The atom leaves the resulting excited state by emission due to internal transitions; however, an atom ionised by loss of a K electron does not necessarily emit a K photon; the energy liberated by the transfer of an outer electron to the K shell can also be used to eject an electron (Auger emission). X-ray line spectra can be completely interpreted through the rules of atomic physics; figure 9.4 shows the energy level diagram and the associated quantum numbers.

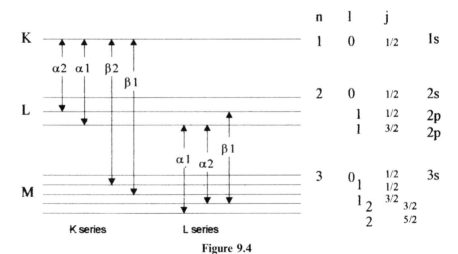

Figure 9.4

---

[1]German for braking radiation.

—The principal quantum number n denotes the shells K, L, M etc.
—The orbital angular quantum number l denotes the orbitals s, p, d etc. ($0 \leqslant l \leqslant n-1$).
—The magnetic quantum number m takes the values $-l \leqslant m \leqslant l$.
—The spin quantum number s takes the values $\pm 1/2$.
—The total angular quantum number j takes the values $j = l + s$.

If the K shell is ionised, the atom is in a state having an energy E_K. The resulting electron hole will be filled by an electron from one of the outer shells of the atom.

A series is named according to the level of the shell to which the electron goes; thus a transition L→K is written K_α, a transition M→K is written K_β and so on.

Orbital energies within the same level are very close, and this gives rise to the formation of multiplets with closely spaced wavelengths:

$$K_{\alpha 1} \Rightarrow K - L_3, \; K_{\alpha 2} \Rightarrow K - L_2; \; K_{\beta 1} \Rightarrow K - M_3; \; K_{\beta 2}^I \Rightarrow K - N_3; \; K_{\beta 2}^{II} \Rightarrow K - N_2 \ldots$$

It can be shown in atomic physics that allowed transitions (those with a probability other than zero) satisfy the following selection rules:

$$\Delta n \geqslant 1, \; \Delta l = \pm 1, \; \Delta j = 0 \pm 1$$

The limit of a series corresponds to an unbound electron jumping to the ionised level; the wavelength of the limit is therefore: $\gamma_K = hc/E_K$.

For a series S (S = K, L, M etc.) to be emitted, the incident electrons must have an energy higher than E_S, i.e. the accelerating potential difference must be greater than the ionisation threshold V_S of level S.

The intensity of a line is proportional to the transition probability for the electron between the initial and final levels. The lines $K_{\alpha 1}$ and $K_{\alpha 2}$ have roughly the same starting energy, but the population of the $2p^{5/2}$ level (4 electrons) is twice that of the $2p^{3/2}$ level; the intensity of the $K_{\alpha 1}$ line is roughly double that of the $K_{\alpha 2}$ line if the atomic number Z lies between 20 and 50. For the same range of values of Z we also have: $I_{K\beta} \approx 0.2 \, I_{K\alpha 1}$.

The wavelengths for the most commonly used anodes in X-ray crystallography are given in table 9.1. (The value for $\gamma_{K\alpha 1 Cu}$ radiation taken as a measurement standard is 1.5405974_{15}Å).

Table 9.1. Wavelengths

Target		Wavelengths/Å			Threshold V_K /volt
Type	Z	$K_{\alpha 2}-K_{\alpha 1}$	K_β	K limit	
Chromium	24	2.2935–2.2896	2.0848	2.070	5 950
Iron	26	1.9399–1.9360	1.7565	1.743	7 100
Cobalt	27	1.7928–1.7889	1.6208	1.608	7 700
Nickel	28	1.6616–1.6578	1.5001	1.488	8 300
Copper	29	1.5443–1.5406	1.3922	1.380	9000
Molybdenum	42	0.7135–0.7093	0.6323	0.6198	20 000
Tungsten	74	0.2138–0.2090	0.1844	0.1783	69 500

NOTE: To a first approximation, X-ray emission lines are not affected by chemical bonds, as the excitation takes place in the inner shells of the atoms. The frequencies depend only on the atomic number Z of the atom according to Moseley's empirical law:

$$\sqrt{v} = A.(Z - B)\ (A \text{ and } B \text{ are constants characteristic of the series}).$$

Wavelengths are chosen according both to the cell parameters of the compound under study and to the chemical elements present. For example, it is not recommended to use a copper target for a compound containing iron since the energy of the $K_{\alpha Cu}$ photons is high enough to ionise the K level of the iron, which will then emit its own characteristic radiation, thereby increasing the intensity of the white spectrum.

3 ABSORPTION OF X-RAYS

Total absorption is the result of two phenomena: scattering and the photoelectric effect; however, the effects of scattering are more or less negligible compared with those of the photoelectric effect. The latter is the result of coherent scattering, in which no change of wavelength occurs (Thomson scattering) and of incoherent scattering (Compton scattering).

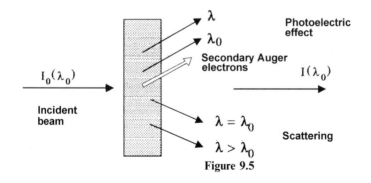

Figure 9.5

3.1   Mass Absorption Coefficient

Consider a monochromatic beam of unit cross-section passing through a homogeneous screen. It will lose energy dI proportional to the mass of the screen per unit surface (dp), and at an incident intensity I:

$$dI = -\mu.I.dp$$

μ is the *mass absorption coefficient* of the screen. On integrating we have:

$$\boxed{\frac{I}{I_0} = e^{-\mu.p} = e^{-\mu.\rho.x}}$$

(x is the thickness of the screen and ρ is the density).

3.2   Variation of the Mass Absorption Coefficient

The mass absorption coefficient depends on both the atomic number of the element and the wavelength.

❐ VARIATION WITH WAVELENGTH

A plot of μ against wavelength presents discontinuities due to the photoelectric effect, i.e., absorption of a photon by the atom with expulsion of an electron. Secondary radiation is emitted (so-called 'fluorescence') and this may be accompanied by Auger electrons and secondary electrons.

For a shell to be ionised, the energy $h\nu$ of the primary proton must be greater than the binding energy of the electron. A given shell, for instance the K shell, will only be ionised by radiation of frequency ν higher than ν_K such that $h\nu_K = W_K = hc/\lambda_K$. The wavelength must be lower than:

$$\lambda_K \text{Å} = \frac{h.c}{W_K} = \frac{h.c}{eV_K} = \frac{12394}{V_K(\text{volt})}$$

As soon as λ is lower than λ_K, the K shell becomes ionised and absorption by this shell is at a maximum which then decreases with λ. The same phenomenon occurs with the L shell but the relative amplitude of the discontinuities is smaller.

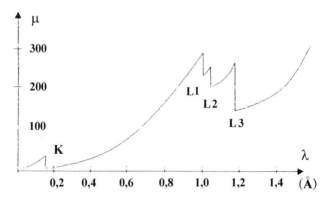

Figure 9.6. Variation of the coefficient μ for a tungsten screen

In the regions between the discontinuous zones, the mass absorption coefficient of an element varies in close accordance with $C.Z^3.\lambda^3$ (The Bragg–Pierce law).

❏ VARIATION WITH THE TYPE OF ELEMENT
Absorption increases with the atomic number of the element; light elements are weak absorbers whereas heavy elements are strong absorbers. μ does not increase continuously with Z and the discontinuities have the same origin as above.

If, for example, we consider the K_α line of copper ($\lambda = 1.542$Å), elements with an atomic number equal to or lower than 27 (that of cobalt) have a critical wavelength λ_K higher than $\lambda(\lambda_K Co = 1.608$Å$)$. The reverse is true for the elements after cobalt (nickel: $\lambda_K Ni = 1.489$Å) and ionisation of the K shell then becomes impossible. For $\lambda_K Cu$ radiation, there is a sharp decrease in the coefficient μ between cobalt and nickel.

3.3 Applications

❏ WINDOWS AND SCREENS
Problems of absorption govern the choice of material used in X-ray crystallography. Windows in tubes and detectors are made in materials of low absorbency which therefore have low atomic numbers. As organic materials do not support a vacuum sufficiently well, beryllium, despite its difficult workability, remains the principal material here. Ordinary glass absorbs highly, and so special glasses (Lindemann glass) have to be used for sample containers. On the other hand lead with its high absorbency is the most commonly used material for screens, either in the form of sheet metal or lead glass windows.

□ FILTERS

A screen with a discontinuity at λ_K is a strong absorber of radiation at wavelengths shorter than λ_K.

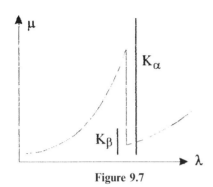

Figure 9.7

The K_α doublet is close to the K_β line whose relative intensity is quite high. The diffraction phenomena due to the K_α radiation have those due to K_β superimposed on them, thus complicating the interpretation of diagrams. Since the K_β line has a shorter wavelength than that of K_α, a filter can be provided to absorb much of the K_β and little of the K_α.

To make the filter, we require an element whose K discontinuity occurs between the two lines. Note that although such a filter enables the K_β line to be eliminated, it does not eliminate the white spectrum, nor the separation between the $K_{\alpha 1}$ and $K_{\alpha 2}$ lines.

To obtain a true monochromatic beam, crystal monochromators must be used (see chapter 12, section 8).

Table 9.2 gives the types of filters used with the most common anodes for eliminating the K_β line. Thicknesses are calculated to give a ratio of $1:100$ between the intensities of the K_β and K_α lines. As can be seen from the transmissions, filtering reduces the incident intensity by a factor of about 2.

Table 9.2. Filters for the most common anodes

Type of anode	$K_\alpha/\text{Å}$	Filter Type	Filter thickness/μm	% transmission K_α	% transmission K_β
Cr	2.291	V	11	58	3
Fe	1.937	Mn	11	59	3
Co	1.791	Fe	12	57	3
Cu	1.542	Ni	15	52	2
Mo	0.710	Zr	81	44	1

In the case of the chromium anode, as it is impossible to produce very thin vanadium sheet, the filter is made from vanadium oxide with a binder.

4   X-RAY DETECTORS

4.1   Fluorescent Screens

Although X-rays are invisible, they may be transformed into visible radiation: they have the property of making certain substances such as zinc sulphide fluoresce; the greater the intensity of the X-ray beam, the brighter the light emitted from such a screen (this is the principle of clinical radiography). Screens are now only used to locate beams during adjustments.

4.2   Photographic Films

For many years photographic film was used for accurate determination of the position and intensity of lines in diffraction patterns, but this practice has now been abandoned in favour of other more precise techniques. The photographic emulsions used were of large-grain silver bromide; an X-ray photon acting on an Ag^+ ion transforms this into an atom of Ag^0, producing a latent image of the pattern in the emulsion. The film is then developed and all the Ag^+ ions in a grain of emulsion containing an atom of silver are transformed into Ag^0. At the same time, a few non-activated grains develop spontaneously, giving rise to a clouding of the image. The undeveloped grains are then removed by a fixer. For medium intensities, the darkening of the film is proportional to the exposure time. However, intensity measurements are of poor precision, even if the films are processed with great care, and the use of film is now limited to techniques where it is not required to measure the intensities of diffraction patterns.

4.3   Gas Counters

☐ THE GEIGER–MULLER COUNTER
When an X-ray photon interacts with an atom of an inert gas (e.g. Xenon), the atom can be ionised into a 'positive–negative ion' pair. The energy required is of the order of 20–30 eV (20.8 eV for xenon). A $K_{\alpha Cu}$ photon possesses an energy of 8.04 keV, and it is thus capable of creating about 350 ion pairs in the gaseous medium.

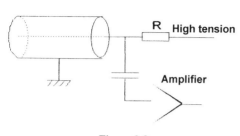

Figure 9.8

The Geiger–Müller (G-M) counter is made up of an earthed metal tube through which passes a wire (the anode) held at a potential of around 1500–2000 V. The tube is filled with a mixture of gases and has a window transparent to X-rays. The electrons produced by the ionisation process in the gas are attracted to the anode, the positive ions to the body of the tube.

The electrons are accelerated by the electric field around the anode and acquire sufficient energy to ionise any neutral gas atoms encountered. There is an avalanche effect (the amplification factor is between 10^4 and 10^7 and depends on the intensity of the electric field). The burst of electrons arriving at the anode causes a drop in potential at the capacitor connected to the anode. The resulting pulse is amplified, shaped and sent to a counter. The positive ions produced take a certain time to disappear (the dead time), and if a new photon enters the tube during this time, it is not detected. At high intensities, the counter becomes saturated and its response is not linear. Manufacturers of G-M tubes therefore use gaseous mixtures designed to give the shortest possible dead times ($\approx 10^{-4}$ s).

❐ PROPORTIONAL COUNTERS

If a weaker electric field and a gain lower than 10^5 are used, the amplitude of the output voltage pulse from the tube becomes proportional to the energy of the photon. The signal has to be amplified much more than with a conventional G-M counter, but on the other hand the dead time is much lower. Moreover, since the amplitude of the pulses is proportional to the energy of the photons, discrimination can be made between the 'right' photons, i.e. those that correspond to the right wavelength, and the others. Only those pulses corresponding to photons with energies between certain well-defined limits (the 'window') are transmitted.

This technique of 'pulse height selection' has enabled a considerable improvement in the signal-to-noise ratio by eliminating the white spectrum and fluorescence of the sample.

4.4 Scintillation Counters

In this type of detector, the energy of the photon is converted into electrical energy in a two-stage process. In the first stage, the X-ray photon is transformed into a visible photon (phosphorescence). A thallium-doped sodium iodide crystal is generally used; this re-emits at around 4100 Å. In the second stage, the visible photon energy is transformed into electrical energy by a photomultiplier. Photomultipliers give very high gain, are easy to adjust and have a very rapid response. The dead time is very short and the response is practically linear over a wide range of intensities.

In sophisticated work, use is also made of position detectors; these can either be gas counters with metal grid electrodes or scintillation counters coupled to a charge-transfer image amplification system. Also used are matrices of semiconductor detectors (reverse-biased diodes in which X-ray photons create electron-hole pairs) coupled with charge-transfer devices.

5   COUNTING ERRORS

X-ray emission is a random phenomenon; all measurements involved are therefore subject to statistical laws. The emission distribution obeys the Poisson law which, for a suitably large number of events N, can be approximated by a Gaussian law. The distribution is then symmetrical about the mean N_0. If $\sigma = \sqrt{N_0}$ is the standard deviation, then with N sufficiently large we have: $\sigma \approx \sqrt{N}$. For a Gaussian distribution, there is a probability of 68.3% that the value of N will lie in the interval $N \pm \sigma$, a probability of 95.4% for an interval $N \pm 2\sigma$ and a probability of 99.7% for an interval of $N \pm 3\sigma$.

Counting errors are random errors, and the accuracy of the measurement is a function of σ. Another indicator of probable error is the relative standard deviation ε:

$$\varepsilon = \frac{\sigma}{N} = \frac{1}{\sqrt{N}} \Rightarrow \varepsilon(\%) = \frac{100}{\sqrt{N}}$$

When measuring the intensity of a line, the background noise is superimposed on the signal, thereby increasing the uncertainty in the measurement. Let N be the total count, N_b the background and $N_c = N - N_b$ the real count; to evaluate N_b two measurements are generally carried out, one on either side of the line. The standard deviations on N and N_b are σ and σ_b, and the standard deviation on N_c is:

$$\sigma_c = \sqrt{\sigma^2 + \sigma_b^2} = \sqrt{N + N_b}$$

and the relative uncertainty (background of 68.3%) is:

$$\frac{\Delta N_c}{N_c} = \sqrt{\frac{N + N_b}{N - N_b}}$$

For accurate intensity measurements, counts as large as possible should be used, with a sufficiently high signal-to-noise ratio; however, extremely long counts are not advantageous owing to long-term fluctuations in the power supplies.

6 X-RAY OPTICS

6.1 Geometrical Optics

According to the classical theory of electromagnetic waves, the refractive index of a pure substance tends to 1 at high frequencies according to the relation:

$$n = 1 - \varepsilon = 1 - \frac{N.r_e.}{2\pi} \frac{Z}{M} \rho.\gamma^2$$

(Where: N is the Avogadro number, $r_e = (1/4\pi\varepsilon_0)(e^2/m.c^2) = 2.82.10^{-15}$ m is the classical radius of the atom, Z is the atomic number of the medium, M is the atomic mass, ρ is the density and λ is the wavelength).

With ρ expressed in MKSA units, λ in Å, and Z/M close to $1/2$, we have:

$$\varepsilon \approx 1.3 \times 10^{-6} \rho \lambda^2$$

The refractive index of the medium for X-rays is thus very slightly less than 1 and optical systems for X-rays cannot therefore be constructed easily; the X-ray telescope mirrors used in astrophysics have complex profiles.

If a beam travelling through a vacuum meets a medium of lower refractive index, then if the angle of incidence exceeds the limiting angle of reflection there is total reflection.

For X-rays, this angle is several minutes, and this property has been used for the absolute measurement of wavelengths of 'soft' X-rays ($\lambda > 50$Å), using metal gratings and very small angles of incidence. The property has also been used in the study of surfaces by X-rays, and also in synchrotron beam focusing.

6.2 Physical Optics

For interference and diffraction to occur, the structures of the objects viewed must have dimensions of the same order of magnitude as the wavelength of the incident radiation.

In crystals, lattice distances vary from a few tenths of angstroms to several dozen angstroms, and these distances are compatible with diffraction of radiation of the order of an angstrom.

In the following chapters we shall only deal with elastic scattering of X-rays, where the diffracted rays have the same wavelength as the incident rays.

Like all ionising radiation, X-rays are potentially dangerous for the operator.

The X-rays used in crystallography are not very penetrative and can cause serious burns to the dermis (radio-necrosis).

X-ray generators are periodically given safety inspections.

It is important not to 'bypass' any of the safety measures.

Chapter 10
Fundamentals of Diffraction Theory

1 REVIEW OF DIFFRACTION

1.1 Fraunhofer Diffraction

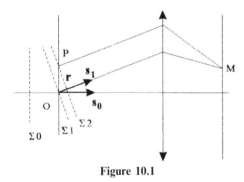

Consider a diffractor illuminated by a plane wave $\Sigma 0$ of wave vector s_0. The diffraction pattern is observed *ad infinitum* in a direction characterised by the wave vector s_1 (this is equivalent to making the observation in the focal plane of a convex lens).

Figure 10.1

The point O of the diffractor is chosen as the origin of the phases. Let $\Sigma 1$ and $\Sigma 2$ be the wave planes passing through O and through P characterised by the vector $\mathbf{r}$. The phase difference between the two wave planes is equal to:

$$\varphi = \frac{2.\pi.\mathbf{r}.(\mathbf{s}_1 - \mathbf{s}_0)}{\lambda} = 2.\pi.\mathbf{r}.\mathbf{S} \qquad \left(\mathbf{S} = \frac{\mathbf{s}_1 - \mathbf{s}_0}{\lambda}\right)$$

We denote by $A(\mathbf{r})$ the scattering aperture (for a screen with slits, the transparency is equal to one for the slits and zero for the opaque regions). We assume that the amplitude dA scattered by the diffracting element around point P is proportional to the length $d\mathbf{r}$ of this element. With these assumptions, the scattered amplitude at M is:

$$A_s = \int\limits_{Diffractor} A(\mathbf{r}).e^{j2.\pi.\mathbf{r}.S}.d\mathbf{r} \tag{1}$$

In Fraunhofer diffraction, the diffraction pattern is the Fourier transform of the diffractor aperture function.

1.2 Diffraction by a Plane Lattice

Consider an optical grating with N slits of width a, and a distance b apart, whose height is large compared with a and b. The grating is illuminated with coherent light of wavelength λ, and the light transmitted in a plane parallel to the grating at a distance D from it is observed. We let x be the distance from the point of observation P from the optical axis of the system and we write:

$$u = \frac{x.b}{\lambda.D}, \quad k = \frac{b}{a}$$

It can be shown that the luminous intensity at P is equal to:

$$I_{(P)} = C \left(\frac{\sin\dfrac{\pi.u}{k}}{\dfrac{\pi.u}{k}} \frac{\sin N.\pi.u}{\sin \pi.u} \right)^2 = C.D.A_N^2 \tag{2}$$

This formula is quite general:

☞ | *The diffraction pattern of a periodic structure is equal to the product of the diffraction pattern of the repeating unit, D, and the function A_N^2 characteristic of the periodicity of the structure.*

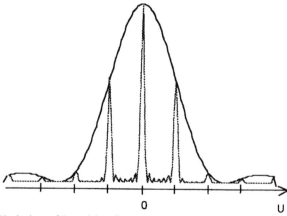

Figure 10.2. Variation of $I_{(P)}$ with u for a seven-slit grating with $k = 2.5$ (in the domain $-4 \leqslant u \leqslant +4$) and D envelope curve

The function A_N^2 has the following properties:

◆ It is a periodic function: $A_N^2(u+n) = A_N^2(u)$ if n is an integer.
◆ For N integer, it has principal maxima at intensity N^2

$$\lim_{u\to 0} \frac{\sin^2 N.\pi.u}{\sin^2 \pi.u} = \lim_{u\to 0} \frac{\sin^2 N.\pi.u}{(\pi.u)^2} = \lim_{u\to 0} N^2 . \frac{\sin^2 N.u}{(N.\pi.u)^2} = N^2$$

◆ It has $N-1$ minima between two principal maxima:

$$A_N^2(u) = 0 \text{ if } \sin \pi.N.u = 0 \text{ i.e. } u = \frac{m}{N} \ (m \text{ integer})$$

The intensity of the first subsidiary maximum is:

$$\frac{1}{\sin^2(3.\pi/2.N)} \approx \frac{1}{(3.\pi/2.N)^2} = \frac{4.N^2}{9.\pi^2} \approx 0.04.N^2$$

We can draw several important conclusions from this study of optical gratings and the relation for the diffracted intensity at P: $I_{(P)} = CDA_N^2$:

❏ The position of the maxima in the diffraction pattern of a periodic structure (the term A_N^2) is a function only of the periodicity of the structure.
❏ The intensity of the maxima in the diffraction pattern is a function of the repeating unit of the structure.
❏ In a diffraction photograph, the quantity which we can determine is the luminous intensity, and this is proportional to the square of the amplitude of the diffracted waves. The phases of the waves, on the other hand, are impossible to determine.
❏ From the interference pattern, the periodicity of the structure can readily be determined, but the aperture function of the repeating unit is not easy to determine.

2 SCATTERING OF X-RAYS BY AN ELECTRON

Apart from fluorescence radiation, matter illuminated by X-rays emits radiation of wavelength equal to or longer than that of the incident radiation; this is scattered radiation. Its intensity is very low, but radiation scattered without change of wavelength enables interference phenomena to be studied provided the medium in question is periodic.

2.1   Incoherent Scattering, or Compton Scattering

This was discovered by A.H. Compton in 1926. Scattering is the result of the impact between an incident photon and an electron. We assume the electron to be initially at rest (energy mc^2). Its velocity after the impact is v; we let $\beta = v/c$. The momentum of the incident photon is $p = hv/c$. From the conservation of energy and momentum we obtain the following system of equations:

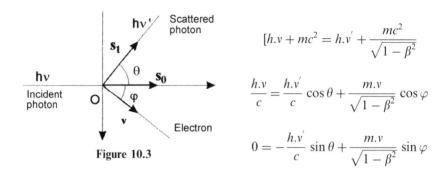

$$[h.v + mc^2 = h.v' + \frac{mc^2}{\sqrt{1 - \beta^2}}$$

$$\frac{h.v}{c} = \frac{h.v'}{c} \cos \theta + \frac{m.v}{\sqrt{1 - \beta^2}} \cos \varphi$$

$$0 = -\frac{h.v'}{c} \sin \theta + \frac{m.v}{\sqrt{1 - \beta^2}} \sin \varphi$$

Figure 10.3

On solving the equations we have:

$$\lambda' - \lambda = \frac{h}{m.c} (1 - \cos \theta) \tag{3}$$

The wavelength of the scattered photon depends on the direction of observation. There is no phase relation between the incident and scattered waves. The waves scattered by the various electrons never interfere—their intensities simply add. In X-ray crystallography, Compton scattering is a parasitic phenomenon which increases the background noise.

2.2   Coherent or Thomson Scattering

Under incident radiation with a wave vector s_0 and an electric field E_0, the electron undergoes an acceleration γ and emits secondary radiation which, far from the source, has the structure of a plane wave polarised in the plane E_0, s_0. The amplitude of the scattered electromagnetic field is proportional to the acceleration.

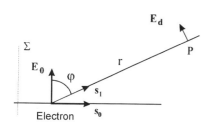

Figure 10.4

The motion of the electron is given by the relation:

$$m\frac{\mathrm{d}^2 x}{\mathrm{d}t^2} + F\frac{\mathrm{d}x}{\mathrm{d}t} + m.\omega_0^2.x = -e.E_0.e^{j\omega t}$$

If the electron is weakly bound, damping and restraining forces are negligible; the acceleration of the electron is then:

$$\gamma = \frac{e}{m}E_0.e^{j\omega t}$$

The scattered amplitude at P is equal to[1]

$$E_p = \frac{\mu_0}{4.\pi m}\frac{e^2}{r}\frac{\sin\varphi}{r}E_0 = \frac{1}{4.\pi.\varepsilon_0}\frac{e^2}{m.c^2}\frac{\sin\varphi}{r}E_0$$

NOTE: Electrons of light atoms and external electrons of heavy atoms behave towards X-rays as free electrons, since their binding energy with the nucleus corresponds to natural frequencies very much lower than those of the incident radiation. For inner shell electrons of heavy atoms, the binding energy is comparable to that of the radiation, and coupling, giving rise to scattering, may result.

2.3 Polarisation Factor

We take a coordinate frame in which the Ox axis coincides with the incident wave vector s_0 and such that the plane xOy contains the point of observation P. The incident electric field can be written $\mathbf{E} = \mathbf{E}_y + \mathbf{E}_z$. The incident radiation is the sum of a large number of incoherent vibrations the effect of which is obtained by taking the sum of the intensities.

If the incident radiation is unpolarised, the mean values of $\mathbf{E}_y$ and $\mathbf{E}_z$ are equal and hence: $I_y = I_z = I/2$. The contribution to the scattered wave of the component along Oy of the incident wave is:

---

[1]For a proof of this, refer to an account of radiation by antennae.

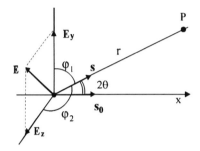

Figure 10.5

$$I_y^P \left(\frac{\mu_0}{4.\pi} \frac{e^2}{m} \frac{\sin \varphi_1}{r} \right)^2 . \frac{I}{2}$$

similarly:

$$I_z^P = \left(\frac{\mu_0}{4.\pi} \frac{e^2}{m} \frac{\sin \varphi_2}{r} \right)^2 . \frac{I}{2}$$

Now, $\varphi_2 = \pi/2$ and $2\theta + \varphi_1 = \pi/2$. By summing the intensities we then deduce Thomson's formula:

$$\frac{I_{diffused}}{I_{incident}} = \left(\frac{\mu_0}{4\pi} \right)^2 \frac{e^4}{m^2.r^2} \frac{1 + \cos^2 \theta}{2} \qquad (4)$$

Writing:

$$r_e = \frac{1}{4.\pi.\varepsilon_0} \frac{e^2}{m.c^2} = 2.818 10^{-15} m \text{ and } P(\theta) = \frac{1 + \cos^2 2\theta}{2}$$

(Where r_e is the classical radius of the atom and $P(\theta)$ the polarisation factor), the intensity scattered by an electron can be written in the form:

$$I_{el} = I_0 \left(\frac{r_e}{r} \right)^2 .P(\theta) \qquad (5)$$

❐ *The scattered radiation is partially polarised.*
On emerging from a crystal monochromator, the radiation, unlike that from the anode of an X-ray tube, is not isotropic.
❐ In Thomson's formula, the mass occurs in the denominator as its square. The same calculation can be applied to the nucleus; the proton-scattered intensity is $(1840)^2$ times smaller than the electron-scattered intensity.

☞ | *In the scattering of X-rays by matter, only the contribution of electrons is significant.*

3 SCATTERING OF X-RAYS BY MATTER

3.1 Electron Density Function

The contribution of the atomic nucleus to X-ray diffraction is negligible, and the degree of transparency to X-rays is thus determined by the electron density. In quantum mechanics the single point electron of classical theory is replaced by a charge density $\rho(\mathbf{r}) = |\Psi(\mathbf{r})|^2$, where $\Psi(\mathbf{r})$ is the wave function of the electron. The volume dv is assumed to contain a charge $\Psi(\mathbf{r})\,dv$ and to scatter a wave whose amplitude is that scattered by the electron multiplied by $\rho(\mathbf{r})dv$. The electron density function has maxima at the centres of the atoms and minima between the atoms. The total amplitude diffracted by the sample in a direction defined by the vector $\mathbf{S} = (\mathbf{s}_1 - \mathbf{s}_0)/\lambda$ ($\mathbf{s}_0$ is the incident wave vector, $\mathbf{s}_1$ the diffracted wave vector) is then:

$$A_s = A_{el}. \int_{Sample} \rho(\mathbf{r}).e^{j2.\pi.\mathbf{r}.\mathbf{S}}.dv_{\mathbf{r}}$$

A_s is the Fourier transform of the electron density. Through the inverse transform we can determine:

$$\rho(\mathbf{r}) = \int A_s.e^{j2.\pi.\mathbf{r}.\mathbf{S}} dV_S$$

The function A_S is a complex function:

$$A_s = ||A_S||.e^{j2.\pi.\phi_{(S)}}$$

The intensity, which is observable, is proportional to the square of the amplitude:

$$I_S = ||A_S||^2$$

If the function $\rho(\mathbf{r})$ is known, it is possible to determine $||A_S||$ and hence the intensity. On the other hand, measurement of the intensity does not enable the phase of A_S to be determined. It is therefore impossible to calculate $\rho(\mathbf{r})$ *a priori* from experimental measurements alone. For structure determination, the crystallographer has to use models whose validity has been tested by comparing values of intensities calculated from the model with experimental ones.

3.2   Atomic Scattering Factor

❑   CALCULATION PRINCIPLES

X-ray diffraction can be studied, to a first approximation, by assuming that matter is composed of independent atoms and that the effect of chemical bonds on the electron distribution can be ignored. To determine the intensity scattered by an isolated atom, the shell model can be used: the electron density is a function only of r. If z is the number of electrons, we can write:

$$\int_0^\infty \rho(\mathbf{r}).4\pi.r^2.dr = z \tag{6}$$

If A_{el} is the amplitude scattered by an isolated electron, the volume element dv containing $\rho(\mathbf{r})dv$ electrons will scatter an amplitude of $dA_S = A_{el}\rho(\mathbf{r})dv$. The phase of the wave scattered by the volume element dv is φ. (The nucleus is taken as phase origin).

The intensity scattered by the atom is thus:

$$A_S = \int\int\int A_{el}.\rho(\mathbf{r}).e^{j\varphi}.dv = A_{el}\int\int\int \rho(\mathbf{r}).e^{j\varphi}.dv$$

Let $\mathbf{s}_0$ and $\mathbf{s}_1$ be the incident and diffracted wave vectors, as before. We set $2\theta = \{\mathbf{s}_0, \mathbf{s}_1\}$ and we denote by $\mathbf{ON}$ the vector normal to the bisector of $\{\mathbf{s}_0, \mathbf{s}_1\}$ and α the angle $\{OP, ON\}$.

The path difference for the point P is then:

$$\delta = \mathbf{r}.\mathbf{s}_1 - \mathbf{r}.\mathbf{s}_0 = \mathbf{r}.(\mathbf{s}_1 - \mathbf{s}_0).$$

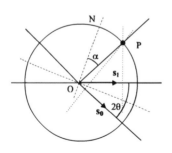

Figure 10.6

The phase difference between the points P and O is thus:

$$\varphi = \frac{2\pi}{\lambda}\mathbf{r}.(\mathbf{s}_1 - \mathbf{s}_0) = \frac{2\pi}{\lambda}r.||\mathbf{s}_1 - \mathbf{s}_0||.\cos\alpha$$

If we set $t = \dfrac{4.\pi}{\lambda}\sin\theta$, we can write:

$$\varphi = t.r.\cos\alpha$$

All points P on a circular crown of axis ON, thickness dr and width $r.d\alpha$ will have the same values of r and of the angle α.

The volume of this elementary crown is:

$$d\upsilon = 2\pi r \sin \alpha . r . d\alpha dr = 2\pi r^2 \sin \alpha d\alpha dr$$

Since $d\varphi / d\alpha = -t . r . \sin \alpha$, we can write $d\upsilon$ in the form: $d\upsilon = -(2.\pi/r)r.d\varphi.dr$

$$A_s = -A_{el} \int \int \int \rho(\mathbf{r}).e^{j\varphi} . \frac{2.\pi.r}{t} .d\varphi.dr$$

We calculate the integral by varying α from 0 to π and r from 0 to infinity. (In the shell model hypothesis we can separate the variables):

$$A_s = A_{el} \frac{2.\pi}{t} \int_0^\infty \rho(\mathbf{r}).r.dr \int_{-tr}^{+tr} e^{j\varphi} d\varphi$$

We finally obtain the following expression for the diffracted amplitude:

$$A_s = A_{el} \int_0^\infty \rho(\mathbf{r}). \frac{\sin(4.\pi.r. \sin \theta/\lambda)}{4.\pi.r. \sin \theta/\lambda} 4.\pi.r^2.dr \qquad (7)$$

The function $\rho(\mathbf{r})$ must be known to carry out the calculation. We let:

$$A_S = A_{el}.f(\sin \theta/\lambda)$$

The term $f(\sin \theta/\lambda)$, which is the Fourier transform of the electron density, is called the **atomic scattering factor**.

❐ PROPERTIES OF THE ATOMIC SCATTERING FACTOR

◆ It is a decreasing function of $\sin \theta/\lambda$.
◆ Much work has been devoted to calculating its values. Reliable data are tabulated in the 'International Tables for Crystallography'.
◆ According to relation (7), *if θ is zero, f is equal to z*, the number of electrons in the atom or ion. e.g. if θ is zero, f is equal to 18 for K^+, Ar and Cl^-.
◆ The electrons which are most involved in coherent diffraction are those in the outer shells; the scattering factors of an atom and its ions only differ for small values of θ. Similarly, the contribution of bonding electrons, which are diffusely spread out, decreases very rapidly with θ.
◆ When a wave is scattered by an atom, there is generally a phase shift of π and the scattering factor is real. If the frequency of the incident wave is close to the discontinuity in the K absorption of the atom, damping and

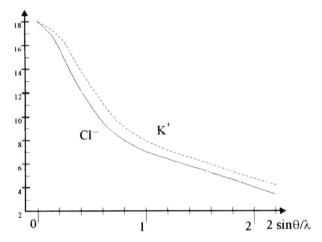

Figure 10.7. Variation of atomic scattering factors of K and Cl^- with $2\sin\theta/\lambda$ (λ in Å)

restraining terms in the equation of motion of the electron can no longer be neglected, (cf § 2.2). Coupling of the frequencies results in scattering which is referred to as anomalous scattering. The atomic scattering factor becomes:

$$f_t = f + \Delta f' + j \cdot \Delta f'' \tag{8}$$

The phase shift is then no longer equal to π and the atomic scattering factor includes an imaginary part. To a first approximation, it can be assumed that the correction terms are independent of the diffraction angle θ. $\Delta f''$ is always positive, $\Delta f'$ is negative if ω is less than ω_0 and positive otherwise. The values of the correction terms are also tabulated for the common values of λ.

3.3   Diffraction of X-rays by a Crystal

In a crystal, the electron density $\rho(\mathbf{r})$ is assumed to be the overlap of individual electron densities $\rho_i(\mathbf{r}')$, centred on the points $\mathbf{r}_i$, of the atoms making up the crystal:

$$\rho(\mathbf{r}) = \sum_i^{crystal} \rho_i(\mathbf{r} - \mathbf{r}_i)$$

The expression for the diffracted amplitude:

$$A_S = A_{el}. \int_{Sample} \rho(\mathbf{r}).e^{j2.\pi.\mathbf{r}.\mathbf{S}}.dv_{\mathbf{r}}$$

then becomes ($V = (\mathbf{a}, \mathbf{b}, \mathbf{c}) = (V^*)^{-1}$):

$$A_s = A_{el} \int \sum_i \rho_i(\mathbf{r} - \mathbf{r}_i).e^{j2.\pi.\mathbf{r}.\mathbf{S}}.dv_{\mathbf{r}} = \frac{A_{el}}{V} \sum_i \left(\int \rho_i(\mathbf{R}_i).e^{j2.\pi.\mathbf{R}_i\mathbf{S}}.dv\mathbf{R}_i \right).e^{j2.\pi.\mathbf{r}_i\mathbf{S}}$$

In a crystal the electron density is three-dimensional. If $\mathbf{a}$, $\mathbf{b}$ and $\mathbf{c}$ are the base vectors of the lattice, we have:

$$\rho(\mathbf{r}) = \rho(\mathbf{r} + u.\mathbf{a} + v.\mathbf{b} + w.\mathbf{c}) \text{ where } \mathbf{r} = x.\mathbf{a} + y.\mathbf{b} + z.\mathbf{c}$$

$$(u, v, w \text{ integers and } 0 \leqslant x, y, z < 1)$$

The expression for A_S can also be written:

$$A_s = \frac{A_{el}}{V} \sum_{crystal} \left(\int_{lattice} \rho(\mathbf{r}).e^{j2.\pi.\mathbf{r}\mathbf{S}}.d_{\mathbf{r}} \right).e^{j2.\pi(u.\mathbf{a}+v.\mathbf{b}+w.\mathbf{c})\mathbf{S}} \qquad (9)$$

If we write:

$$F_S = \frac{A_{el}}{V} \int_{lattice} \rho(\mathbf{r}).e^{j2.\pi.\mathbf{r}\mathbf{S}}.dv_{\mathbf{r}} \qquad (10)$$

and if m, n, p are the number of cells along Ox, Oy, Oz, we conclude:

$$A_S = F_S. \sum_{u=1}^{m} e^{j2.\pi.u.\mathbf{a}\mathbf{S}} . \sum_{v=1}^{n} e^{j2.\pi.v.\mathbf{b}\mathbf{S}} . \sum_{w=1}^{p} e^{j2.\pi.w.\mathbf{c}\mathbf{S}} \qquad (11)$$

Each sum is equal to:

$$A_q(\mathbf{a}_i.\mathbf{S}) = \frac{\sin \pi.q.\mathbf{a}_i.\mathbf{S}}{\sin \pi \mathbf{a}_i.\mathbf{S}}$$

The function F_S is the Fourier transform of the electron density of a cell, and is known as the **structure factor.**

The diffracted intensity is then:

$$I = ||F_S||^2.A_m^2(\mathbf{a}.\mathbf{S}).A_n^2(\mathbf{b}.\mathbf{S}).A2_p^2(\mathbf{c}.\mathbf{S}) = ||F_S||^2.L^2 \qquad (12)$$

We thus obtain the general result stated in paragraph 1.2:

> The diffraction pattern of a periodic structure is equal to the product of the diffraction pattern of the repeating unit and a function dependent only on the periodicity of the structure.

4   DIFFRACTION BY A THREE-DIMENSIONAL LATTICE

4.1   Laue Conditions

Since the number of cells in a crystal is extremely large, the scattered intensity (expression 12) is zero throughout (see § 1.2) unless the three functions $A_q(\mathbf{a}_i.\mathbf{S})$ are at a simultaneous maximum, i.e. the products $\mathbf{a}_i\mathbf{S}$ are integers.

The vectors $\mathbf{S}$ then obey the Laue conditions:

$$\left.\begin{array}{l} \mathbf{a}.\mathbf{S} = h \\ \mathbf{b}.\mathbf{S} = k \\ \mathbf{c}.\mathbf{S} = l \end{array}\right\} \quad h,\,k,\,l \text{ integers}$$

NOTE: The diffraction directions for a three-dimensional lattice can readily be found using the following method:

Consider a row of point diffractors separated by a distance, a, illuminated by a plane wave of wave vector $\mathbf{s}_0$, which we observe to infinity in the direction $\mathbf{s}_1$.

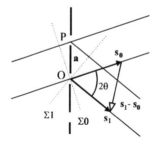

Figure 10.8

The phase difference between the incident wave and the diffracted wave is between the points O and P, equal to:

$$\varphi = \frac{2.\pi.\mathbf{a}.(\mathbf{s}_i - \mathbf{s}_0)}{\lambda}$$

Putting:

$$\mathbf{S} = \frac{(\mathbf{s}_1 - \mathbf{s}_0)}{\lambda},$$

we obtain: $\varphi = 2.\pi.\mathbf{a}.\mathbf{S}.$

Constructive interference occurs if the phase difference between two successive nodes is equal to an integral number times 2π, i.e. if $\mathbf{a}.\mathbf{S} = h$ (integer). The result can be generalised to three dimensions.

❐ THE NATURE OF THE VECTOR S

Let the angle between the wave vectors $\mathbf{s}_1$ and $\mathbf{s}_0$ be 2θ. The modulus of the vector $\mathbf{S}$ is equal to: $||\mathbf{S}|| = (2\sin\theta)/\lambda$. The Laue conditions can be written:

$$\frac{\mathbf{a}}{h}\mathbf{S} = 1 \quad \frac{\mathbf{b}}{k}\mathbf{S} = 1 \quad \frac{\mathbf{c}}{l}\mathbf{S} = 1$$

Hence: $(\mathbf{a}/h - \mathbf{b}/k).\mathbf{S} = 0$ $(\mathbf{a}/h - \mathbf{c}/l).\mathbf{S} = 0$

The vectors $\mathbf{a}/h - \mathbf{b}/k$ and $\mathbf{a}/h - \mathbf{c}/l$ belong to the plane $h(x/a) + k(y/b) + l(z/c) = 1$ in the family of lattice planes (hkl). The vector $\mathbf{S}$ is normal to two vectors contained in the first lattice plane of the family of indices h, k and l which does not pass through the origin, and it is therefore normal to the family of planes (hkl).

The distance d_{hkl} between two planes in the family is equal to the projection of the vector $\mathbf{a}/h$ onto the unit vector normal to these planes:

$$d_{hkl} = \frac{\mathbf{a}}{h} \cdot \frac{\mathbf{S}}{||\mathbf{S}||} = \frac{1}{||\mathbf{S}||}$$

The vector $\mathbf{S}$ is thus equivalent to the reciprocal vector $\mathbf{N}^*_{hkl}$

☞ | *The vectors $\mathbf{S}$ are vectors of the reciprocal lattice.*

The allowed diffraction directions in the lattice are those defined by the rows of the reciprocal lattice.

❐ DIFFRACTION DOMAINS

From relation (12) the diffracted intensity can be written in the form of the product of a **form-factor** L^2 and a structure factor F^2. As $\mathbf{S}$ is a reciprocal vector equal to: $\mathbf{N}^*_{hkl} = h.\mathbf{A}^* + k.\mathbf{B}^* + l.\mathbf{C}^*$, the form factor L can be written as the product:

$$L_S = \sum_{u=1}^{m} e^{j2.\pi.u.h} \cdot \sum_{v=1}^{n} e^{j2.\pi.v.k} \sum_{w=1}^{p} e^{j2.\pi.w.l}$$

This is the product of three geometrical progressions that can be written:

$$L_S = \frac{\sin \pi.m.h}{\sin \pi.h} \frac{\sin \pi.n.k}{\sin \pi.k} \frac{\sin \pi.p.l}{\sin \pi.l}$$

The **interference function** corresponding to the diffracted intensity is equal to L_S^2.

For an *infinite crystal* ($m = \infty$, $n = \infty$, $p = \infty$), this function is zero throughout except for integral values of h, k and l, when it is infinite. This is a Dirac peak distribution over the lattice nodes.

For a finite crystal, there will be a three-dimensional function with principal maxima for integral values of h, k and l and secondary maxima separated by minima of zero. Only the principal maxima have substantial intensity, and in reciprocal space they form a three-dimensional distribution of diffraction domains with dimensions $2A^*/m$, $2B^*/n$ and $2C^*/p$ (the distances between the first zero minima); for a finite crystal, the diffracted intensity is not immediately cancelled on departing from the Laue conditions.

Diffraction occurs when the end of a diffracted ray remains inside the diffraction domain of the node in question. There is then said to be *relaxation of the diffraction conditions.*

Two other formulations equivalent to the Laue conditions are often used: Ewald's construction and Bragg's law, which are based on simple geometrical constructions.

4.2   Ewald's Construction

A wave vector beam s_0 strikes the crystal diffractor at O.

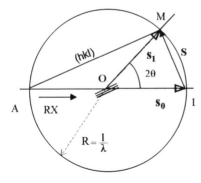

Figure 10.9

The so-called Ewald sphere has its centre at O and radius $R = 1/\lambda$. The incident beam AO crosses the sphere through I. If the vector $\mathbf{IM} = \mathbf{S} = (\mathbf{s}_1 - \mathbf{s}_0)/\lambda$ is such that OM is a diffraction direction, then M is a node in the reciprocal lattice constructed with the point I as origin (the node 000). The straight line AM is parallel to the lattice planes giving rise to diffraction.

Conversely, the diffraction directions which are possible are those defined by straight lines joining the origin O to nodes of the reciprocal lattice which lie on the Ewald sphere. With a randomly oriented crystal there is in general no diffracted ray; the crystal must be rotated about O to bring a reciprocal lattice node onto the sphere.

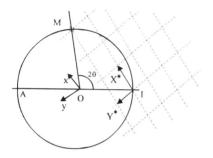

Figure 10.10

When the crystal is rotated about O, the reciprocal lattice rotates about the point I. Figure 10.10 shows the intersection of the Ewald sphere with a lattice plane (001)* in the reciprocal lattice. Since the node M lies on the sphere, it defines the direction of OM. In the example shown in this figure, diffraction by the (310) planes is occurring.

NOTE: If the Ewald sphere is constructed with a radius equal to R_0, the reciprocal lattice must be constructed to the scale $\sigma^2 = R_0\lambda(\mathbf{a}.\mathbf{A}^* = \sigma^2$, $\mathbf{b}.\mathbf{A}^* = 0$ etc.)

4.3 The Bragg Equation

❏ BRAGG'S LAW

Ewald's construction (figure 10.10) enables us to write:

$$\mathbf{IM} = \mathbf{S} = \mathbf{N}_{hkl} = h.\mathbf{A}^* + k.\mathbf{B}^* + l.\mathbf{C}^*$$

The modulus of the reciprocal vector is $\|\mathbf{N}_{hkl}\| = \frac{2.\sin\theta}{\lambda}$

This is related to the interplanar spacing (hkl) by $\|\mathbf{N}_{hkl}\|.d_{hkl} = 1$. From this we deduce the following relation, which is Bragg's law:

$$\boxed{2.d_{hkl}.\sin\theta = \lambda} \qquad (13)$$

❐ Some Remarks on Bragg's Law

◆ For diffraction by a family of lattice planes (which contains all the lattice nodes) to occur, the condition $\lambda < 2d_{hkl}$ must hold; however, for diffracted rays to be observed, θ must not be too small:

In X-ray diffraction experiments, the wavelength λ used must be of the same order of magnitude as the lattice interplanar spacings d_{hkl} in the crystal.

◆ Consider the family of planes $(H\,K\,L)$ such that H, K, L are relatively prime and the family $(h\,k\,l) = (nH\,nK\,nL)$ with n an integer. We then have: $d_{hkl} = d_{HKL}/n$. Bragg's law can therefore also be written:

$$2d_{HKL}\sin\theta = n\lambda \qquad (14)$$

The nth order reflection (with path difference δ between two consecutive rays equal to $n\lambda$ from the planes $(H\,K\,L)$ can be interpreted as a first order reflection $(\delta = \lambda)$ from the fictitious lattice planes $(nH\,nK\,nL)$ separated by a distance d_{HKL}/n. (In a family $(nH\,nK\,nL)$, only one in n planes contains nodes).

❐ Conventional Interpretation of Bragg's Law

The lattice is represented by a succession of parallel equidistant lattice planes. For nodes in the plane, there is phase agreement between the scattered rays if the diffracted beam obeys the Snell–Descartes laws. The angles of incidence and diffraction are equal and it can be verified from figure 10.11 that the optical path lengths for nodes N0 and N1 are equal when this condition is fulfilled.

There must also be phase agreement between waves from different planes. The path difference between nodes N1 and N2, which is $2d\sin\theta$, must be equal to $n\lambda$ with n an integer:

$$2d_{hkl}\sin\theta = n\lambda$$

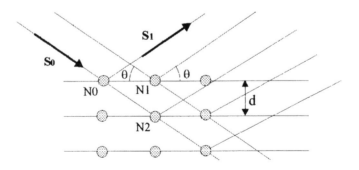

Figure 10.11

If the Bragg condition is satisfied, reflection of the incident ray from the lattice planes will occur according to the Snell–Descartes laws.

NOTE: For this proof to hold, it is not necessary for the nodes to be ordered within the lattice planes, and the proof does not account for all the phenomena involved.

4.4 Conclusions

☞ | *The Laue conditions, the Bragg equation and Ewald's construction are three equivalent representations of the same phenomenon: the directions of diffraction in a lattice are determined by its reciprocal lattice.*

The nature of the repeating unit only influences the diffracted intensity, not the directions. Measurements of X-ray diffraction angles in a crystal only give information about the translational lattice; diffraction directions are determined using one of the above methods, according to the nature of the problem and the diffraction technique actually used. To determine the position of atoms within the unit-cell, intensities within the diffraction pattern must be also be measured.

5 DIFFRACTION INTENSITIES

5.1 The Temperature Factor

In a crystal, one atom is bound to the others by various forces; its equilibrium position is that which minimises its energy. Any perturbation will result in an oscillation of that atom about its equilibrium position, and in particular, thermal agitation will change the diffracting power of the atoms. A complete analysis of these phenomena is rather long and complex, and here we shall limit ourselves to an explanation of the calculation principles.

We assume that the equilibrium position of the atom is chosen as origin, that the probability of finding the centre of that atom at $\mathbf{r}'$, is $p(\mathbf{r}')$ and that the electron density at $\mathbf{r}$ when the centre is at $\mathbf{r}'$, is $\rho_a(\mathbf{r}-\mathbf{r}')$. The electron density modified by thermal motion (the average obtained by integrating over all displacements) becomes:

$$\rho_t(\mathbf{r}) = \int \rho_a(\mathbf{r} - \mathbf{r}').p(\mathbf{r}').\mathrm{d}\mathbf{r}'$$

Here we assume that the shape of the electron cloud is unchanged by movements of the nucleus. The mean atomic scattering factor is the Fourier transform of $\rho_t(\mathbf{r})$. According to the above expression, $\rho_t(\mathbf{r})$ is a convolution product, and its Fourier transform is therefore equal to the product of the Fourier transform of the convoluted functions. The Fourier transform of the probability function is called the *temperature factor*, or *Debye–Waller factor*.

$$q(\mathbf{S}) = \int p(\mathbf{r}')e^{2j\pi.\mathbf{r}'\mathbf{S}].}\,\mathrm{d}\mathbf{r}' \tag{15}$$

If thermal agitation is taken into account, the atomic scattering factor f_t to be considered is thus equal to the product of the classical atomic scattering factor f, which is the Fourier transform of $\rho(\mathbf{r})$, and the function $q(\mathbf{S})$. Assuming that the thermal agitation movement has spherical symmetry, $p(\mathbf{r}')$ is isotropic and can be described by a Gaussian function:

$$p(\mathbf{r}') = p(\mathbf{r}') = \frac{1}{\sqrt{2\pi}}\,\frac{1}{\sqrt{U}}.e^{-r'^2/2U}$$

$U = \langle \mathbf{r}'2\rangle$ is the deviation from the equilibrium position.
The Fourier transform of $p(\mathbf{r}')$ is also a Gaussian distribution:

$$q(\mathbf{S}) = e^{-2\pi^2.US^2} = e^{-8\pi^2.U.(\sin^2\theta)/\lambda^2} = e^{-B.(\sin^2\theta)/\lambda^2} \tag{16}$$

$B = 8\pi^2 U$ is the *atomic temperature factor*.
Thermal agitation has the effect of making the electron density more diffuse (the lattice planes have a 'thickness'), and of lowering the atomic scattering factor: the closer the lattice planes (and hence the higher the diffraction angle), the lower the atomic scattering factor. At ambient temperatures, values of U are typically 0.01–0.1 Å^2, and for harmonic vibrations U is a substantially linear function of temperature. Thermal agitation is, in general, anisotropic. If we assume that $p(\mathbf{r}')$ is represented by a three-dimensional Gaussian distribution, the surfaces of equal probability are ellipsoids centred on the mean positions of the atoms in the crystal. The temperature factor representing the ellipsoid of thermal agitation in the reciprocal lattice becomes:

$$q(\mathbf{S}) = \exp[-2\pi^2(U_{11}.X^{*2} + U_{22}.Y^{*2} + U_{33}.Z^{*2}$$
$$+ 2U_{12}.X^*.Y^* + 2.U_{13}.X^*.Z^* + 2.U_{23}.Y^*.Z^*)]$$

The six parameters U_{ij} define the directions and lengths of the axes of the thermal ellipsoid.

5.2   Structure Factor

In the calculations of the intensity scattered by the crystal, we derived the term:

$$F_S = A_{el} \int\limits_{\text{cell}} \rho(\mathbf{r}).e^{j2.\pi.\mathbf{r}.\mathbf{S}}.dv_{\mathbf{r}}$$

which is the *structure factor*. The following equivalent expression (making use of the atomic scattering factors) is more commonly used:

$$F_S = F_{hkl} = \sum_{i=1}^{n}(f_i)_t.e^{2j\pi.\mathbf{r}_i.\mathbf{S}} \sum_{i=1}^{n}(f_i)_t.e^{2j\pi(h.x_i+k.y_i+l.z_i)} \qquad (17)$$

The term A_{el} is the same for all the atoms in the cell, and is therefore omitted in this expression for the structure factor. The effect of chemical bonds is neglected and the summation is carried out over the n atoms in the cell. The atomic scattering factors of each atom must be corrected for thermal agitation effects. The scattered intensity in a direction defined by the vector $\mathbf{S}$ is proportional to the product of F_{hkl} and its complex conjugate F_{hkl}^{*}.

$$I_{hkl} \propto F_{hkl}.F_{hkl}^{*}$$

5.3   Example of Structure Factor Determination

We consider cesium chloride, CsCl. This has a simple cubic lattice with unit-cell composed of a Cl^- ion and a Cs^+ ion. If we set the origin on the chloride ion $(0, 0, 0)$, the reduced coordinates of the cesium ion are $1/2$, $1/2$, $1/2$. The structure factor is therefore:

$$F_{hkl} = \sum_{m=1}^{2} f_m.e^{2j\pi.(h.x_m+k.y_m+l.z_m)} = f_{Cl^-} + f_{Cs^+}.e^{j\pi(h+k+l)}$$

$$\left. \begin{array}{l} \text{If } h+k+l \text{ is even } F_{hkl} = f_{Cl^-} + f_{Cs^+} \\ \text{If } h+k+l \text{ is odd } F_{hkl} = f_{Cl^-} - f_{Cs^+} \end{array} \right\} = f_{Cl^-}, f_{Cs^+} = f\left(\frac{\sin\theta}{\lambda}\right)$$

For this compound, if the sum of the indices of the diffracted line is even, then the line is strong; if the sum is odd, the line is weak.

5.4   The Relation Between the Structure Factor and the Reciprocal Lattice

If the intensity scattered in a direction S is zero, we can consider that the corresponding node in the reciprocal lattice does not exist. This will enable us easily to establish a number of properties of the reciprocal lattice.

EXAMPLE: What is the reciprocal lattice of a cubic I-lattice?

An atom with coordinates x, y, z corresponds to one with coordinates:

$$x + 1/2, \; y + 1/2, \; z + 1/2;$$

Gathering together the n atoms of the cell in pairs, we can express the structure factor in the form:

$$F_{hkl} = \sum_{m=1}^{n/2} f_m.(e^{2j\pi.(h.x_m+k.y_m+l.z_m)} + e^{2j\pi.(h.(x_m+1/2)+k.(y_m+1/2)+l.(z_m+1/2))})$$

$$F_{hkl} = \sum_{m=1}^{n/2} f_m.e^{2j\pi.(h.x_m+k.y_m+l.z_m)}.(1 + e^{j\pi.(h+k+l)})$$

If $h + k + l$ is odd, F_{hkl} is *always* zero.

We see immediately that a cubic lattice with cell parameter a, and having no nodes with an odd sum of indices, is actually an F-lattice with cell parameter $2a$. We have thus established that *the reciprocal lattice of an I-lattice is an F-lattice*.

5.5   Friedel's Law

Consider a reflection from a family of planes (hkl), characterised by a vector S, and one from a family $(\bar{h}\bar{k}\bar{l})$, with diffraction vector $-$S. The structure factor for the family $(\bar{h}\bar{k}\bar{l})$ is:

$$F_{\bar{h}\bar{k}\bar{l}} = \sum_{m=1}^{n} f_m.e^{-2j\pi.\mathbf{r}_m\mathbf{S}}$$

If the atomic scattering factors f_m of all the atoms in the cell are real, the structure factor of the family $(\bar{h}\bar{k}\bar{l})$ is the complex conjugate of the structure factor of the family (hkl): $F_{\bar{h}\bar{k}\bar{l}} = F^*_{hkl}$. From this we deduce **Friedel's law**:

$$\boxed{I_{\bar{h}\bar{k}\bar{l}} = I_{hkl} \propto F_{hkl}.F^*_{hkl}} \tag{18}$$

☞ | *The intensities of the reflections (h k l) and ($\bar{h}\,\bar{k}\,\bar{l}$) are equal even if the crystal is non-centrosymmetric.*

The diffraction pattern will always have a centre of symmetry, even if the crystal itself is non-centrosymmetric. Diffraction methods enable the Laue class to be established, but not its point group. In fact the law is approximate since it assumes that the atomic scattering factors are real. If we choose a wavelength at which *at least* one atom presents anomalous dispersion, then Friedel's law is no longer applicable. In certain cases this method is used to distinguish between (*h k l*) and ($\bar{h}\,\bar{k}\,\bar{l}$) reflections.

5.6 Lorentz Correction

In an ideal infinite crystal, the nodes of the reciprocal lattice are points. In a real crystal of finite dimensions with defects, the reciprocal nodes occupy a non-negligible volume of space in the reciprocal lattice. The longer a node remains in the diffraction position, the stronger the corresponding reflection. If all the nodes scattered in the same time period during an experiment, this would not matter, but in classical diffraction methods, the times for the various reciprocal nodes to cross the sphere are different. The diffraction time depends on the position of the node in the lattice and the speed with which it crosses the sphere. Consider, for example, a crystal rotating with constant angular velocity ω about an axis of rotation normal to the direction of diffraction of the node in question.

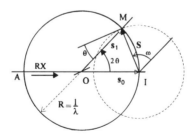

Figure 10.12

The reciprocal lattice rotates about I at the same velocity ω. If V_N is the component of the linear velocity of the node in the direction of diffraction, the **Lorentz factor** is defined by:

$$L(\theta) = \frac{\omega}{V_N.\lambda}$$

This factor is proportional to the time which the node takes to cross the Ewald sphere ($R = 1/\lambda$). The linear velocity of M is: $V = ||\mathbf{S}^*||.\omega = S^*.\omega$. The projection of $\mathbf{V}$ onto the direction of the refracted ray $\mathbf{s}_1$ is $V_N = S^*.\cos\theta$, Using the Bragg equation we can write:

$$S^* = \frac{1}{d} = \frac{2.\sin\theta}{\lambda} \Rightarrow V_N = \frac{\omega}{\lambda} 2.\sin\theta = \frac{\omega}{\lambda}\sin 2\theta.$$

In this example, we find:

$$L(\theta) = 1/\sin 2\theta$$

$L(\theta)$ depends on the diffraction technique used:

For powder methods we find:

$$L(\theta) = \frac{1}{\sin^2\theta.\cos\theta},$$

and for a rotating crystal (radius normal to the axis of rotation):

$$L(\theta) = \frac{1}{\sin 2\theta}$$

6   REFLECTING POWER OF A CRYSTAL

In our study of diffraction by a crystal, certain phenomena have been neglected: some of the primary and secondary radiation is absorbed by the sample, and the secondary radiation from the sample can be re-diffracted. These effects are taken into account by an absorption correction term denoted by A. The value of A can only be calculated if the sample has a simple shape such as a sphere or a cylinder.

The intensity diffracted by a crystal can now be written in the final form:

$$\boxed{I_{hkl} = C.m.L(\theta).P(\theta).A\,\frac{\Omega}{V^2}\,||F_{hkl}||^2}$$

◆ C is a constant which includes A_{el}^2 (the intensity scattered by an isolated electron) and the intensity of the primary radiation.

◆ m is the multiplicity of the line and corresponds to the number of equivalent families of lattice planes which give the same diffraction line.

◆ $L(\theta)$ is the Lorentz factor corresponding to the velocity with which the reciprocal node under consideration crosses the Ewald sphere.

◆ $P(\theta)$ is the polarisation factor. This factor is equal to $(1 + \cos^2 2\theta)/2$ for a non-polarised incident beam.

◆ *A* is the sample absorption correction.

◆ Ω is the sample volume and *V* the unit-cell volume.

◆ F_{hkl} is the structure factor which involves:

—the atomic scattering factors of the atoms in the repeating unit, corrected if necessary for abnormal scattering;

—the Debye factors, which depend on temperature, on the nature of the atoms in the unit-cell and their environments;

—the relative positions of the atoms in the unit-cell.

Given the approximations made in the determination of some of the parameters, the accuracy obtained for the diffracted intensities is of the order of a few percent. Intensity measurements are always relative since it is difficult to determine precisely the value of the constant *C*, which depends on the incident beam.

We conclude this chapter with the following diagram which summarises the steps followed in this study of diffraction by crystalline structures.

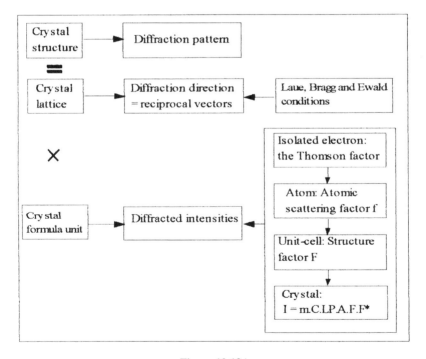

Figure 10.12A

Chapter 11

The Laue Method

1 PRINCIPLE OF THE METHOD

A single crystal placed at random in a beam of X-rays will not in general emit diffracted radiation. The Bragg equation $2d_{hkl} \sin \theta = \lambda$ must be satisfied for diffraction to be observed. With a single crystal, this can be done in two ways:

—using monochromatic radiation and rotating the crystal with respect to the beam; this is the *rotating crystal method*;
—using polychromatic radiation and a stationary crystal; this is the *Laue method*, and historically was the first to be used. (The first experiment was carried out in 1912 by W. Friedrich and P. Knipping after suggestions by M.von Laue).

Laue diagrams exhibit the following characteristics:

◆ *One* spot on the diagram corresponds to *one* family of lattice planes.
◆ For a given spot, the wavelength of the incident beam is unknown; it is not therefore possible to deduce information on the dimensions of the diffracting unit from the diagram.
◆ The intensity of emission from an X-ray tube anode is not constant with wavelength, and it is therefore not possible to obtain information from the intensity of the diffraction spots.
◆ The diagrams show the relative positions of the various lattice planes and therefore enable internal symmetries of the specimen to be revealed.

2 EXPERIMENTAL METHOD

The specimen (Figure 11.1) is generally stuck to a goniometer head (see Figure 11.10) which allows the crystal to be precisely oriented with respect to the incident beam. The beam is obtained by placing a collimator perforated with pinholes against the X-ray tube window. This collimator limits the divergence of the beam. With a conventional tube, usable wavelengths lie between $\lambda_{min} \approx 12\,400/V$ (λ in ångströms and V, the difference in potential between the filament and the anode, in volts) and λ_{max} of the order of 3 Å. The generator has to supply a 'white' radiation as intense as possible, and a tungsten anode is used; this is held at as high a voltage as possible without exciting the K series. The diffraction spots are usually recorded on flat photographic film place a few centimetres from the crystal, perpendicular to the incident beam.

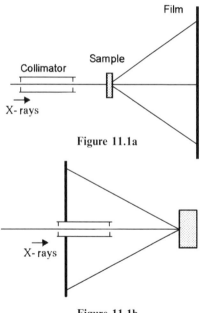

Figure 11.1a

Figure 11.1b

The diffraction patterns are either recorded by transmission (for thin or weakly absorbing specimens), as in figure 11.1a, or by reflection (for large specimens, giving 'back patterns') as in figure 11.1b.

The direction of the incident beam remains fixed in relation to the specimen. A family of lattice planes (*hkl*) of interplanar spacing d_{hkl} making an angle θ with the direct beam will diffract the wavelength λ_θ when the Bragg condition $n\lambda_\theta = 2d_{hkl}\sin\theta$ is satisfied.

Each spot on a Laue diagram corresponds to a family of lattice planes whose orientation with respect to the incident beam can be deduced from the reflection conditions.

3 CONSTRUCTION OF THE LAUE DIAGRAM

Diffraction directions are defined by the intersections with the Ewald sphere of nodes in the reciprocal lattice constructed around the origin I.

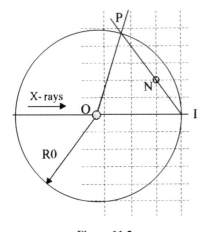

Figure 11.2

Since the wavelength is unknown, the radius of the Ewald sphere is arbitrarily taken as equal to R_0. The reciprocal lattice should be constructed to the scale $\sigma^2 = R_0\lambda$; it is in fact drawn to an arbitrary scale. If the scale is made to vary with λ, each node N of the lattice moves along the row IN. For a certain value of λ, the node will lie on the Ewald sphere at P. The straight line OP then defines the direction of the diffracted beam corresponding to the node N.

The intersection of this line with the plane of the film defines the position of the diffraction spot.

The reciprocal vector **IP**, which is normal to the family of lattice planes that diffract in the direction OP, has the modulus:

$$\| \mathbf{IP} \| = IP = 2R_0 \sin \theta = R_0\lambda/d_{hkl} \tag{1}$$

REMARKS:

◆ All the nodes of a given reciprocal row IN give the same point P, and hence a single diffraction spot: the 'harmonics' ($nh\ nk\ nl$) of a family of planes ($h\ k\ l$) with h, k, l relatively prime, all reflect under the same angle of incidence θ. We could equally well consider that the family ($h\ k\ l$) which gives first-order diffraction at wavelength λ under the angle of incidence θ, also gives second, third, fourth order, etc. diffraction at wavelengths $\lambda/2$, $\lambda/3$, $\lambda/4$ etc.

◆ Not all reciprocal lattice nodes can give rise to diffraction, since the range of wavelengths used is limited.

We consider the reciprocal lattice and the spheres of radius $1/\lambda_{min}$ and $1/\lambda_{max}$.

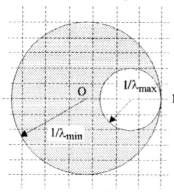

Figure 11.3

Only nodes in the grey area defined by the two spheres can give rise to diffraction.

The low wavelength limit is effectively determined by the anode supply voltage. The high wavelength limit depends in general on the crystal.

If d_M is the largest interplanar spacing in the direct lattice, the longest wavelength capable of giving a diffraction spot is, according to Bragg's law: $\lambda_{max} = 2d_M$.

◆ Figure 11.2 is constructed with the incident beam parallel to a reciprocal row. In this particular case, the construction of diffraction directions shows that families of lattice planes symmetrical with respect to the beam give symmetrical diffraction spots.

Although the Laue technique gives no useful information on the unit-cell parameters, it does show the relative arrangement of the lattice planes and hence the internal symmetries of the specimen. If the beam is oriented parallel to a symmetry element of the crystal, the diffraction pattern shows the same symmetry.

The method does not, however, enable distinction to be made between centrosymmetric and non-centrosymmetric crystals; this is due to Friedel's law.

The applications of the Laue method are thus:

—*identifying symmetry elements in unknown specimens;*
—*orienting crystals of known symmetry.*

4 FEATURES OF LAUE DIAGRAMS

◻ BLIND REGION

The minimum value of IP (cf relation 1) is $R_0\lambda_{min}/d_M$; in the centre of Laue diagrams there is an area with no diffraction spots, called the 'blind region'.

◻ ZONE CURVES

We remember that planes $(h_ik_il_i)$ are said to be in a 'zone' if they all contain the same row $[uvw]$ called the 'zone axis'. For all planes $(h_ik_il_i)$ in the zone, the reciprocal directions $[h_ik_il_i]^*$ are perpendicular to the zone axis and are contained in the reciprocal plane $(uvw)^*$. Each plane in the zone therefore satisfies the relation: $h_iu + k_iv + l_iw = 0$. The direction of the zone axis is determined by the intersection of two planes in the zone.

In Laue diagrams, the diffraction spots are seen to lie on ellipses (figure 11.5) or hyperbolas. These curves are the loci of spots corresponding to families of lattice planes with the same zone axis. Consider for example the planes $(h_ik_il_i)$ having the row $[uvw]$ as zone axis (Figure 11.4). The reciprocal plane $(uvw)^*$, normal to the zone axis, intersects the Ewald sphere in a circle. This plane contains the reciprocal rows $[h_ik_il_i]^* = IA_i$.

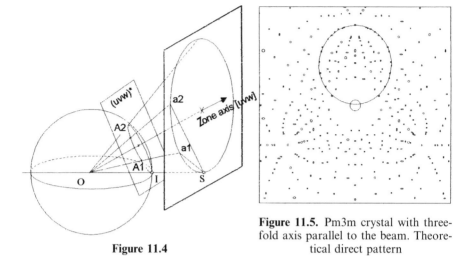

Figure 11.4

Figure 11.5. Pm3m crystal with three-fold axis parallel to the beam. Theoretical direct pattern

There corresponds to each reciprocal node A_i a diffraction spot a_i on the film. The diffracted beams OA_ia_i are in general generators of a cone whose axis is the zone axis. The zone curve of axis $[uvw]$ is thus the section of the cone through the plane of the film. Let α be the angle between the zone axis and the incident beam. If the Laue diagram is obtained by transmission, α is less than $45°$; the zone curves are then ellipses. For back patterns ($\alpha > 45°$), they are hyperbolas.

This feature is used to orient crystal by the technique of back-pattern Laue photography. If the incident beam is parallel to a symmetry axis, the zone curves are symmetrical about the centre of the pattern.

The so-called Greninger diagrams, on which are drawn networks of hyperbolae, enable this symmetry to be identified and facilitate orientation of the crystal with respect to the beam. If it is possible to determine the indices of the zone axis from those of certain spots in the zone, indexing the other spots on the zone curve becomes easier.

5   INDEXING A LAUE PHOTOGRAPH

A Laue photograph cannot generally be interpreted by inspection since the relation between the diffraction pattern and the reciprocal lattice is not simple. The **gnomonic projection**[1] method can be used for photographs obtained with a symmetry element of the crystal parallel, or nearly so, to the direction of the incident beam. Under these conditions this method enables diffraction spots in

---

[1]From the Greek Gnomon meaning sundial.

the photograph to be indexed. If, however, the crystal is randomly oriented with respect to the beam, the stereographic projection method has to be used. The stereogram obtained generally enables symmetry elements to be identified and also enables calculation of the angles through which the specimen must be rotated in order to bring a symmetry element into line with the incident beam.

5.1 Gnomonic Projection

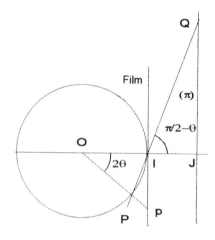

Figure 11.6

The normal IP to the planes (hkl) which gives rise to a diffraction spot p on the film makes an angle ($\pi/2 - \theta$) with the direction of the incident beam OI. Let π be the plane parallel to the plane of the film and lying at a distance IJ=1 (arbitrary unit) from it.

R is the film-specimen distance.

The point Q at the intersection of IP with π is the gnomonic projection of the plane (hkl).

From Figure 11.6 we have:

$$JQ = r = IJ.\cot\theta$$

Hence:

$$r = IJ.\cot(\tfrac{1}{2}\arctan Ip/R) \qquad (2)$$

If the distance $r = f(\text{Ip})$ has previously been calculated, the gnomonic projection can be drawn directly from the Laue photograph.

The gnomonic projection of the reciprocal lattice of the specimen is thus constructed; it is closely analogous to a 'shadow-graph' projection, since the pattern obtained is easy to identify, providing the object is practically parallel to the plane of projection. For this method to give usable results it is therefore necessary to have a significant lattice plane parallel to the plane of the film.

To enable the gnomonic projection to be drawn rapidly, a rule (Mauguin's rule) is graduated to one side of the centre in Ip (linear graduation) and along the other in $r = JQ$. The centre of the rule is set at the centre of the photograph (at the point of impact of the incident beam); the rule is then rotated so that its edge touches the centre of a spot. The distance from the centre is then measured in centimetres; the gnomonic transform lies at the corresponding graduation on the other side of the centre (see Figure 11.7).

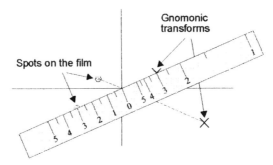

Figure 11.7

The projection cannot be interpreted by inspection. Consider for example a cubic crystal in a beam parallel to the direction of a fourfold axis. Figure 11.8 shows the projection onto (010)* of the reciprocal lattice, the film and the plane of the gnomonic projection. A rotation is also carried out of the gnomonic projection onto the plane of the pattern (the (001)* plane).

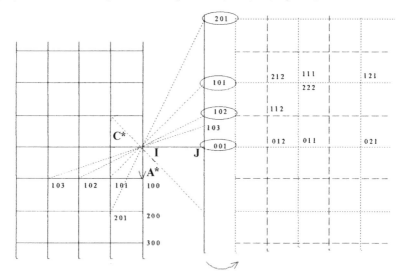

Figure 11.8

The plane π of the gnomonic projection is parallel to the reciprocal plane (001)*. The transforms of the (*hk*0) planes are at infinity. The normals to the (*hkl*) planes meet π in a network of squares of side IJ (dotted lines).

The normals to the (*hk*2) planes meet π in a network of squares of side IJ/2 (dashed lines). In general, planes with indices *h, k l* have normals passing through the points: *h*/*l*; *k*/*l*; *l*: the gnomonic projections have the coordinates *h*/*l* and k/l (in units of IJ). A spot which gives a transformed point of indices 5/3, 3/2—i.e. 10/6, 9/6—has indices of 10, 9 and 6.

In this example the diffraction spots resulting from 'harmonic' planes, can be seen superimposed both on the film and on the gnomonic projection.

This construction, then, gives quite an accurate representation of the reciprocal lattice. Its drawback is that all the reciprocal planes are superimposed on the projection. It enables diffraction spots in the photograph to be indexed provided the direction of the incident beam is very close to that of a symmetry element.

5.2 Stereographic Projection

If the crystal is randomly oriented with respect to the incident beam, gnomonic projections cannot be used. A method of stereographic projection in which the angles are conserved is then used. Consider the Ewald sphere of radius R and the diffraction spot p corresponding to the reciprocal node P defined by an angle 2θ.

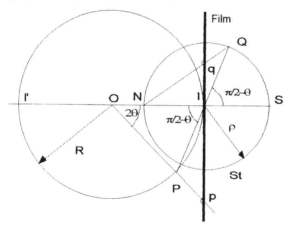

Figure 11.9

We construct a sphere of centre I and radius ρ. The equatorial plane of this sphere, which is also the plane of the film, is the plane of the stereographic projection. The straight line IP intersects the sphere St at Q. If the point N is taken as the centre of inversion, the straight line NQ intersects the plane of projection at q which is the stereographic transform of the reciprocal node P. In figure 11.9, the angle {SNQ} is half the angle {SIQ} and hence:

$$Iq = \rho.\tan\left(\frac{1}{2}\left(\frac{\pi}{2} - \theta\right)\right) \Rightarrow 2.\theta = \arctan\frac{Ip}{R} = \arctan\frac{d}{R}$$

$$Iq = \rho.\tan\left(\frac{\pi}{4} - \frac{1}{4}\arctan\frac{d}{R}\right)$$

Using this relation, a rule can be constructed and used to plot the stereographic projection by the same method as described for the gnomonic projection. In this way we obtain the stereographic projection of the reciprocal lattice of the crystal. Indexing the photograph is equivalent to indexing the stereogram; the latter is easier if we note that the planes in a zone with the row [*uvw*] give spots on the film which are distributed along a zone curve (ellipse or hyperbola), the stereographic projections of whose reciprocal nodes lie on a circle with the row [*uvw*] as zone axis. As angles are conserved in the stereographic projection, it is generally possible to identify the position of symmetry elements on the stereogram and to determine their orientation with respect to the incident beam. (The latter is normal to the plane of projection). This method can also be used for back-projections; the plane of the film then intersects the axis of the beam at I' but the origin of the reciprocal lattice remains at I.

5.3 Conclusions

The Laue method reveals the symmetry elements of a crystal. The *Laue class* can in principle be determined by taking several photographs with different orientations of the specimen; however, more sophisticated techniques are now used, including Buerger's method and the four-circle goniometer, to establish the symmetry elements. Practically the only use for Laue diagrams now is for the orientation of large specimens.

With a classical X-ray tube and normal film, an exposure time of one hour is generally sufficient to obtain a usable photograph. With ultra-sensitive films, the exposure time can be reduced to a few minutes. The accuracy of orientation of the crystal will be of the order of 10–20 minutes if a good quality goniometric head is used.

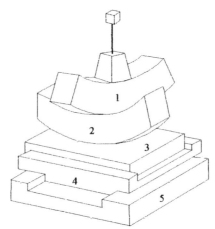

Figure 11.10. A goniometer head. 1 and 2: rotation cradles (concentric with the centre of the specimen); 3 and 4: sliding tables. 5: base

Chapter 12
The Rotating Crystal Method

1 PRINCIPLE OF THE METHOD

When a crystal is bathed in a beam of monochromatic X-rays, diffraction only occurs if a node in the reciprocal lattice lies on the surface of the reflection sphere. In order to bring reciprocal lattice nodes onto the Ewald sphere, the crystal is rotated about an axis normal to the incident beam. Rotation of the crystal results in rotation of the reciprocal lattice.

If the crystal rotation axis is randomly oriented with respect to the crystal lattice, the diffraction pattern will generally be highly complex and unusable. If on the other hand the crystal rotates about a row $\mathbf{n}_{uvw}$, the diffraction pattern will be particularly simple: the family of lattice planes $(uvw)^*$ of spacing D^*_{uvw} in the reciprocal lattice will be normal to the rotation axis, and during rotation these planes will intersect the Ewald sphere in circles S_0, S_1, $S_2 \ldots$ of spacing D^*_{uvw}.

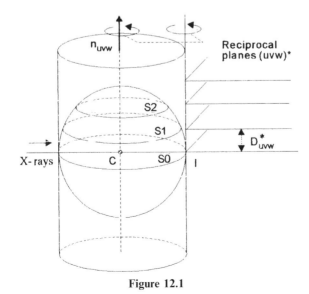

Figure 12.1

The diffracted rays will thus be spread out on a series of cones of revolution with apex at C bearing on the circles S_0, S_1, S_2.

2 CYLINDRICAL FILM ROTATION CAMERA

The crystal C is glued to a goniometer head GH and rotates about the axis of a cylindrical cassette of radius R inside which a film is held.

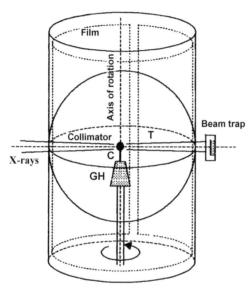

Figure 12.2

The circumference of the cassette is usually 180 mm; 1 mm of unrolled film then corresponds to 2°.

A collimator directs the incident beam onto the crystal; the beam then exits along a tube T to a lead glass beam trap; the collimator and tube together eliminate the rays diffracted by the air in the cassette.

With this apparatus, the diffraction cones intersect the film in a series of non-equidistant circles, and once the film is unwound, the diffraction spots lie on straight lines called layers.

3 DETERMINING THE PARAMETER OF THE ROTATION AXIS ROW

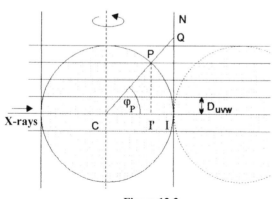

Figure 12.3

There exists a simple relation between the period along the rotation axis n_{uvw} and the distance between the layers on the film. Consider a Ewald sphere with radius R equal to that of the cassette. The reciprocal lattice must be constructed to the scale: $R\lambda = \sigma^2$. The distance between two reciprocal planes $(uvw)^*$ is D^*_{uvw}.

For the layer of order p, we have:

$$\sin \varphi_p = I'P/R = p.D^*_{uvw}/R \quad \text{or} \quad D^*_{uvw}.||\mathbf{n}_{uvw}|| = R\lambda = \sigma^2$$

$$\mathbf{n}_{uvw} = p.\lambda/\sin \varphi_p$$

On the film we measure $IQ = y_p$. From this we deduce $\varphi_p = \arctan y_p/R$ and the value of the parameter of the rotation axis row:

$$\boxed{\mathbf{n}_{uvw} = p.\lambda.\sqrt{\frac{(R^2 + y_P^2)}{y_P}}} \tag{1}$$

The accuracy can be improved by taking the two symmetrical layers which are the furthest from the origin layer (the equatorial layer).

This method enables absolute measurement of the parameters of the rows in the direct lattice. It therefore enables the unit cell parameters to be calculated.

4 INDEXING THE PATTERN

4.1 The Blind Region

Rotating the crystal about $\mathbf{n}_{uvw}$ results in rotation of the reciprocal lattice about IN (figure 12.3). During this rotation, only those nodes within the toroid generated by the rotation of the circle of centre C and radius R about IN will penetrate the Ewald sphere. The nodes lying outside this toroid will lie in the blind region of the cassette.

4.2 Relation Between the Rotation Row Indices and Indices of the p-layer Spots

Consider a spot in the layer p corresponding to a reflection from the lattice planes $(h\ k\ l)$. The reciprocal lattice node $h\ k\ l$ will lie on the p-th reciprocal plane $(uvw)^*$ above the origin. The vector normal to this family $(uvw)^*$ is the direct lattice vector: $\mathbf{n}_{uvw} = u.\mathbf{a} + v.\mathbf{b} + w.\mathbf{c}$

From the relation $D^*_{uvw}||\mathbf{n}_{uvw}|| = 1$, we deduce that the unit vector normal to the (uvw) planes is:

$$\frac{\mathbf{n}_{uvw}}{||\mathbf{n}_{uvw}||} = (u.\mathbf{a} + v.\mathbf{b} + w.\mathbf{c}).D^*_{uvw}$$

The projection of the row $[h\ k\ l]^*$ onto the normal to the planes $(uvw)^*$ is equal to D^*_{uvw}:

$$(h.\mathbf{A}^* + k.\mathbf{B}^* + l.\mathbf{C}^*)/.(u.\mathbf{a} + v.\mathbf{b} + w.\mathbf{c}).D^*_{uvw} = p.D^*_{uvw}$$

The indices h, k, l of the diffraction spots of the p-th layer are related to the indices u, v, w of the rotation row:

$$h.u + k.v + l.w = p \qquad (2)$$

If, for example, the rotation axis is [110], the spots on the equatorial layer ($p = 0$) will have the indices h, $-h$, l, and those of the first layer ($p = +1$) will have the indices $h, -h+1, l$, and so on.

4.3   Zero-layer Indexing

The distance of a diffraction spot from the centre of the pattern is $2R\theta$, The angle θ is defined by the Bragg equation:

$$n\lambda = 2d_{hkl}\sin\theta \qquad (3)$$

If the specimen makes complete rotations, a single node will penetrate the sphere twice, giving rise to two spots symmetrically disposed about the origin of the pattern. Nodes equidistant from the origin I of the reciprocal lattice will give the same diffraction spots; in particular, nodes symmetrical about the origin I, will give superimposed diffraction spots. If the pattern is not symmetrical (as in the case of incomplete rotations), the position of the direct beam in relation to the film must be established. From measurements of the angle θ, the values of the d_{hkl} spacings can be deduced.

Indexing the zero-layer is analogous to indexing a Debye–Scherrer photograph, but in the former case, only those diffraction spots corresponding to nodes in the reciprocal lattice plane containing the origin, appear on the photograph. In practice, to index the zero-layer spots, the reciprocal lattice is constructed to the scale $\sigma^2 = R\lambda$ and the positions of the spots are marked on a circle of radius R. The angle 2θ is deduced from the distance x between the spot and the centre of the pattern. The reciprocal lattice is rotated about I (figure 12.4) to locate the spots on nodes of the reciprocal lattice, and from this are deduced the indices of the spots in this layer.

NOTE: For a cubic or tetragonal crystal rotating about [001], the lattice planes corresponding to spots in the zero-layer have indices ($hk0$) and their spacings are equal to $a/\sqrt{(h^2 + k^2)}$.

The distances between the layers enable the rotation row parameters to be calculated. For a tetragonal crystal, a single photograph will enable the two unit-cell parameters to be determined.

4.4 Indexing Spots in Other Layers

As shown in figure 12.3, reciprocal nodes in the plane of order p intersect the Ewald sphere along a small circle of radius R_p such that:

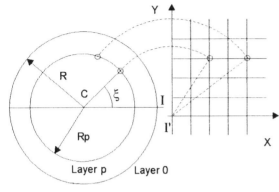

Figure 12.4

$$R_p = \sqrt{R^2 - p^2 . D_{uvw}^2}$$

The diffraction spots are transferred onto this circle; the angles ξ are deduced from the arcs x' (see figure 5) by:

$$x' = R\xi$$

The reciprocal lattice is then rotated about I′, the projection of the origin of the p-th reciprocal plane in the plane of the pattern (figure 12.4).

To determine the position of I′, the position of the axes of the unit-cell of the reciprocal lattice must be calculated in terms of the indices of the rotation row.

4.5 Coordinates of a Spot on the Photograph

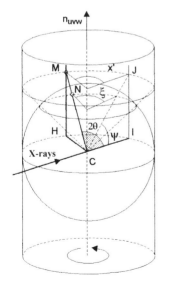

Figure 12.5

Consider a reciprocal node lying on the Ewald sphere. There will be a corresponding diffraction spot M from the p layer. The diffracted ray CM will make an angle 2θ with the incident ray CI. We also have:

$$MH \perp HC \quad \text{and} \quad MH \perp CI.$$

The distance $MH = y_p$ between the layer p and the zero layer is characterised by the angle $\psi(\tan \psi = y_p/R)$.

On the film the distance x' is equal to $R\xi$. From the vector equation:

$$\mathbf{MC} = \mathbf{MH} + \mathbf{HC}$$

we have:

$$\mathbf{MC}.\mathbf{CI} = \mathbf{MH}.\mathbf{CI} + \mathbf{HC}.\mathbf{CI}$$
$$\|\mathbf{MC}\|.R\cos 2\theta = R^2 \cos \xi$$

$$\boxed{\cos 2\theta = \cos \psi . \cos \xi}$$

The diffraction spot of indices h, k, l belongs to the p-th layer; from this we deduce y_p and ψ. From the value of d_{hkl} of this spot, relation (4) can be used to determine ξ and the value of x', and hence the position of the spot on the film.

4.6   The Advantages of the Method

The rotating crystal method enables the *unit-cell parameters to be determined*. On the other hand, determining the angles between the base vectors using this method alone is not always simple, as during complete rotations, diffraction spots corresponding to different reciprocal nodes are superimposed on the photograph. Moreover, constructing the reciprocal lattice from diffraction photographs requires geometrical constructions which can be quite complex. For this reason, other methods have been developed.

5   THE WEISSENBERG CAMERA

❐   PRINCIPLE

The Weissenberg camera differs from the Bragg camera in two ways:

◆ Translation of the film is coupled with rotation of the specimen.
◆ The angle ξ between the incident beam and the axis of the camera can be modified (figure 12.8).

With this camera, a single reciprocal plane is recorded on the film. A metal screen shields the film from reflections arising from any other reciprocal planes. To determine the correct position of the screen, a photograph is first produced without translation, with $\xi = 90°$, in order to determine the distance D^*_{uvw} between the reciprocal planes. If x and z are the coordinates of the spot on the film, ω the rotation speed of the specimen and R_f the radius of the film, we have:

$$\frac{2\theta}{2\pi} = \frac{x}{2\pi.R_f} = K_1.x$$

Cameras are usually constructed so that $K_1 = 2°\,\mathrm{mm}^{-1}$.
Since the speeds of rotation and translation are both uniform we have $\omega = K_2.z$. The gearing is arranged so that $K_1 = K_2$.

❐   THE ZERO-LAYER PHOTOGRAPH

For a row IM' of the zero layer passing through the origin I of the reciprocal lattice (figure 12.6 left) we have $\omega = \theta$. The indices of this row will be, for example, h, 0, 0 or 0, k, 0. During rotation of the crystal, the nodes of this row penetrate the Ewald sphere one by one, diffracting.

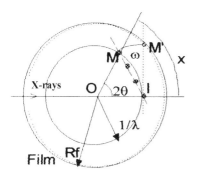 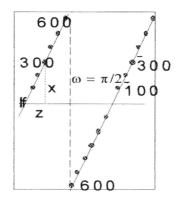

Figure 12.6

The coordinates x and z of a spot on the film are related by the expressions:

$$\theta = \frac{1}{2}K_1.x = K_2.z \Rightarrow x = 2\frac{K_2}{K_1}z$$

If $K_1 = K_2$ then this is the expression for a straight line of slope 2. The diffraction spots corresponding to the reciprocal row IM′ will be spread out along a straight line passing through the origin I_f. When $\omega = \pi/2$ the row is parallel to the incident beam and at the edge of the film. For $\omega > \pi/2$ the spots lie on a new straight line parallel to the first (figure 12.6 right). For $\omega = \omega + \pi$, there is a change of sign of the index which varies during rotation. The same analysis must be carried out for the other base vector of the reciprocal lattice. If ζ^* is the angle between the two reciprocal base vectors, two parallel systems of straight lines are in fact observed (indexed 1 and 2 on figure 12.7), separated by a distance z corresponding to a rotation $\omega = \zeta^*$.

Thus, from a zero layer photograph, the parameters of the base vectors of the reciprocal plane normal to the rotation row, and the angle between these reciprocal vectors, can be determined.

For a row PJ which does not pass through the origin (figure 12.7 left) the relation $\omega = \theta$ no longer holds: the diffraction spots are no longer spread along a straight line, but lie on festoons as shown in figure 12.7 right.

If ϕ is the angle between IP and the reciprocal axis coincident with the beam for $\omega = 0$, we have: $\omega = \pi - (\pi/2 - \theta) - \phi$.

The equation for the festoons is therefore: $x = 2(K_2/K_1)z - (\pi/2 - \phi).K_2$

Each festoon corresponds to a reciprocal row whose nodes have a common index. On the film there are two families of festoons corresponding to the two axes of the reciprocal plane being studied. The intersection of the festoons sets the position of the diffraction spots.

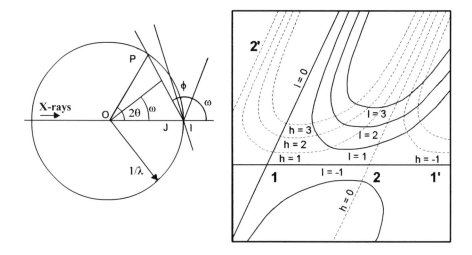

Figure 12.7

The spots can be indexed rapidly with the aid of charts.

◻ OTHER LAYERS

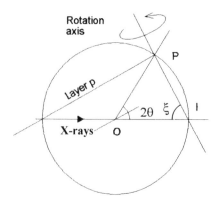

Figure 12.8

Like the Bragg camera, the Weissenberg camera has a blind region. All the nodes of the plane can be studied by tilting the axis of rotation and translation with respect to the incident beam.

In figure 12.8, the point P is the origin of the plane being studied. During rotation about IP all the nodes in the plane penetrate the Ewald sphere. A screen enables the diffraction spots of this plane alone to be selected.

In this method, a diffraction spot corresponds to a single node in the reciprocal lattice, simplifying measurement of the intensities of the diffraction spots. On the other hand, indexing cannot be carried out by inspection, but requires the use of charts. Weissenberg photographs are now only used to determine unit-cell parameters. Only the zero layer is studied, intensities now being measured using other techniques.

6 THE BUERGER PRECESSION CAMERA

6.1 Description of the Method

This method enables an undistorted representation of the planes of the reciprocal lattice of the diffracting crystal to be obtained directly, without the use of additional constructions. This in turn enables the unit-cell parameters to be determined immediately. Indexing of spots is also very simple, and the systematic absences in the plane being studied, appear clearly.

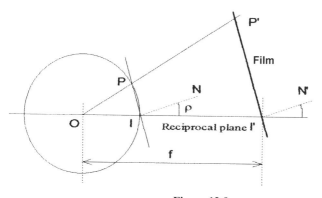

Figure 12.9

In this method, the planes of the reciprocal lattice and of the film are kept parallel. Under these conditions, we have from figure 12.9: $I'P' = IP.f/R$ and putting $R = OI = 1$, we obtain the relation: $I'P' = f.IP$.

In Buerger's method, the diffraction pattern on the film is homothetic with the corresponding plane of the reciprocal lattice.

It is because the planes of the film and of the reciprocal lattice are kept parallel that the proportionality factor, f, between the reciprocal lattice and the pattern on the film is constant. To obtain a diffracted ray, a reciprocal node must lie in a reflection position, i.e. on the Ewald sphere. In this method, this is done by having the plane of the reciprocal lattice precess about the normal which passes through the origin.

6.2   The Zero-Level Plane

While the crystal is rotating, the normal IN to the reciprocal plane containing the origin I describes a cone of angle ρ whose axis is the incident beam (figure 12.9). During this movement (figure 12.10), the plane of the origin intersects the Ewald sphere along a circle C_0. If a node P in the reciprocal lattice penetrates the Ewald sphere, diffraction occurs and the ray OP strikes the plane of the film at Pf. The diffraction spots thus lie on the projection of the circle C_0 onto the film, i.e. on a circle C'_0, since the plane of the film is parallel to C_0. The observable diffraction spots lie inside the circle of centre I' and radius equal to $2\sin\rho/\lambda$.

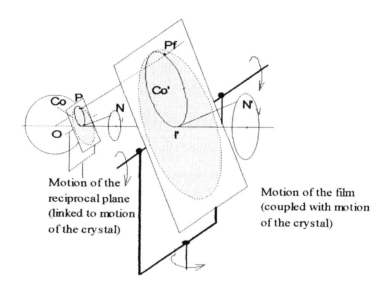

Motion of the reciprocal plane (linked to motion of the crystal)

Motion of the film (coupled with motion of the crystal)

Figure 12.10

These spots give a homothetic image of the reciprocal lattice. The lattice parameters can thus be measured directly on the photograph, corresponding to zero level.

6.3 The Other Planes

If another plane of the reciprocal lattice, situated at level n, is to be observed without distortion and to the same scale, the plane of the film must be moved a distance: $h = n.D^*_{hkl}.f/R$.

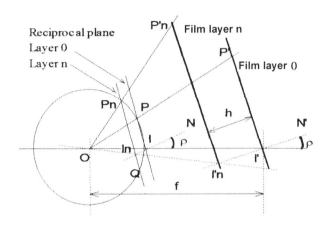

Figure 12.11

Recordings of planes at level n are especially important for the study of systematic absences in the specimen. Figure 12.11 shows that blind regions are present in the photograph and so account must be taken of the fact that nodes lying inside the circle of radius I_nQ cannot penetrate the sphere and are therefore invisible in diffraction.

6.4 The Use of Screens

To isolate diffraction spots from the level n plane, a perforated circular screen is used; this only allows through diffracted rays making an angle (α with the axis of revolution of the system. The spacing between the reciprocal planes being studied must first therefore be determined.

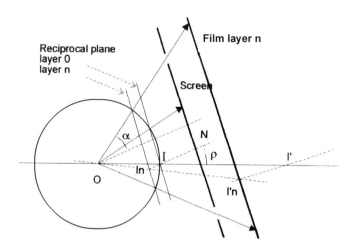

Figure 12.12

6.5   The Advantage of the Method

Each photograph is a homothetic image of a reciprocal lattice plane; from the zero layer, the unit-cell parameters can be obtained (these are the moduli of the base vectors and the angle between them). Absent spots on the photograph enable systematic absences to be identified unambiguously and to determine the space group. Other methods (e.g. the Rimsky and De Jong-Bouman retigraphs) enable undeformed images of the reciprocal lattice to be obtained. However, the relative ease of use of the Buerger camera makes it practically the only one actually used.

7   THE FOUR-CIRCLE GONIOMETER

Measuring intensities on the film is tricky and rather inaccurate, and use is now made of single-crystal diffractometers with electronic detectors (proportional or scintillation counters). The crystal is positioned in the beam using a goniometer, the most widely used being the four-circle model with Euler cradle. The Euler cradle (χ circle) drives a goniometric head TG on which the crystal is fixed. The cradle turns about the principle axis AP of the system, this axis being normal to the incident beam RX. The angle ω is defined by the rotation of the cradle about AP and the angle Φ by the rotation about the axis of the goniometric head (figure 12.13). The detector rotates about AP in the equatorial plane. The angle between the incident beam and the axis of the detector is 2θ.

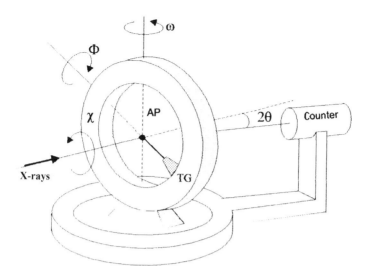

Figure 12.13

The angle 2θ is zero when the detector is aligned with the primary beam; χ is zero when the axis of the goniometer head is parallel to the principle axis; ω is zero when the plane of the cradle is perpendicular to the beam. The origin of the angle Φ is arbitrary. In principle, the angles χ and Φ are sufficient to place a reciprocal node in the equatorial plane in the diffraction position, but steric hindrance problems make ω necessary too. The four movements are achieved with computer-controlled motors.

Another type of goniometer, the kappa goniostat, is also used. This is simpler in construction and there is more room available for temperature control of the specimen.

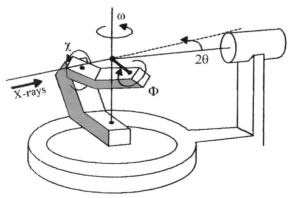

Figure 12.14

To study a crystal using these devices, the following steps are followed:

◆ The specimen is stuck to the goniometer head and optically centred on the beam.

◆ A random search for diffraction spots is carried out. From the data collected, the orientation of the crystal in the frame of the laboratory is determined (the orientation matrix)[1] and the unit-cell parameters are estimated.

◆ The unit cell parameters are refined. The values calculated in the previous step enable the diffraction directions to be defined. The directions calculated for high values of θ and for a suitable number of spots are tested and refined. At the end of this operation, accurate values for the unit-cell parameters and the orientation matrix are available.

◆ The intensities of diffraction spots are recorded.

Once the unit-cell parameters and orientation matrix are known, the values of Φ, χ, ω, and θ for which a given node hkl is in the diffraction position can be calculated. The intensities of several thousand spots are recorded.

Using this technique, intensities are recorded spot by spot; it is therefore not suitable for use with substances such as proteins which are degraded by X-rays.

8   CRYSTAL MONOCHROMATOR

In diffraction methods using monochromatic radiation, the use of a filter for eliminating K_β radiation is often insufficient, and there remain in the spectrum both high and low wavelengths which can excite fluorescent radiation in the specimen. In addition, overlapping of the $K_{\alpha 1}$ and $K_{\alpha 2}$ lines complicates the interpretation of spectra.

The solution is to use a monochromator which isolates the chosen radiation. A crystal reflection from a family of lattice planes can be used, such that the Bragg equation: $n\lambda = 2d_{hkl}\sin\theta$ is satisfied for the chosen $K_{\alpha 1}$ radiation. The harmonics $\lambda/2$, $\lambda/3$, λ/n etc. are white background radiation, and hence of much lower intensities than the $K_{\alpha 1}$ line. The drawback is that exposure times are much longer with a crystal monochromator than with a filter. To increase the useful size (and energy) of the beam, a mechanically deformed crystal can be used, in which the diffracting lattice planes have the form of a cylinder of revolution. Various types of monochromators are available and we shall describe the most widely used model, that of Johansson.

---

[1]For a detailed study, consult the International Tables.

The Johansson monochromator

A parallel-faced cylindrical disc is cut from a block of crystal; this disc has a radius 2R and its cylindrical generators are parallel to the lattice planes.

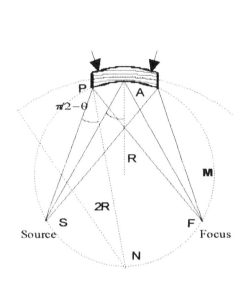

Figure 12.15

Using a press, the disc is applied to a cylinder M of radius R; the radius of curvature of the lattice planes is then 2R. All rays coming from the source S will make the same angle θ with the lattice planes. The normals to the lattice planes pass through the centre of curvature N of the disc. All angles SPN and NPF are equal to $\pi/2 - \theta$.

All the diffracted rays converge on F giving a monochromatic and stigmatic image of S.

Now, $SA = H = 2R \sin \theta$ and $n\lambda = 2d_{hkl} \sin \theta$; hence the distance between the source and the centre of the disc must be:

$$H = R \frac{\lambda}{d_{hkl}}$$

If the source S is at the focus of the tube, the energy concentrated at F will be very large. The discs used must be able to withstand the processes of cutting and elastic bending and have high reflecting power. Quartz, graphite and silicon are the most commonly used substances. The crystal must be cut and the press machined to very high accuracy to ensure constant curvature of the disc.

Monochromators are supplied in the form of compact units which fit directly onto the cathode of the X-ray tube (front monochromator); alternatively, if space is insufficient, they are placed between the crystal and the detector (back monochromator). When making intensity measurements, it should be remembered that the radiation from the monochromator will be polarised.

In the simplest case, the initial beam, the beam from the monochromator and the diffracted beam will be co-planar. If θ_M is the angle of reflection from the monochromator, the polarisation factor is:

$$P(\theta) = \frac{\cos^2 2\theta.|\cos 2\theta_M| + 1}{1 + |\cos 2\theta_M|}$$

Chapter 13

X-ray Diffraction of Polycrystalline Materials

Powder diffraction methods are widely used in the study of crystalline materials; they enable materials to be characterised both quantitatively and qualitatively without the need for single crystal specimens.

Qualitatively, diffraction techniques on powder materials enable, among other things:

—the chemical composition of the powder to be determined by comparing
—the spectrum of the sample with those contained in data base;
—the presence of impurities to be detected;
—the crystallinity of the material to be assessed.

Quantitatively, these methods allow a study of:

—the crystal parameters a, b, c, α, β, γ;
—the atom positions and space group for the simplest cases;
—powder mixtures and solid solutions;
—the presence of any structural disorder;
—the dependence on temperature of the sample parameters.

1 PRINCIPLE OF THE METHOD

The method, invented by P. Debye and P. Scherrer, involves Diffraction of a *monochromatic* beam of X-rays by a sample made up of a large number of randomly oriented single crystals. The size of the crystals is of the order of 0.01–0.001 mm, and given the very large number (from 10^7 to 10^{13}) contained in the sample, there will always be a large number for which a family of lattice planes (hkl) makes an angle θ with the incident beam to satisfy the Bragg equation $n\lambda = 2d_{hkl}\sin\theta$.

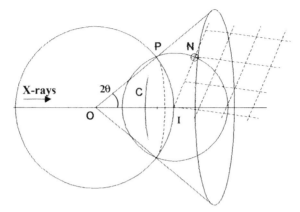

Figure 13.1

Each suitably oriented microcrystal thus gives a diffracted beam deflected by 2θ from the primary beam. The set of reflected beams forms a cone of angle 2θ with the incident beam as axis.

The problem may also be approached through the Ewald construction: the set of microcrystals can be replaced by a single crystal rotating about O; the reciprocal lattice then rotates about I and each reciprocal node N describes a sphere centred on the node at the origin I (000). Each of these spheres intersects the Ewald sphere in a circle C normal to the primary beam (figure 13.1).

The intersection of these cones with a flat film normal to the incident beam produces circular rings; if there are insufficient microcrystals, the rings appear incomplete. Each value of d_{hkl} corresponds to a diffraction cone and hence to a line on the film, a study of which enables a list of lattice spacings to be made for the sample.

2   THE DEBYE–SCHERRER CAMERA

The specimen is inside a cylindrical cassette (figure 13.2); all the diffraction lines for planes where $d_{hkl} > \lambda/2$ are obtained. The apparatus is comprised of:

—A collimator to limit the aperture and control the direction of the incident beam.

—A beam trap to absorb the primary beam as close as possible to the specimen, in order to reduce diffusion by air, which tends to fog the film and reduce contrast.

—An excentrically mounted specimen holder to centre the specimen in the beam.

—The film, wrapped round the inside of the cassette; the circumference of the latter is either 360 mm (1° per mm) or 180 mm.

The film may be positioned in three ways (figure 13.2):

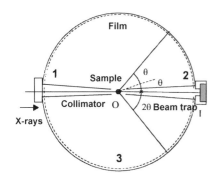

Figure 13.2

1—Normal mounting, enabling all lines for which $\theta < \pi/4$ to be observed (transmission).

2—Van Arkel mounting, enabling all lines for which $\theta > \pi/4$ to be observed (reflection).

3—Stroumanis mounting, enabling both transmission and reflection lines to be observed; this is the most common system.

The incident beam is controlled by the collimator and made monochromatic either by a filter, in which case it will consist of $\lambda_{K\alpha1}$ and $\lambda_{K\alpha2}$ radiation, or by a crystal monochromator.

The specimen is rod-shaped with a diameter of 0.3 to 0.5 mm. The powder is obtained by grinding and sieving and is either stuck to an amorphous whisker or contained in an X-ray transparent capillary (the Lindemann tube).

NOTE: Each correctly oriented microcrystal will diffract in a single direction θ, giving a spot on the film; the lines are the result of the average effect of all the microcrystals. This averaging effect can be enhanced by rotating the specimen about an axis normal to the beam with a motor fitted to the cassette.

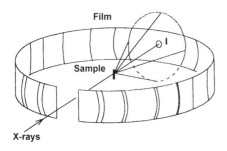

Figure 13.3.

With a cylindrical cassette, the diffraction cones form elliptical lines on the film; these have a minor axis of $4R\theta$, and it is unnecessary to record the whole ring. If a K_β filter is used, this will allow through $\lambda_{K\alpha1}$ and $\lambda_{K\alpha2}$, giving rise to two superimposed diffraction patterns.

For a given family of planes we have:

$$2d_{hkl} = \frac{\lambda_{K\alpha1}}{\sin\theta_1} = \frac{\lambda_{K\alpha2}}{\sin\theta_2}$$

For $\lambda_{K\alpha Cu}$ we have:

$$\frac{\lambda_{K\alpha1}}{\lambda_{K\alpha2}} = \frac{1.5405}{1.5443} = \frac{\sin\theta_1}{\sin\theta_2} = 0.9975.$$

The separation between lines is perceptible from angles $\theta > 15°$, and for $\theta = 80°$ the gap $\Delta\theta$ is $0.8°$. When making measurements it is advisable to sight the inner edge of the line for transmission spectra (for an unresolved doublet) and the outer edge of the line or the outer line when the doublet is resolved. Under these conditions the wavelength will be taken as $\lambda_{K\alpha1}$.

When the doublet is unresolved, the centre of the line can be sighted and the wavelength taken as $\lambda_{K\alpha} = 1/3(2\lambda_{K\alpha1} + \lambda_{K\alpha2})$. This weighting takes account of the relative intensities of the two components of the doublet.

Figure 13.4

Using Straumanis mounting, for a transmission line $(0 \leqslant \theta \leqslant \pi/4)$ of 'diameter' D, the angle θ is given by:

$$D/2\pi R = 4/2\pi.$$

For reflection $(\pi/4 < \theta < \pi/2)$, $D'/2\pi R = (2\pi - 4\theta)/2\pi$.

3 INDEXING PATTERN LINES

3.1 Measurement of Values of d_{hkl}

Diffraction line diameters must be measured as accurately as possible, care being taken to minimise systematic errors. Special care must be taken when centring the specimen and sighting the lines. For very precise measurements, a calibration compound may be mixed with the sample under study, and measurements of d_{hkl} refined by interpolations. Neglecting the error in λ, the relative uncertainty is:

$$\frac{\delta(d_{hkl})}{d_{hkl}} = \frac{1}{d_{hkl}} \delta\left(\frac{\lambda}{2.\sin\theta}\right) = -\cot\theta.\delta\theta$$

Thus the higher the value of θ, the smaller the uncertainty; the most precise measurements are in principle made on the outermost reflection lines, but the width of these (related to the natural width of the K_α line) makes precise sighting less easy. With care and a methodical approach, it is possible to obtain values of d_{hkl} to a precision of 0.002 Å.

3.2 Indexing Pattern Lines

From the Bragg equation, an equation of the type:

$$\frac{1}{d_{hkl}^2} = \frac{4.\sin^2\theta}{\lambda^2} = h^2.A^{*2} + k^2.B^{*2} + l^2.C^{*2} + 2.h.k.\mathbf{A}^*.\mathbf{B}^* + 2.h.l.\mathbf{A}^*.\mathbf{C}^*$$
$$+ 2.k.l.\mathbf{B}^*.\mathbf{C}^*$$

is obtained for each line. The reciprocal vector parameters are the same for all the equations and the indices h, k and l are integers characteristic of each line. The resulting system of equations with six unknowns is capable of solution and several computer programs are available which can deal even with compounds of low symmetry.

When searching a solution 'manually', various types of possible lattice can be tried, starting with *cubic lattices*. For these:

$$d_{hkl} = \frac{a}{\sqrt{h^2 + k^2 + l^2}} = \frac{a}{\sqrt{s}}$$

($s = h^2 + k^2 + l^2$ is a number which can be any positive integer except $s = (8p + 7)^{4q}$).

Rings can be indexed rapidly by the following method using a slide-rule. The sliding scale is removed and slid back so that its figures are upside down, its x^2-scale being in contact with the x-scale on the fixed part of the rule. If now the 1 of the sliding x^2-scale is placed opposite a number a on the fixed scale, any number n on the sliding x^2-scale will be opposite $a/\sqrt{n}$ on the fixed x-scale. The measured lattice spacings are now marked off on the fixed x-scale and the sliding scale is moved until all the lattice spacings are opposite an integer; the value of the lattice parameter a will now be found opposite 1 on the slide rule.

A graphical method can also be used: the abscissa is graduated in values $Ox = 1/d^2$ and the ordinate is graduated in the possible values of s. The value of a^{-2} is obtained directly from the straight line passing through the points so obtained and the origin.

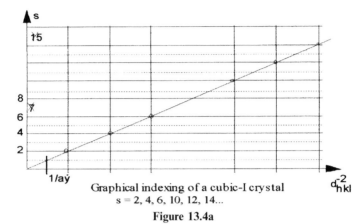

Graphical indexing of a cubic-I crystal
s = 2, 4, 6, 10, 12, 14...

Figure 13.4a

If the indices are indifferent, the lattice is type **P**. If for all the lines the values of hkl are such that the sum $h + k + l$ is even, the lattice is type **I** (as in the case of the figure above). If for all the lines the values of hkl are such that h, k, l are simultaneously even or odd, the lattice is type **F**.

Using a calculator the sequence of values $K_i = 1/d^2_{i(hkl)}$ can be calculated, and the sequence $S_i = Ki/Kl$ deduced from this. If the crystal is simple cubic, S_i is the sequence of integers (except 7, 15, 23 . . .). If $2S_i$ is a sequence of even integers, the lattice is cubic I. If $3S_i$ produces the sequence 3, 4, 8, 11, 12, 16 . . ., the lattice is cubic F.

NOTE:

—An examination of table 13.1 reveals that to distinguish a cubic P type lattice with parameter a from a cubic I type lattice with parameter $a/\sqrt{2}$, at least seven lines are necessary.

—Rays may be missing from the pattern because of systematic absences.

If the search for a cubic lattice fails, lattices with principal axes (tetragonal, trigonal and hexagonal) are tried. Charts (the Hull chart and the Bunn chart) are available to facilitate the search. Manual searching of lattices of lower symmetry is very unreliable.

It should be noted that the lower the symmetry, the greater the number of lines; e.g. the six reflections 100, 010, 001, $\bar{1}$00, 0$\bar{1}$0, 00$\bar{1}$ are superimposed for a cubic compound but give two distinct lines with a tetragonal compound and three with an orthorhombic compound. *This degeneracy in hkl (line multiplicity) must be taken into account when measuring the intensities of the lines.*

☞ | *The classical Debye–Scherrer method thus enables the lattice dimensions to be established but not the symmetry.*

Table 13.1. Possible values of s according to lattice type.

hkl	P	I $h + k + l = 2n$	F $h\,k\,l$ same parity
100	1		
110	2	2	
111	3		3
200	4	4	4
210	5		
211	6	6	
220	8	8	8
300, 221	9		
310	10	10	
311	11		11
222	12	12	12
320	13		
321	14	14	
400	16	16	16
410, 322	17		
411, 330	18	18	
331	19		19
420	20	20	
421	21		
332	22	22	
422	24	24	24

4 SPECIAL CAMERAS

4.1 Variable temperature cameras

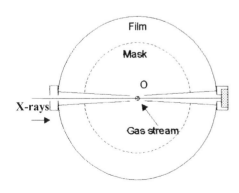

Figure 13.5

To enable the specimen temperature to be controlled, the upper and lower covers of the camera are perforated, a black paper mask protecting the film from ambient light. Temperature control of the powder specimen is achieved with a gas blower. Using this type of camera, the variation of unit-cell parameters with temperature may be followed.

4.2 Focusing Cameras

One of the major drawbacks of the Bragg camera is that the lines are often excessively wide, and this limits the accuracy of sighting and of intensity measurements. One solution is to use focusing cameras, which give very narrow lines.

❐ THE GUINIER CAMERA

The specimen is in the form of an arc of a circle C; it is bathed in a beam from a crystal monochromator converging at F on the parafocusing circle C. The diffracted beam also converges on the circle C. This type of camera can be used in transmission but only for small diffraction angles.

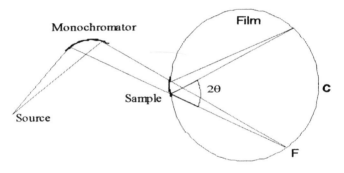

Figure 13.6

❐ THE SEEMAN–BOHLIN CAMERA

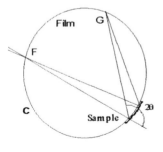

Figure 13.7

The specimen is in the form of an arc of a circle C; it is bathed in a divergent beam from the point F of the parafocusing circle C. The diffracted beam also converges on the circle C at the point G. This camera can be used in reflection with large diffraction angles. With these focusing cameras, exposure times are much shorter than with standard cameras and the lines are very narrow.

5 AUTOMATIC DIFFRACTOMETERS

The classical Debye–Scherrer camera, often involving very long exposure times, and requiring development of the film and analysis of the image, is now hardly ever used. Instead, there are now diffractometers; these are quicker to use and give results in the form of data which can be interpreted automatically.

☐ PROPORTIONAL COUNTER DIFFRACTOMETER

The specimen (figure 13.8a) is placed on a flat support which can rotate about a vertical or horizontal axis, according to the type of apparatus. A proportional counter can move about the same axis of rotation. S is the image of the source from the monochromator (using the *front* monochromator system). When the specimen support rotates through an angle θ, a system of gears causes the arm carrying the detector to rotate through 2θ; this means that when the Bragg condition is satisfied for a given position of the specimen, the detector is positioned correctly to receive the diffracted photons. To eliminate the $\lambda_{K\alpha2}$ line, a crystal monochromator can be used, placed this time between the specimen and the detector (*back* monochromator).

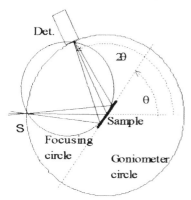

Figure 13.8a

A system of vertical slits (F1, F2, FD) and horizontal slits (Soller slits FS) enables a beam of large height (1 cm), and hence of high energy, to be used. The rotation speed of the specimen can be set between approximately $2°$ min^{-1} and $0.125°$ min^{-1}. Angles are read off the goniometer to a precision of $0.01°$; using goniometers fitted with optical coders, a resolution of one thousandth of a degree may be obtained.

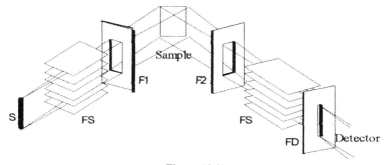

Figure 13.8b

With the specimen rotating continuously, the output of the counter is fed to an integrator; this has a time constant which smooths out fluctuations in the recorded signal, but which distorts the lines. With stepwise rotation, however, the integrator is unnecessary and the line profiles can be recorded undistorted. In both cases, the rotation speed must be slow enough for random fluctuations in the count rate to be negligible, even for weak lines. At the slowest speed, the complete recording of a spectrum requires about ten hours. This is comparable with the time for recording a spectrum on film.

A computer data acquisition system controls the motor driving the diffractometer, records the intensities of the diffracted beams, determines the positions of the lines and calculates values of d_{hkl}. For routine analyses an automatic specimen changer can be fitted.

With vertical axis goniometers it is difficult to make the powder stick to the support; the powder has to be mixed with a binder and pressed. Preferential orientation of the powder grains is then very difficult to avoid; the microcrystals are no longer randomly oriented and measurements of line intensities are erroneous. With horizontal axis goniometers, on the other hand, the powder can simply be sprinkled onto the support and the risk of preferential orientation will be smaller.

Figure 13.9 shows an example of a powder spectrum recorded with an automatic diffractometer using monochromatic radiation. The recording was made in stepwise mode with steps of $0.03°$. The compound examined is orthorhombic, which accounts for the rather large number of lines.

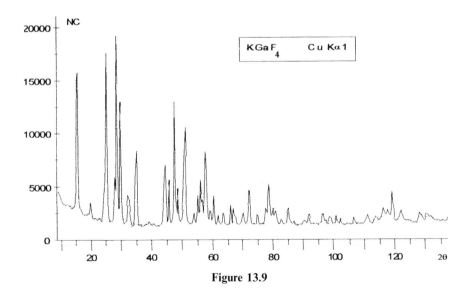

Figure 13.9

◻ LINEAR DETECTOR DIFFRACTOMETER

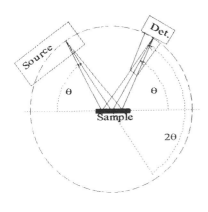

In this type of diffractometer, the specimen is immobile and horizontal. The anode, in front of which is a crystal monochromator, is placed on a moving arm rotating abut a horizontal axis. The detector is also fixed to an arm rotating about the same axis. The motions of the two arms are linked so that the angle between the incident beam and the diffracted beam is 2θ. The detector, which has an aperture angle of about 10°, is connected to a multichannel recorder.

Figure 13.10

The detector is a proportional counter with a resistive grid as cathode; it has a very linear response with angle and has a maximum resolution of the order or 0.005°. The specimen support is a temperature-controlled platinum leaf. The apparatus can be used both as a standard diffractometer and as a static mode diffractometer (without rotation); in this mode the continuous behaviour of the small detected zone in the diffraction spectrum can be studied. This is particularly useful in the study of kinetics, of the temperature variation of unit-cell parameters and of phase transitions.

◻ CURVED DETECTOR DIFFRACTOMETER

A specimen contained in a capillary tube or placed on a holder is interposed between the beam and the detector D. The radiation from a focusing monochromator is used.

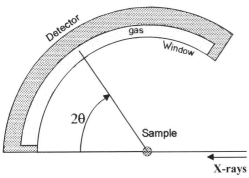

The detector is a curved counter with a 120° aperture, fitted with a continuous metal blade. This blade detects the electrons produced by conversion triggered in an exchanger gas by the photons diffracted by the specimen. The electrons produce an electric current on the blade, and this current separates into two currents i_1 and i_2 which take times t_1 and t_2 to reach the ends of the detector. $(T = t_1 + t_2)$

Figure 13.11

As T is known, measurement of $t_2 - t_1$ enables t_1 and t_2 to be determined and hence the position of the diffracted photon on the blade (figure 13.12). The memory of the corresponding channel is then incremented. Using a 120° detector and a 4096-channel memory, a precision of the order of 0.03° in the position can be achieved. The dead time of the detector is comparable to that of a proportional counter.

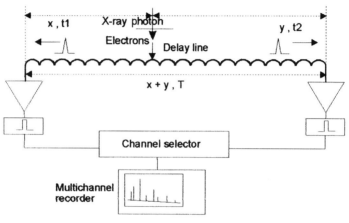

Figure 13.12

The control program of the analyser enables the contents of the memories to be displayed permanently on the screen, so that the spectrum can be watched as it builds up. The advantage of the curved detector is that a spectrum over 120° can be obtained very quickly (less than ten minutes), compared with at least ten hours using a Debye–Scherrer camera or standard detector. However, the angular linearity is not perfect and careful calibration is necessary.

6   APPLICATIONS OF POWDER METHODS

6.1   Identification of Crystallised Compounds

Each crystalline compound gives a unique powder photograph, a sort of 'signature'.

The analysis of powder diffraction photographs is a powerful method of identification. During the 1930s a data file, the Hanawalt System, was started. This was taken up again around 1940 and developed by the American Society for Testing and materials (ASTM) who published it, first in bound volumes, then as file cards, and finally in microfiche form.

27—1085   i

Cs_2NaAlF_{12}		$2CsF \cdot NaF.3AlF_3$	dÅ	Int.	*hkl*	dÅ	Int.	*hkl*
			5.8	35	101			
Cesium Sodium Aluminium Fluoride			3.66	80	104			
			3.52	70	110			
			3.14	55	015			
			3.05	100	113			
Rad. $CuK\alpha_1 \lambda$ 1.54051	**Filter**	**d-sp**	3.01	85	021			
			2.887	16	202			
Cutoff **Int.**		**I/Icor**	2.533	50	024			
Ref Courbion, G. et al., *Mater. Res. Bull.*, **9** 425 (1974)			2.397	7	107			
Sys. Rhombohedral	S.G.R#m (166)		2.340	40	205			
a 7.026(3) **b**	**c** 18.244(5) **A**	**C** 2.5966	2.284	25	211			
α β	γ **Z**3	**mp**	2.137	16	018			
Ref	5		2.054	18	214			
D_x D_m	SS/FOM $F_{18} = 19(.040,24)$		2.028	30	300			
			1.980	3	027			
Prepared by heating CsF, NaF and AlF_3 in a closed system			1.946	10	125			
under Ar between 600 and 800C for 12 hours, qunched to			1.924	50	303			
room temperature. Rhombohedral parameters: a = 7.310,			1.824	30	208			
$\alpha = 57.43°$, PSC: hR18.								

Figure 13.13. A JCPDS file card

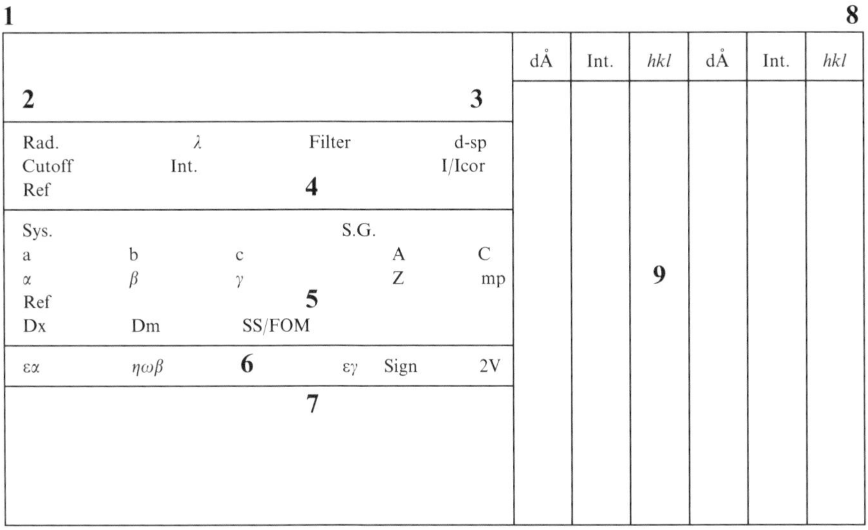

1   8

	dÅ	Int.	*hkl*	dÅ	Int.	*hkl*
2 **3**						
Rad. λ Filter d-sp						
Cutoff Int. I/Icor						
Ref **4**						
Sys. S.G.						
a b c A C						
α β γ Z mp			**9**			
Ref **5**						
Dx Dm SS/FOM						
$\varepsilon\alpha$ $\eta\omega\beta$ **6** $\varepsilon\gamma$ Sign 2V						
7						

Figure 13.14. Zones of a JCPDS file card

By 1970 the database held around 30 000 entries and in 1986, 44 000; there are now close to 60 000 entries. An international organisation, the Joint Committee for Powder Diffraction Standards (JCPDS), updates and distributes the database and user software. A CD-ROM version is now available, affording compactness and fast and easy access to data.

Classification of the database is based on the lattice spacings of the three families of planes giving the most intense diffraction lines on the photograph. Intensities are expressed as a percentage of the intensity of the strongest line which is conventionally given an intensity of 100. Figure 13.13 shows a JCPDS file card, and figure 14.14 shows the various data zones:

1—Code number (serial number followed by the compound number in the series 1–1500 for inorganic compounds and 1501–2000 for organic compounds).

2—The chemical formula, chemical name and mineralogical name.

3—The structural formula ('dot' formula).

4—Experimental conditions: Rad = source, λ = wavelength, d-sp = method, Cutoff = maximum measurable d_{hkl}, Int = method, I/Cor = ratio between the intensities of the strongest lines for the specimen and for corundum (50-50 mixture by weight).

5—Physical data for the specimen: Sys = crystal system, S.G. = group symbol, a, b, c, α, β, γ = lattice parameters, A = a/b, C = c/b, Z = number of units per cell, mp = melting point, Dx = calculated density, Dm = measured density, SS/FOM = Smith–Snyder factor of merit.

6—Optical data: $\varepsilon\alpha$, $\eta\omega\beta$, $\varepsilon\gamma$ = refractive indices, Sign = optical sign, 2V = angle between the optical axes.

7—Additional information (chemical analysis, method of synthesis etc.)

8—Quality mark. A star ★ denotes very accurate data, an **i** fairly accurate, a circle ○ unreliable, a **C**, data calculated from the structure and an **R** refinement by the Rietveld method.

9—List of all the d_{hkl}, with intensities and Miller indices.

With modern software tools it is very easy to compare powder spectra recorded on an automatic diffractometer with ones on the database and thus to identify a compound or mixture of compounds.

6.2   Quantitative Analysis of Crystalline Compounds

Consider a mixture of crystalline species whose mass concentrations c_i we wish to determine. For each species i we measure the intensity I_i of a strong line and compare this with the intensity I_i^0 of the same line measured in a mixture of known concentration c_i^0. The concentration ratio c_i^0/c_i should in principle be equal to the intensity ratio I_i^0/I_i; in practice the relation does not in general hold owing to absorption by the specimen. A reference calibration mixture, for which the relative intensities of the lines are known, must be added to the specimen and absorption effects then corrected. The composition of a mixture can be determined to within a few percent.

6.3 Determination of the Unit-cell Parameters

The principles governing the determination of unit-cell parameters are given in paragraph 3.2. The method is rapid and the measurements of the parameters can reach a precision of 10^{-5}. When such precision is required, the specimen is mixed with calibration powders (silicon, diamond etc.) whose d_{hkl} are known to a precision of 10^{-6}. The positions of the lines of the compound being studied are refined by interpolation with those of the standards.

If the specimen holder is fitted with temperature control, the technique can be used to study the temperature variation of the parameters (thermodilatometry); it is the most precise method available for determining coefficients of thermal expansion of materials.

6.4 Textures

In some materials, microcrystal orientations are not random, certain orientations predominating. This preferential orientation, or **texture**, may arise from the geometry of the microcrystals or from treatments the material has undergone.

In the case of *fibre textures*, the crystallites have one of their rows [uvw] oriented in a common direction (the fibre axis); the diffraction pattern obtained is intermediate between that for a crystal rotating about the row [uvw] and a powder photograph (the lines having uniform intensity): at the points where spots would be recorded for a rotating crystal there are lines with reinforcements in the form of centred arcs.

With *layer textures*, the crystallites tend to have normals to the plane of layers oriented in the same direction: only those reflections which correspond to the planes of the layers appear on the photograph (00l lines for the (001) planes).

6.5 Phase Transitions

With a temperature controlled specimen holder, structural phase transitions can be studied. The appearance in the crystalline medium of a new periodicity which is a multiple of the initial periodicity results in the appearance of new diffraction lines in the photograph; these are called *superstructure lines*. If the lattice spacing d_{hkl} for a family of lattice planes (hkl) becomes nd_{hkl}, the reciprocal parameter N^*_{hkl} becomes N^*_{hkl}/n.

If the phase transition is accompanied by a lowering of symmetry, the degeneracy of certain diffraction lines may be lifted. For example, during a cubic ⇔ tetragonal transition, a cubic line (100) of multiplicity 6 splits into two components: a (001) line of multiplicity 2 and a (100) line of multiplicity 4. Analysis of the splitting enables the relationship between the groups of the different phases to be established.

The method can also be used to study order–disorder transitions. Consider for example the alloy $AuCu_3$ which exhibits such a transition. In the disordered phase, obtained by quenching the compound at a temperature higher than 425°C, the atoms are distributed randomly. The diffraction pattern is identical to that of a crystal having the same lattice (P in this case) and a single type of atom C. If f_A and f_B denote the atomic diffusion factors of the components A and B, present in the proportions p_A and p_B ($p_A + p_B = 1$), then the atomic diffusion factor of the imaginary single atom C is ($p_A \cdot f_A + p_B \cdot f_B$).

In the disordered phase, the imaginary atoms C = [1/4Au + 3/4Cu] can be considered to occupy the sites 0, 0, 0; $\frac{1}{2}, \frac{1}{2}, 0$; $\frac{1}{2}, 0, \frac{1}{2}$ and 0, $\frac{1}{2}, \frac{1}{2}$. In the ordered phase, obtained by annealing, the atomic positions will be:

Au at 0, 0, 0; Cu at $\frac{1}{2}, \frac{1}{2}, 0$; $\frac{1}{2}, 0, \frac{1}{2}$ and 0, $\frac{1}{2}, \frac{1}{2}$.

The structure factors for the diffraction lines are thus:

for h, k, l of the same parity:

$$(F_{hkl})_{\text{ord}} = f_{Au} + 3f_{Cu}$$
$$(F_{hkl})_{\text{dis}} = f_{Au} + 3f_{Cu}$$

and for mixed parities:

$$(F_{hkl})_{\text{ord}} = f_{Au} - f_{Cu} \text{ (superstructure lines)}$$
$$(F_{hkl})_{\text{dis}} = 0$$

The diffraction pattern of the ordered phase has the same lines as for the disordered phase with, in addition, the so-called 'superstructure lines' which are much weaker than the 'normal' lines.

6.6   Structure Determination

Certain very simple structures such as NaCl, CsCl and rutile, depend on only a few parameters, and can be solved very rapidly by the Debye–Scherrer method. In general, however, crystal structure determination involves solving a system having nine unknowns for each atom in the repeating unit: the three position-coordinates and six thermal agitation parameters; a powder photograph will only provide us with between 20 and 40 spectral line intensities as data.

When large enough single crystals are not available, however, the powder method is the only resort. In 1969, Rietveld proposed a method enabling structures of medium complexity to be solved using powder diagrams. This is based on *profile refinement* of the diffraction lines. An initial model for the structure is proposed, and this model is then refined by point-by-point comparison with calculated and measured profiles. The spectrum is recorded in stepwise mode. From the positions of the lines, the unit-cell parameters are deduced; from the indexing and possible systematic absences, a space group is proposed; and from physico-chemical considerations or by comparison with other compounds similar to the one studied, a structural model is proposed.

For each step *i*, the intensity I_i^c is calculated and compared with the measured intensity I_i^{ob}. The least-squares method is used to minimise the quantity:

$$S = \sum_i \omega_i . |I_i^{ob} - I_i^c|^2$$

(ω_i is a weighting factor which depends on the quality of the measurement and I_i^c is the sum of the contributions of the Bragg lines close to the step *i* under consideration).

$$I_i^c = s. \sum_k m_k . LP_k . |F_k|^2 . G(\Delta\theta_{ik}) + I_i^b$$

I_i^b is the white background intensity, *s* a scaling factor, m_K the line multiplicity factor, Lp_k the Lorentz polarisation correction, F_k the structure factor, $\Delta\theta_{ik} = 2(\theta_i - \theta_k)$ and $G(\Delta\theta_{ik})$ is the reflection profile function.

There exists a large number of possible analytical functions G: Lorentzian, Gaussian or mixtures of Lorientzian and Gaussian (pseudo-Voigt) functions can all be used. In the latter case, if L_k is the width at half-maximum, then after normalisation ($0 \leqslant x \leqslant 1$) we have:

$$G = x \frac{2}{\pi . L_k} (1 + 4.X_{ik}^2)^{-1} + (1 - x).2 \sqrt{\frac{\ln 2}{\pi}} \frac{1}{L_k} e^{-4.\ln 2.X_{ik}^2}$$

The convolution of a Lorentzian by a Gaussian (pure Voigt) can also be used for G.

In the Rietveld method, the parameters to be adjusted are those for the unit-cell and atomic positions, thermal parameters, the function G and the background intensity. Several powerful computer programs are available for routine use of this method, but the user should first become familiar with the working of the apparatus by testing standard specimens. Care also should be taken to obtain a perfectly random distribution of microcrystals in the specimen.

The method is widely used in neutron diffraction, where it is often impossible to grow large enough single crystals.

This list of applications of powder methods is by no means exhaustive and shows how useful the technique is in routine laboratory work.

Chapter 14

Neutron and Electron Diffraction

Techniques using the diffraction by crystals of neutrons and electrons are complementary to X-ray diffraction. Here we give the reader a general idea of these special methods; for a fuller account, specialised works should be consulted.

1   NEUTRON DIFFRACTION

1.1   Production and Detection of Neutrons

During fission reactions in nuclear reactors, very fast high-energy neutrons are produced; the associated de Broglie wavelength $\lambda = h/mv$ is very small and of little use in diffraction experiments. The flux of neutrons is therefore passed through a moderator (heavy water or graphite) to 'thermalise' them through collisions. Once the neutrons have undergone a large number of collisions with the atoms in the moderator, they are in thermal equilibrium with these atoms and their average kinetic energy is related to the temperature in the moderator medium by:

$$\frac{1}{2}m.v^2 = \frac{3}{2}k.T$$

The average wavelength is therefore:

$$\lambda = \frac{h}{mv} = \sqrt{\frac{h^2}{3m.k.T}} = \frac{25.14}{\sqrt{T}} \quad (\lambda \text{ in Å}, T \text{ in Kelvin})$$

At a temperature of $0\,°\mathrm{C}$, the wavelength is $0.55\,\text{Å}$ and therefore suitable for diffraction by crystals. Since the speeds obey the Maxwell distribution law, the radiation is polychromatic. A crystal monochromator (Ge, Cu, Zn or Pb) is used to select a particular wavelength. Figure 14.1 shows how a neutron diffractometer functions. A cadmium collimator directs the incident neutron beam onto the monochromator crystal; a second collimator is used to select the useful radiation. The neutrons are detected by detecting the charged particles which are created during a nuclear reaction in the detector with boron, which has a sufficiently large capture cross-section for thermal neutrons.

$$n_0^1 + B_5^{10} \Rightarrow Li_3^7 + He_2^4$$

The diffracted rays are analysed by proportional counters filled with boron trifluoride or by scintillators enriched with B^{10} which detects the charged particles formed during ionisation of the light atoms produced.

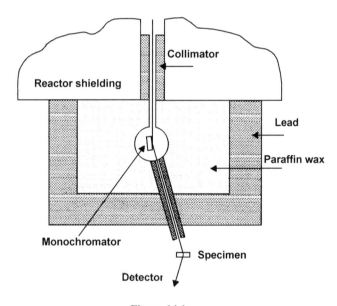

Figure 14.1

Neutrons can also be produced by a process known as spallation: pulses of high energy protons ($\approx 1 MeV$) are directed onto a uranium target; each proton generates about 25 high energy neutrons which are emitted over a very short time (about $0.4 \mu s$).

1.2 Neutron Scattering

Neutrons interact with matter on two levels: they interact with the nuclei, and their magnetic moment, associated with their spin, interacts with the magnetic moments of the target atoms.

◆ **Neutron-nucleus** interactions depend on short-range nuclear forces. Nuclear dimensions ($\approx 10^{-15}$cm) are negligible compared with the wavelengths associated with the incident neutrons; the nucleus behaves as a point, and the nuclear scattering factor b_0 is independent of the angle of diffraction. The neutron-nucleus interaction results in the formation of an unstable nucleus which returns to its ground state by emitting a neutron. For certain energies a resonance effect can occur and the scattering factor can be negative (for H^1, Ti^{48}, Mn^{55}) or may contain an imaginary part (Cd^{113}). Calculation of scattering factors is complex, and the values actually used are empirical.

The capture cross-section depends on the configuration of the nucleus, and is not related to the atomic number Z of the atoms; it is, however, very sensitive to the isotopic configuration of the nucleus. Diffraction amplitudes of neutrons and X-rays (for $\theta = 0$) are compared in table 14.1 (units are 10^{-12} cm).

Table 14.1. Scattering coefficients of X-rays and neutrons

Element	Z	b neutrons	f ($\theta=0$) X-rays
H^1	1	−0.38	0.28
H^2	1	0.65	0.28
O	8	0.58	2.25
Si	14	0.40	3.95
Fe^{54}	26	0.42	7.30
Fe^{56}	26	1.01	7.30
Fe^{57}	26	0.23	7.30
Pb	82	0.96	23.1

◆ The magnetic moment I of the nucleus can affect neutron scattering. The neutron spin can couple with I in parallel or antiparallel mode to give a total spin $J = I \pm 1/2$ and scattering coefficients b^+ and b^-. There are a total of $\{2(I+1/2)+1\} + \{2(I-1/2)+1\} = 2(2I+1)$ possible states of which there is a fraction

$$\omega^+ = \frac{2(I+1/2)+1}{2(2I+1)} = \frac{I+1}{2I+1}$$

for parallel spins with a factor b^+ and a fraction

$$\omega^- = \frac{2(I + 1/2) + 1}{2(2I + 1)} = \frac{I}{2I + 1}$$

for antiparallel spins with a factor b^-. The scattering factor $b_M = \omega^+ b^+ + \omega^- b^-$ is responsible for the coherent contribution to the nuclear moments.

♦ Atoms having an orbital magnetic moment arising from unpaired electrons interact with the magnetic moment of the neutron, giving an additional scattering factor which depends on $\sin\theta/\lambda$.

Magnetic and nuclear scattering factors have comparable orders of magnitude; when calculating the values of scattering factors, account must be taken of isotope abundances in the specimen nuclei.

1.3   Special Features of Neutron Diffraction Methods

❐ Interactions are 10^3–10^4 times weaker than with X-rays; high fluxes and large crystals are needed if a reasonable signal to noise ratio is to be achieved.

❐ b and Z are not related: it is possible to distinguish atoms with closely similar atomic numbers (e.g. $b_{Mn} = 0.36$ and $b_{Fe} = 0.96$) and to localise light atoms accurately. Hydrogen atoms, which are practically invisible in X-ray diffraction, have a scattering factor which makes them easily located with neutrons.

❐ When determining structures by Fourier analysis, positions of nuclei are obtained from nuclear coefficients and spin density distributions from magnetic coefficients.

❐ Nuclear coefficients are independent of $\sin\theta/\lambda$: with diffraction at large angles precisions greater than those obtained with X-rays can be achieved.

❐ The energy of a neutron with a wavelength around 1 Å is of the order of 0.08 eV. This is comparable with the energy of thermal vibration in a crystal and there is **inelastic scattering** of thermal neutrons; the scattered radiation no longer has the frequency Ω of the incident radiation, but has a new frequency $\Omega' = \Omega \pm \omega$, ω being the frequency of the scattering elastic wave. For X-rays, Ω and ω are respectively of the order of 10^{18} and 10^{12} Hz; the change in frequency is therefore undetectable.

On the other hand, the energy difference between the incident and scattered neutrons is readily measurable and represents the phonon $\hbar\omega$ responsible for the scattering. Since it is the geometry of the experimental apparatus which determines the value of the wave vector of the phonon, inelastic scattering of neutrons enables dispersion curves for elastic waves in the crystal to be studied.

1.4 Time-of-Flight Method

Using spallation sources, time analysis of the diffraction pattern can be carried out, instead of the usual angular analysis. For this, the technique of pulsed neutrons, introduced in 1981, is used. The neutrons produced, slowed down by a moderator, are all produced simultaneously but have different energies and speeds. In the time-of-flight method, neutrons of different wavelengths are detected according to their time of arrival at the detector. If L is the total distance travelled before reaching the detector, $mv = mL/t = h\lambda$. The detector receives neutrons scattered at a fixed angle θ_0. A family of planes (hkl) diffracts the wavelength $\lambda_{hkl} = 2d_{hkl}\sin\theta$, and for this family the time of flight will be:

$$t_{hkl} = \frac{m}{h}L.\lambda_{hkl} = \frac{2m}{h}L.d_{hkl}.\sin\theta_0$$

(For a $d_{hkl} \approx 1A$ and $L\sin\theta_0 \approx 14\,m$, the time of flight is of the order of $7\,ms$).

1.5 Magnetic Structures

Neutron diffraction directly reveals the pattern of magnetic moment orientations in crystals. Consider for example the compound MnO. Above 120 K, X-ray and neutron diffraction patterns are identical. The compound has a cubic structure of the NaCl type, with a unit-cell parameter (the chemical cell) $a = 4.43\text{Å}$. At temperatures lower than 120 K (the Neel temperature), the neutron diffraction spectrum shows extra lines and interpretation of the data shows that the unit-cell parameter (the magnetic cell) is equal to $8.86\,\text{Å}$. Analysis of the spectra shows that in one (111) plane, the magnetic moments of the Mn^{2+} ions are all parallel, and that in two successive (111) planes the moments are antiparallel (the compound is antiferromagnetic).

Neutron diffraction is a powerful tool widely used to study complex magnetic structures (ferromagnetic, antiferromagnetic, helimagnetic etc.)

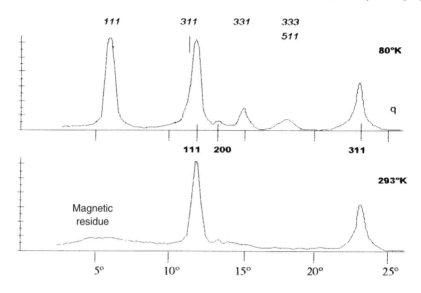

Figure 14.2. After Shull *et al.* Phys. Rev. **83** 1951. Indices in bold type: chemical cell lines. Indices in italics: magnetic cell

1.6   Neutron Absorption

For neutrons, a coefficient of mass absorption can also be defined; values are much smaller than for X-rays and only a few elements (boron, cadmium, gadolinium) have substantial coefficients. Table 14.2 gives some examples of coefficients μ of mass absorption (in $cm^2\ g^{-1}$) together with the thickness of material t (in cm) required to produce an attenuation of 99% of the incident beam for various radiations.

Table 14.2. Coefficients of mass absorption

Radiation	Be	Al	Cu	Pb
X CuKα	μ=1.50	μ=48.6	μ=52.9	μ=232
(8 keV)	t=1.67	t=0.035	t=0.01	t=0.0017
X MoKα	μ=0.298	μ=5.16	μ=50.9	μ=120
(17 keV)	t=8.3	t=0.33	t=0.01	t=0.0034
Neutrons ($\lambda \approx 1.5$ Å)	μ=0.0003	μ=0.003	μ=0.021	μ=0.0003
(0.035 eV)	t=8900	t=600	t=26	t=1430
Electrons (100 keV)	39.10^{-4}	42.10^{-4}	11.10^{-4}	$0.6.10^{-4}$

2 ELECTRON DIFFRACTION

2.1 Production and Detection

Electron beams are obtained by emission from a heated filament and are accelerated by a high potential V. Their kinetic energy is:

$$\frac{1}{2} mv^2 = eV$$

The associated wavelength is therefore:

$$\lambda = \frac{h}{mv} = \frac{h}{\sqrt{2meV}}$$

When the relativistic mass correction is included, the associated wavelength becomes:

$$\lambda = \frac{12.26}{\sqrt{E(1 + 0.979 \times 10^{-6}E}} \quad (\lambda \text{ in Å}, \ E \text{ in } eV)$$

At energies lower than 100 keV, the relativistic correction can be neglected and the simplified relation used:

$$\lambda = \frac{12.26}{\sqrt{E}} \quad (\lambda \text{ in Å}, \ E \text{ in } eV)$$

At a potential of 100 keV the wavelength is 0.037 Å.

A monochromatic beam is thus obtained. Electromagnetic lenses reduce the divergence of the beam to 10^{-3} or 10^{-4} radians. Two energy ranges are used: high energy electrons (V $\approx$ 50–120 kV, corresponding to $\lambda \approx 0.05$ Å) and low energy electrons (V $\approx$ 10–300 V, $\lambda \approx$ 4–1 Å).

Absorption by matter is substantial (see table 2) and transmission diffraction can only be carried out with very thin specimens (t $=10^{-5}$–10^{-7} cm).

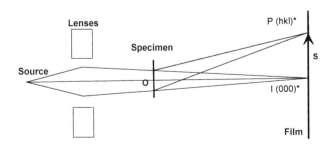

Figure 14.3

Because of the high absorption, detectors must be capable of working in a vacuum with no window. A fluorescent screen enables the diffraction pattern to be observed directly, allowing real-time positioning of the specimen in the beam. Photographic film is highly sensitive to electrons; other detectors include scintillation devices and silicon p-n junctions connected to charge transfer devices.

2.2   Electron Scattering Factor

The interaction of electrons can interact with matter can be classified as follows:

❐ Absence of interaction.
❐ Elastic scattering due to the coulombic potential of the nuclei.
❐ Inelastic scattering through interaction of electrons with the target.

The scattering coefficient f_S^e of electrons can be expressed in terms of the scattering factor f_S^x of X-rays and the atomic number Z by the Mott equation:

$$f_S^e = \frac{1}{4\pi\varepsilon_0} \frac{me^2}{2h^2} \frac{\lambda^2}{sin^2\,\theta} \left(Z - f_S^x\right)$$

Expressing f_S^e and λ in terms of m, we have:

$$f_S^e = 2.40 \times 10^8 \frac{\lambda^2}{\sin^2\theta} \left(Z - f_S^x\right)$$

2.3   Special Features of Electron Diffraction Methods

❐ Electron scattering factors are greater than those of X-rays, and so very small specimens can be used.
❐ The dependence of f on atomic number Z is less pronounced than for X-rays; light atoms in the presence of heavy atoms can therefore be more easily located.
❐ As the wavelength λ is very small in comparison with lattice spacings, $\sin\theta$ can be assimilated to θ. The Bragg equation then becomes:

$$2\theta d_{hkl} = n\lambda$$

Diffraction angles are of a few degrees.

❏ The radius of the Ewald sphere (OI in figure 14.3) is very large compared with the base vectors of the reciprocal lattice; the sphere can be assimilated to its tangent plane and a flat film used as detector.

❏ In transmission, the specimen thickness must be very small because of absorption; the Laue conditions are substantially relaxed in the direction normal to the plane of the specimen. If the beam is parallel to the direct row $[uvw]$, the reciprocal plane $(uvw)^*$ passing through the origin appears on the photograph. The spots in this plane, characterised by the vector $\mathbf{S} = \mathbf{IP}$, correspond to reflections h, k, l such that: $hu + kv + lw = 0$.
The diffraction pattern is the gnomonic projection of the reciprocal plane containing the origin.

❏ Since diffraction patterns can be obtained for microcrystals, this technique can be used for detailed analysis of polycrystalline specimens.

❏ Because of absorption, electron diffraction techniques can only be used to study surfaces.

Parallel beam electron diffraction is usually coupled with an imaging system (in the transmission electron microscope). In the imaging mode, the microcrystals are selected; their diffraction patterns are then analysed (figure 14.4).

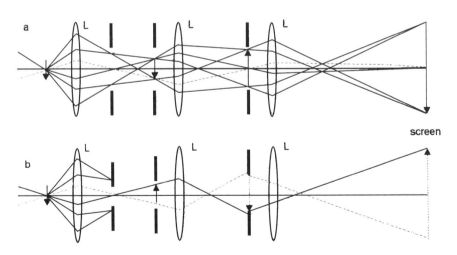

L : electrostatic or magnetic lenses (objective, intermediate, projection).
a : diffraction mode.   b : imaging mode.

Figure 14.4. The transmission electron microscope and diffraction device

Chapter 15

Determination of the Atomic Structure of Crystals

The positions of spots in a diffraction pattern depend only on the unit-cell parameters; the amplitude of the diffracted radiation, on the other hand, depends on the positions of the atoms in the unit-cell. For a given structure it is easy to determine *a priori* the diffraction pattern, but the reverse process, determining the structure from the diffraction pattern, is much more difficult; only the intensity of the diffraction spots, which is proportional to the square of the amplitude of the diffracted wave, is experimentally available. The phase of the diffracted wave can only be arrived at from experimental data by indirect methods. The problem is tricky, but it can be solved more easily with the powerful tools that are now available for numerical calculations.

Before carrying out a structure determination, the crystallographer must first gather the data: the unit-cell parameters, the overall constituents of the unit-cell, the point group and space group of the crystal. Here we shall simply give the principles behind structure determination methods.

1 DETERMINATION OF THE UNIT-CELL

1.1 Determination of the Unit-cell Parameters

Using a two-circle goniometer, optical measurements on a single crystal will give the angles between the base vectors and the ratios of the latter. This facilitates subsequent orientation of the crystal for further study. Several diffraction methods can be used to determine the parameters; the rotating crystal method will give the row parameters unambiguously, but it requires precise orientation of the crystal. The powder method does not require a single crystal, and enables higher precision to be achieved, but it is difficult to apply to compounds of low symmetry. Four-circle diffractometers also give satisfactory precision.

1.2 Constituents of the Unit-cell

Once the chemical formula has been determined by chemical or spectroscopic analysis, the molar mass M can be determined. Knowing the unit-cell parameters enables the volume V of the unit-cell to be calculated. Next the density μ of the compound is measured, and the number of structural units z per unit-cell is calculated from the equation: $z = \mu V N / M$ (N is the Avogadro number), z must be an integer. The measured density is usually lower than the theoretical value owing to inclusions in the specimen.

2 DETERMINATION OF THE POINT AND SPACE GROUPS

2.1 Determination of the Point Group

To determine the crystal class, the data obtained are treated as follows:

❏ MORPHOLOGY

When shapes peculiar to a certain class or associations of shapes are present, the crystal class and the orientation of its axes may be determined directly. To avoid any ambiguities related to shapes not modified by merohedries (e.g. the cube is only one of the shapes possible in all the cubic classes), a large number of crystals grown in various ways should be studied, since the method of growth can have a considerable influence on the crystal habit. When nucleation embryos are viewed under the microscope, high growth-rate forms appear; these embryos subsequently disappear, but their presence can indicate the crystal class.

❏ CORROSION PATTERNS

When a crystal is attacked by a solvent, a negative image of the rapid growth forms is revealed, and the symmetry of these corrosion patterns provides evidence for the crystal class. This technique can be used on crystals with no natural faces.

❏ EXAMINATION UNDER POLARISED LIGHT

Under polarised light, cubic crystals are isotropic, crystals with a principal axis are uniaxial and the rest are biaxial. It should however be borne in mind that accidental birefringence, or alternatively birefringence too weak to be observed, can occur.

❏ LAUE PHOTOGRAPHS

The Laue method enables the Laue class of the specimen to be determined, the symmetry of the photograph indicating those symmetry elements in zone with the incident beam. Because of the Friedel law, it is impossible to say from the photograph alone whether or not the specimen is centrosymmetric; this can only be done after various additional physical studies.

❏ PHYSICAL PROPERTIES[1]

◆ Some crystals become electrically polarised under a change of temperature: this is called *pyroelectricity*, and the effect can only exist in the so-called polar classes in which the symmetry operations leave the pyroelectric vector unchanged. The ten possible polar classes are:

1: the vector can have any direction.

m: the vector is parallel to a mirror plane.

2, mm2, 3, 3m, 4, 4mm, 6, 6mm: the vector is parallel to the unique axis.

◆ The *piezoelectric effect* is the appearance of an electric dipole when the crystal is mechanically stressed (the direct effect) or a deformation of the crystal under the effect of an electric field (the inverse effect). A study of the effect of symmetry operations on the coefficients of the piezoelectric tensor (third rank) shows that the effect is possible in all the non-centrosymmetric classes with the exception of the class 432.

◆ *Optical rotation or optical activity* is the rotation of the plane of polarisation of a beam of light on passing through the crystal. The phenomenon can be represented by a gyration tensor (axial, second rank). Examination of the effect of crystalline symmetry on the components of the tensor shows optical rotation to be possible in the following enantiomorphic classes:

2, 222, 3, 32, 4, 422, 6, 622, 23, 432, together with the classes:

1, m, mm2, $\overline{4}$, and $\overline{4}$2m.

◆ The *electro-optical effect* is the result of non-linear phenomena occurring when intense light passes through a crystal. Non-centrosymmetric crystals can induce light of twice the frequency. This effect (to which there corresponds a third rank tensor) is possible in all the non-centrosymmetric groups except group 432; it is highly sensitive and is now often used to detect non-centrosymmetric crystals.

From a theoretical point of view, the physical properties of a crystal are of obvious interest; however, these theoretically possible phenomena may not be

---

[1]See for example J. F. NYE, *Physical Properties of Crystals*, Clarendon Press, Oxford (1985).

experimentally detectable and in practice few crystals actually exhibit positive effects.

2.2 Determination of the Space Group

This is based on a study of systematic absences. Using an appropriate method such as the Weissenberg or Buerger camera, as many diffraction spots as possible are obtained and then indexed. Systematic absences, which depend on lattice mode and translational symmetry operations of the group, are deduced. If the class is known, the space group can be now be deduced. Three types of extinctions can be distinguished, according to the number of dimensions in the reciprocal lattice which are involved in their periodicity.

❐ THREE-DIMENSIONAL PERIODICITY RELATED TO LATTICE MODE

In table 15.1 are listed the possible reflection conditions for the various types of lattice.

Table 15.1. Extinctions related to lattice mode

Type of cell	Reflection conditions	Translations
Primitive P	None	$\mathbf{a}, \mathbf{b}, \mathbf{c}$
Face centred C	$h + k = 2n$	$\frac{1}{2}(\mathbf{a} + \mathbf{b})$
Face centred A	$k + l = 2n$	$\frac{1}{2}(\mathbf{b} + \mathbf{c})$
Face centred B	$h + l = 2n$	$\frac{1}{2}(\mathbf{a} + \mathbf{c})$
Body entred I	$h + k + l = 2n$	$\frac{1}{2}(\mathbf{a} + \mathbf{b} + \mathbf{c})$
Face centred F	h, k, l all even or all odd	$\frac{1}{2}(\mathbf{a} + \mathbf{b}), \frac{1}{2}(\mathbf{a} + \mathbf{c})$ $\frac{1}{2}(\mathbf{b} + \mathbf{c})$

❐ TWO-DIMENSIONAL PERIODICITY RELATED TO A PLANE OF TRANSLATIONAL SYMMETRY

Consider as an example a type a glide plane parallel to (010). This brings into correspondence an atom with coordinates x, y, z with an atom of coordinates $x + \frac{1}{2}$, $-y$, z. If the unit-cell atoms are grouped in pairs, the structure factor can be expressed in the form:

$$F_{hkl} = \sum_{m=1}^{n/2} f_m \cdot (e^{2j\pi \cdot (h.x_m + k.y_m + l.z_m)} + e^{2j\pi \cdot (h.(x_m + \frac{1}{2}) - k.y_m + l.z_m)})$$

$$\text{Hence}: \quad F_{h0l} = \sum_{m=1}^{n/2} f_m . e^{2j\pi.(h.x_m + l.z_m).(1 + e^{j\pi.h})}$$

F_{h0l} is other than zero only if h is even.

Table 15.2. Extinctions related to glide planes

Type de mirror	Reflection conditions	Translations
Mirror a (001)	$hk0 : h = 2n$	½ **a**
Mirror a (010)	$h0l : h = 2n$	½ **a**
Mirror b (100)	$0kl : k = 2n$	½ **b**
Mirror b (001)	$hk0 : k = 2n$	½ **b**
Mirror c (100)	$0kl : l = 2n$	½ **c**
Mirror c (010)	$h0l : l = 2n$	½ **c**
Mirror n (001)	$hk0 : h + k = 2n$	½ (**a** + **b**)
Mirror d (001)	$hk0 : h + k = 4n$	¼ (**a** + **b**)

❐ ONE-DIMENSIONAL PERIODICITY RELATED TO A SCREW AXIS

Consider a two-fold screw axis parallel to [010] and passing through $x = 1/4$ and $z = 0$. It brings into correspondence an atom with coordinates x, y, z with an atom of coordinates $1/2 - x$, $1/2 + y$, $-z$. If the unit-cell atoms are grouped in pairs, the structure factor can be expressed in the form:

$$F_{hkl} = \sum_{m=1}^{n/2} f_m . (e^{2j\pi.(h.x_m + k.y_m + l.z_m)} + e^{2j\pi.(h.(-x_m + \frac{1}{2}) + k.(y_m + \frac{1}{2}) - l.z_m)})$$

$$\text{Hence}: \quad F_{0k0} = \sum_{m=1}^{n/2} f_m . e^{2j\pi.k.y_m} . (1 + e^{j\pi.k})$$

F_{0k0} is other than zero only if k is even.

Note that the position of the axis in the unit-cell is unimportant; only its direction matters. (As an exercise, repeat the calculation with another position of the axis).

Table 15.3. Extinctions related to screw axes

Type of axis	Reflection conditions	Translation
2_1 axis along [001]	$00l: l = 2n$	$\frac{1}{2}$ **c**
2_1 axis along [010]	$0k0: k = 2n$	$\frac{1}{2}$ **b**
2_1 axis along [100]	$h00: h = 2n$	$\frac{1}{2}$ **a**
4_1 axis along [001]	$00l: l = 4n$	$\frac{1}{4}$ **c**

In the case where all the unit-cell atoms occupy special positions, there may additionally exist special extinctions.

All of the systematic and special extinctions are listed for each group in the International Tables (Volume A).

Unequivocal determination of the space group is not always possible, in which case the calculations are performed for all possible groups, and the most likely solution is adopted.

3 DETERMINATION OF THE POSITIONS OF THE ATOMS IN THE UNIT-CELL

This is a complex problem; for each atom we must determine the three spatial coordinates and the six thermal vibration parameters, making a total of nine parameters per atom in the case of anisotropic thermal vibration or four if the thermal vibration is isotropic.

3.1 Trial and Error Method

For simple structures of high symmetry, it sometimes possible to determine the structure without any calculations. The space group and the number of atoms of each type in the unit-cell may be sufficient to determine the structure. When searching for a suitable structure, the lists of equivalent positions in the International Tables can be used; certain physical considerations should also be taken into account, including typical bond lengths, atomic or ionic radii and isostructural rules (crystals with similar chemical formula often have the same structure).

To confirm a hypothesis, theoretical intensities of diffraction spots are calculated (with suitable corrections according to the experimental technique used) and compared with the measured intensities. Structure types entirely determined by the space group include: CsCl, NaCl, CaF_2 (fluorite), ZnS (zinc blende), diamond, $CaTiO_3$ (perovskite).

Thus diamond has the face centred cubic structure ($h\ k\ l$ of the same parity) with eight atoms per unit-cell; only reflections of the type $h = 2n + 1$ *or* $h + k + l = 4n$ are present in the diffraction pattern. From the International Tables, the only possibility for the space group of diamond is $F4_1/d\ 3\ 2/m$ with the atoms in sites 8a.

If only limited calculating power is available, the trial and error method can be used, provided that the structure depends only on one or two parameters. A classic example is the determination of structures of the rutile (TiO_2) type; the group is P4/mmm and the atom coordinates are:

Ti: $0, 0, 0; 1/2, 1/2, 1/2$ O: $\pm \left(x, x, 0; 1/2 + x, 1/2 - x, 1/2 \right).$

To determine the structure it is only necessary to find the value of x which gives the best agreement between the calculated and measured intensities.

When the trial and error method cannot be used (i.e. when an initial model cannot be proposed), or when it gives incoherent results, Fourier harmonic analysis methods have to be used.

3.2 Methods Using Fourier Synthesis

◻ THE PHASE PROBLEM

When the atom positions and thermal vibration parameters are known, the structure factors and the amplitude of the diffracted waves can be calculated. It has been shown that:

$$A_S = \sum_{\text{crystal}} \left(\int_{\text{cell}} \rho_i(\mathbf{r}).e^{j.2\pi.\mathbf{r}.\mathbf{S}}.dv_{\mathbf{r}} \right).e^{j2.\pi.(u.\mathbf{a}+v.\mathbf{b}+w.\mathbf{c}).\mathbf{S}} = \sum_{\text{crystal}} F_{hkl}.e^{j2.\pi.(h.u+k.v+l.w)}$$

$$\text{where: } F_{hkl} = \sum_{\text{cell}} (f_m)_t.e^{2j\pi.(h.x_m+k.y_m+l.z_m)}$$

If n is the number of atoms in the unit-cell, the structure factor can also be written in the form:

$$F_{hkl} = \sum_{m=1}^{n} (f_m)_t. \cos 2\pi(h.x_m + ky_m + l.z_m)$$

$$\tag{1}$$

$$+j. \sum_{m=1}^{n} (f_m)_t. \sin 2\pi(h.x_m + k.y_m + l.z_m) = V(A_{hkl} + j.B_{hkl})$$

The diffracted intensity I_{hkl} is proportional to $A_{hkl}^2 + B_{hkl}^2$. The electron density can be calculated using the inverse Fourier transform:

$$\rho_{xyz}^t = \int_{V^*} F(\mathbf{S}).e^{-2j\pi.\mathbf{r}.\mathbf{S}}.d\mathbf{S} = \frac{1}{V} \sum_{h=-\infty}^{+\infty} \sum_{k=-\infty}^{+\infty} \sum_{l=-\infty}^{+\infty} F_{hkl}.e^{-2j\pi(h.x+k.y+l.z)} \tag{2}$$

$$\text{now: } F_{hkl} = V.(A_{hkl} + j.B_{hkl}), \ F_{\bar{h}\bar{k}\bar{l}} = V.(A_{hkl} - j.B_{hkl})$$

$$\rho'_{xyz} - \left[A_{000} + 2 \sum_{h=1}^{\infty} \sum_{k=-\infty}^{\infty} \sum_{l=-\infty}^{\infty} (A_{hkl} \cdot \cos \, 2\pi(h.x + k.y + l.z) \right.$$
$$\left. + B_{hkl} \cdot \sin \, 2\pi(h.x + k.y + l.z)) \right] \tag{3}$$

In the series, the coefficient $A_{000} = F_{000}/V$ is equal to the total number of electrons ρ_0 in the unit-cell. B_{000} is always null and in centrosymmetric compounds, all the B_{hkl} are null. In practice, summations of the indices h, k, l are limited to the domain of measured spots. As a result there arises an uncertainty which can be diminished by working with a shorter wavelength, thereby increasing the number of spots in the pattern.

The electron density can also be written in the form:

$$\rho'_{xyz} = \rho_0 + \sum_{h=1}^{\infty} \sum_{k=\infty}^{\infty} \sum_{l=\infty}^{\infty} |C_{hkl}| \cdot \cos[2\pi(h.x + k.y + l.z) + \alpha_{hkl}] \tag{4}$$

☞ | *The modulus of the coefficients can be obtained directly by experiment, but the phase α_{hkl} remains unknown*

If the general solution to the phase problem is not known, various approximate methods are available which give satisfactory results and allow even very complex structures to be determined.

❒ THE PATTERSON FUNCTION

We consider $P(\mathbf{U})$, the self-convolution function of $\rho(\mathbf{r})$, i.e. the product of the convolution of $\rho(\mathbf{r})$ and $\rho(-\mathbf{r})$:

$$P(\mathbf{U}) = \rho(\mathbf{r}) * \rho(-\mathbf{r})$$

$$P(\mathbf{U}) = \int_V \rho(\mathbf{r}).\rho(\mathbf{r} + \mathbf{U}).d\mathbf{r} = V \int_0^1 dx \int_0^1 dy \int_0^1 dz.\rho(x, y, z).\rho(x + u, y + v, z + w)$$

The Fourier transform of a convolution product is equal to the product of the Fourier transforms of the convoluted functions. The Fourier coefficients of $\rho(\mathbf{r})$ are proportional to the F_{hkl} (equation (2)). But since:

$$F_{hkl} = F^*_{h\bar{k}\bar{l}} \rightarrow F^*_{hkl} = F^*_{h\bar{k}\bar{l}}$$

the Fourier coefficients of $\rho(-\mathbf{r})$ are proportional to the F^*_{hkl} and the Fourier coefficients of the Patterson function $P(\mathbf{U})$ are proportional to the intensities

$I_{hkl} = k.F_{hkl}.F^*_{hkl}$, which are known. This function is therefore always centrosymmetric.

To interpret the Patterson function, the structure can be idealised by replacing each atom in the unit-cell by a point charge equal to the number of electrons $z(\mathbf{r}_i)$. The function then becomes:

$$P(\mathbf{U}) = \sum_{i=1}^{n} z(\mathbf{r}_i).z(\mathbf{r}_i + \mathbf{U})$$

This function is null throughout unless $\mathbf{U}$ is an interatomic vector. In a structure containing heavy atoms (p) and light atoms (l), the values of the function P according to the vectors $\mathbf{U}$ will be:

Vector $\mathbf{U}$	Function P
Heavy atom–heavy atom	$z_p.z_p$ high
Heavy atom–light atom	$z_p.z_l$ intermediate
Light atom–light atom	$z_l.z_l$ small

The periodicity of the Patterson function is the same as that of the crystal and its unit-cell has the same dimensions. In contrast, the number of 'peaks' in the function is much greater than the number of atoms n; there are n^2 peaks of which n correspond to vectors $\mathbf{r}_{ii}$ of null length and $n(n-1)$ distributed through the unit-cell, corresponding to vectors $\mathbf{r}_{ij}$.

The figure below shows projections of a structure with three atoms per unit-cell, and the corresponding Patterson function. The peaks in the Patterson function are spread out more than the electron clouds of the atoms and if there is a large number of atoms in the cell, peaks become superimposed.

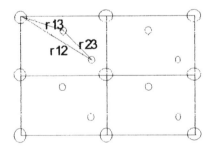
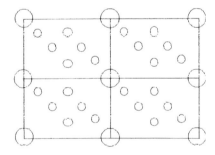

Figure 15.1

As all the vectors are brought to a common origin, the symmetry elements of the cell of the Patterson function must also be translated to this origin, when they lose any translational components. The 230 space groups give rise to only 24 Patterson groups.

❐ THE HEAVY-ATOM METHOD

An atom which is much heavier than the others can be located easily because the corresponding peaks are very strong. Structure factors of atoms of this type are therefore calculated by assuming that it is these which determine the phase of the strongest reflections. This phase is attributed to the corresponding intensity values and the Fourier series is calculated; from this are deduced the approximate positions of a certain number of atoms and the phases of the other reflections. The whole structure is determined by successive iterations. If the compound in question does not contain a heavy atom, synthesis of an isostructural compound which does contain such an atom can be attempted, and the structure then determined. Even if this second compound is not strictly isostructural, important information on atom positions can be obtained.

❐ PATTERSON VECTOR METHOD

The peak positions of the Patterson function of a structure containing the atoms $1, 2 \ldots N$ in the cell can be obtained by superimposing the images M_1, $M_2 \ldots M_N$ obtained when the atoms $1, 2 \ldots N$ are positioned successively on the origin. Solving the inverse problem is much more difficult but it is feasible when the stereochemistry and the structure of fragments assumed to be rigid are known. From this starting point, successive iterations are performed.

The main drawback of the Patterson method is that with increasing numbers of atoms in the unit-cell, overlapping between peaks becomes considerable and the atoms can no longer be identified.

3.3 Direct Methods

These methods are all based on the fact that the electron density is a strictly positive quantity, and this implies a certain number of relations between the structure factors. Through statistical analysis of the amplitude of the structure factors, information on phases can be partially reconstituted and an approximate structure arrived at.

❐ THE BASIS OF THE METHODS

Assuming that all the atoms in the unit-cell vibrate isotropically and identically, the structure factor can be expressed in the form:

$$F_{\mathbf{S}}^{t} = e^{-(B \sin^2 \theta)/\lambda^2} . F_{\mathbf{S}}$$

and if the structure factor is independent of temperature:

$$F_S = \sum_{cell} f_i \cdot e^{2j\pi.(h.x_i+k.y_i+l.z_i)} = \sum_i f_i \cdot e^{j\phi_i}$$

The unit structure factor is defined by: $U_S = F/\Sigma_i f_i$ and the normalised structure factor by:

$$|E_s|^2 = \frac{|U_S|^2}{\langle|U_S|^2\rangle} = \frac{|F_S|^2}{\langle|F_S|^2\rangle}$$

In a domain $\Delta S = S - S'$ in reciprocal space ($\Delta \sin\theta$ in the laboratory frame), we have:

$$\langle F_S . F_S^* \rangle = \sum_p \sum_q f_p . f_q \langle e^{j(\phi_p - \phi_q)}\rangle$$

If the atoms in the unit-cell are positioned randomly with a normal distribution (all positions have the same probability), the phases ϕ_p are also random and therefore $\langle\phi_p - \phi_q\rangle = 0$ if $p \neq q$.

From this we deduce Wilson's relation:

$$\langle|F_S|^2\rangle = \sum_p f_p^2 \ ; \ |E_S|^2 = |F_S|^2 / \sum_p f_p^2$$

From the above hypotheses generalised structure factors can be calculated. The results for centrosymmetric and non-centrosymnmetric structures are different.

Table 15.4

	Centrosym.	Non-centrosym
$\langle|E|^2\rangle$	1	1
$\langle|E|\rangle$	0.798	0.886
$\langle|E^2 - 1|\rangle$	0.968	0.736
$\%|E| > 1$	32	37
$\%|E| > 2$	5	1.8
$\%|E| > 3$	0.3	0.01

As there are considerable differences between the two distributions, statistical analysis of spot intensities should make it possible to decide whether or not there is a centre of symmetry in the structure. It is worth noting that there are only three factors in one thousand for which $|E|$ is greater than 3 in the centrosymmetric case. If the atoms are randomly distributed in the unit-cell, the probability of finding a direction in which a large number of atoms diffuse in phase is very small.

Statistical relations on phases (the inequalities of Harker and Kasper), were first established in 1948. The foundations of statistical analysis of data and the principles of the direct methods were set down between 1950 and 1960 by the mathematician H.Hauptman and the physicist J.Karle.

❒ THE SAYRE RELATION

In 1953, a statistical relation was established by Sayre between the phases and amplitudes of strong reflections. The relation is based on the following observation: for a structure composed of atoms whose electron densities do not overlap, the electron density function and its square are two similar functions. For these two functions, the positions of the maxima (on the atoms) and the minima (between the atoms) are identical. We can write:

$$\rho_s = \frac{1}{V}\sum_{i=1}^{N} f_s^i . \, e^{2j\pi S.r_i} \quad ; \quad \rho_s^2 = \frac{1}{V}\sum_{i=1}^{N} g_s^i . \, e^{2j\pi S.r_i}$$

The Fourier transform of ρ_S is F_s/V where:

$$F_s = \sum_{i=1}^{N} f_s^i . \, e^{2j\pi S.r_i}$$

Assuming all the atoms in the cell to be identical we have:

$$F_s = f_s \sum_{i=1}^{N} e^{2j\pi S.r_i} \; ; \; G_s = g_s \sum_{i=1}^{[N} e^{2j\pi S.r_i} \Rightarrow F_s = \frac{f_s}{g_s} G_s$$

When the atoms are different, these relations become:

$$F_s \approx \frac{\langle f_s \rangle}{\langle g_s \rangle} G_s = \gamma_s . \, G_s$$

The Fourier transform of ρ_s^2 is the convolution product $\frac{1}{V}F_s * \frac{1}{V}F_{s'} . F_s$. Since F_s is defined only on the nodes of the reciprocal lattice, the convolution integral reduces to the summation:

$$G_s = \frac{1}{V}\sum_{s'} F_{s'} . F_{s-s'}$$

From this we deduce the Sayre relation:

$$F_s = \frac{\gamma_s}{V} \sum_{s'} F_{s'} \cdot F_{s-s'}$$

For large values of S (large angles of diffraction), values of F tend to zero; it is therefore preferable to work with unitary structure factors.

If U_s is large, the signs of the large terms in the sum $U_{s'} \cdot U_{s-s'}$ must mostly be that of U_s. We can therefore write:

$$\text{sig}(U_s) = \text{sig}\left(\sum_{s'} U_{s'} \cdot U_{s-s'} \right)$$

In this deterministic form, the relation is of little use since the signs of every term must be known to obtain a single one. The probabilistic version is much more fruitful; if U_s is large, it is *likely* (but not certain) that the signs of the terms in the summation are correct. The larger the products $U_{s'} \cdot U_{s-s'}$, the greater this probability. If we denote the sign of the reflection vector S by $\text{sig}(S)$, the latter observation can be expressed by the relation:

$$\text{sig}(S) \approx (\text{sig}(S') \cdot \text{sig}(S - S')$$

or by the equivalent relation:

$$\text{sig}(S) \cdot \text{sig}(S') \cdot \text{sig}(S - S') \approx +1$$

The sign $\approx$ indicates that the relation is no more than probable.

For non-centrosymmetric structures, the Sayre relation can, after rendering the phases explicit, be written:

$$|F_s| e^{j\phi_s} = \frac{\gamma_s}{V} \sum_{s'} |F_{s'}| \cdot |F_{s-s'}| e^{j(\phi_{s'} + \phi_{s-s'})}$$

On comparing the real and imaginary parts, we obtain the tangent formula from which the value of ϕ_s is obtained:

$$\tan \phi_s = \frac{\sum_{s'} |F_{s'}| \cdot |F_{s-s'}| \sin(\phi_{s'} + \phi_{s-s'})}{\sum_{s'} |F_{s'}| \cdot |F_{s-s'}| \cos(\phi_{s'} + \phi_{s-s'})} :$$

Statistical analysis of the diffracted intensities also gives information on the symmetry elements, allowing detection of elements not revealed through systematic absences.

❐ DIRECT METHODS

Once numerical calculation tools had been developed, the probabilistic approach of Hauptman and Karle to the phase problem proved extremely fruitful.

The direct methods now in use are derived from the so-called symbolic addition method of Karle. From an initial set of *known signs* (known by the choice of origin or through inequalities) and *unknown signs*, to which are given symbols, the greatest possible number of signs are generated for the centrosymmetric structure. All the signs of the strongest terms can be obtained by a series of iterations.

Experience has shown that only a small number of symbols need be introduced (fewer than six), and it is therefore possible to make an exhaustive study of all the possibilities. This is because in a structure, the number of directions in which many atoms diffuse in phase is small.

Various computer programs (SHELX, XTAL, NRCVAX, MULTAN, CRYSTALS etc.), based on complex iterative algorithms, are now available to crystallographers for the determination of structures by successive approximations.

Before the development of modern numerical methods, use was made of an analogue method based on a photographic summation of terms in the Fourier series. Using exposure times proportional to the amplitudes F_{hkl}, a film was exposed, with sinusoidal fringes of pitch and orientation dependent on the values of h, k and l. The method is limited to the first terms in the Fourier series, but it is these which are the most important.

3.4 Structure Refinement

Because of the limited number of spots which are taken into consideration, and the various approximations used, the results obtained are imperfect and the structures must be refined so as to reduce the discrepancies between the measured and calculated intensities of all the diffraction spots.

Before this can be done, the results must be analysed and in particular checked for plausible interatomic distances and bond angles which conform to stereochemical requirements. Similarly, ellipsoids of thermal vibration must have volumes compatible with those of neighbouring atoms.

Powerful software is available for drawing the structures obtained. Stereoscopic images can be produced to allow three-dimensional views, the two images being calculated for the angle of vision of each eye.

Confidence in a hypothetical structure is measured by the *reliability index R*, or *R factor*, defined by:

$$R = \frac{\sum_{hkl} |\sqrt{I_m} - k.\sqrt{I_c}|}{\sum_{hkl} \sqrt{I_m}} \quad \text{where} \quad k = \frac{\sum_{hkl} \sqrt{I_m}}{\sum_{hkl} \sqrt{I_c}}$$

To carry out structure refinement, crystallographers make use of programs (generally part of a structure determination package) which use the least mean squares method to find the best values for the parameters of each atom in the unit-cell.

If any doubt remains about the space group of the compound, the structure and reliability index for each of the possible groups are determined for a maximum number of independent spots. The structure factor giving the lowest R factor is adopted. In practice, R factors lower than 0.05 are seldom obtained.

The quality of a structure determination depends on the quality of the crystal which has been used for the measurements; the greatest care must be taken in choosing a suitable specimen.

Chapter 16

Structural Chemistry

1 INTRODUCTION

From a structural point of view we can consider two types of crystal: molecular crystals, in which the molecules remain individual, and macromolecular crystals formed from periodic three-dimensional networks. In molecular crystals the molecules are held together by Van der Waals forces or by hydrogen bonds, both of which are weak. In macromolecular crystals there may be three-, two- or one-dimensional networks; in the latter two cases sheets or fibres are held together by Van der Waals forces or hydrogen bonds. The strong bonds in these crystals are either localised and covalent, or ionic, or delocalised as in metals. Structural chemistry attempts to predict structures *a priori*; this is a difficult exercise and very often an explanation of the observed facts *a posteriori* is all that can be given.

1.1 Chemical Bonds in Crystals

For strongly electronegative atoms, it is possible to predict the kind of bonds set up between them.

❐ HIGHLY ELECTRONEGATIVE ELEMENT WITH HIGHLY
  ELECTROPOSITIVE ELEMENT

The electropositive element (e.g. Na) tends to lose its electron, which is weakly bound and the electronegative element tends to gain an electron; the result is the formation of ions (Na^+ and Cl^-) and an ionically bonded (heteropolar) structure with cations and anions in electrostatic equilibrium.

❐ Association Of Highly Electronegative Elements

The two atoms are both 'hungry' for electrons; with non-metals there is a sharing of valence electrons through the creation of molecular orbitals and covalent or homopolar bonding.

❐ Association Of Highly Electropositive Elements

This is the case for metals. The valence electrons are weakly bound and circulate freely throughout the structure; a simple model is that of a lattice of nuclei bathing in a sea of conduction electrons.

❐ Actual Bonding

The extreme cases described above are rare, and bonds are usually of an intermediate type. Other phenomena must be taken into account, including long-range forces and polarisability; for example, small highly charged cations can deform the electron cloud of large anions, causing a bond which should be ionic to acquire a marked covalent character.

❐ Relations Between Structure And Properties

The physical properties of a material are strongly influenced by the nature and strength of the bonds and the arrangement of the electrons in the solid. The bond energy affects the hardness, melting point, plasticity and many other properties, while the arrangement of the electrons affects the electrical and optical properties of the material.

1.2 The Ionic Bond

The energy of an ionic crystal arises from two terms: a negative term due to coulombic interaction between the ions and a positive term due to the repulsion occurring when the electron clouds from different ions begin to overlap; it is this which confers on the ion a certain radius. If the Van der Waals forces are neglected, the energy can be written in the form:

$$E = \frac{1}{4\pi\varepsilon_0} \sum_j \frac{q_i \cdot q_j \cdot e^2}{r_{ij}} + \sum_j B_{ij} e^{-\alpha_{ij} \cdot r_{ij}}$$

The repulsive term can be written in the form: $B'.r^{-n}$ ($n \approx 9$).

For crystals made up of monatomic ions, such as NaCl, it is quite easy to calculate the electrostatic energy; if R is the shortest distance $Na^+ - Cl^-$ in the crystal, the Coulombic energy is written:

$$E = -\frac{1}{4.\pi.\varepsilon_0}\sum_j \frac{q_i \cdot q_j \cdot e^2}{r_{ij}} = \frac{e^2}{4.\pi.\varepsilon_0 \cdot R}\left(-\frac{6}{1}+\frac{12}{\sqrt{2}}-\frac{8}{\sqrt{3}}+\frac{6}{2}-\cdots\right) = \frac{e^2}{4.\pi.\varepsilon_0 \cdot R}M$$

(There are 6 Cl^- at R, 12 Na^+ at $R\sqrt{2}$ etc.)

The sum of the series is called the Madelung constant M, and it is a function of the type of structure. Values of this constant are given in table 1 for several structures:

Table 16.1. The Madelung constant for some typical structures

Type	M	Type	M
CsCl	1.76267	CaF_2	2.5194
NaCl	1.74756	TiO_2 (rutile)	2.408
ZnS (wurtzite)	1.64132	$CdCl_2$	2.2445
ZnS (blende)	1.63806	CdI_2	2.1915

In an ionic structure, the ions appear as undeformable charged spheres surrounded by the greatest possible number of oppositely charged neighbouring ions compatible with the overall electrical neutrality of the structure.

1.3 The Covalent Bond

Here it is the sharing rather than the transfer of electrons which gives rise to the forces of attraction between the ions in a crystal. A complete account of covalent bonding is beyond the scope of this introduction; suffice it to say that the rule of $8-N$ can be used to determine the number of covalent bonds formed by an atom: an element in the Nth column of the periodic classification will acquire an inert gas configuration by establishing $8-N$ covalent bonds ($4 \leqslant N \leqslant 7$) with neighbouring ligands. These bonds have well-defined directions in space which are determined, according to the molecular orbital models in general use, by orbital hybridisation.

1.4 Other Types of Bonds

❒ METALLIC BONDING

The electron-sea model is too simplistic, and band theory (the Bloch model) should be used to obtain a correct interpretation of the behaviour of metals.

❐ VAN DER WAALS BONDING

Van der Waals forces arise from interactions between intrinsic or induced dipole moments (The Keesom and Debye forces) or higher order induced moments (London forces). To a first approximation, these forces together produce an attractive force varying as r^{-7}. At very short range, the London force becomes very large and repulsive, resulting in an impenetrable region. For atoms, this domain is a sphere whose radius is the Van der Waals radius.

❐ HYDROGEN BONDING

This results from the association between a molecule H−A (A = O, N, S, C etc.) and a group B (O, N, Cl, F etc.) possessing an electron pair. The bond is symbolised A–H . . . B, and its stability arises from the electrostatic attraction between the polar bond A–H and the lone pair on B, and also from the polarisation of this pair through the action of the dipole A–H. The energies involved are small.

1.5   Rigid Sphere Models

In a structural array, the positions occupied by the atoms are a result of the equilibrium between the forces of attraction and repulsion; this conveys the idea that the atoms are rigid spheres, and in many cases the atoms can indeed be considered as rigid incompressible spheres. Since the forces of attraction between atoms depend on the nature of the bonding, for a given atom several radii must be considered: a Van der Waals radius, a metallic radius, several ionic radii depending on the charge on the ion, and covalent radii dependent on the type of bond.

Table 16.2. Ionic radii

Li^+	0.74	Mg^{2+}	0.72	Al^{3+}	0.53	F^-	1.33
Na^+	1.02	Ca^{2+}	1.00	Ga^{3+}	0.62	Cl^-	1.81
K^+	1.38	Ba^{2+}	1.36	Cr^{3+}	0.61	Br^-	1.96
Rb^+	1.49	Zn^{2+}	0.75	Fe^{3+}	0.64	I^-	2.20
Cs^+	1.70	Cu^{2+}	0.73	Ti^{4+}	0.60	O^{2-}	1.40

[1]Schannon R.D., Prewitt C.T.—*Acta Cryst.*, **B25**, 925 (1969) & **B26**, 1046 (1970)
Schannon R.D. —*Acta Cryst.*, **A32**, 751 (1976).

Various tables giving these radii are available. As an example, table 16.2 gives values of ionic radii selected from the table of Schannon and Prewitt,[1] and table 16.3 gives several values of metallic radii.

Values in Å, based on an O^{2-} radius of 1.40Å and for 6-coordination. These values are affected by the coordination, the spin state and the polarizability of the atom.

Metallic radii given for 12-coordination or 8 for the alkali metals (Values in Å)

Table 16.3. Metallic radii

Li	1.52	Mg	1.60	Al	1.43	Ag	1.44
Na	1.86	Ca	1.97	Ga	1.35	Au	1.44
K	2.30	Ba	2.22	Cr	1.28	Cd	1.51
Rb	2.47	Zn	1.34	Fe	1.26	Hg	1.51
Cs	2.67	Cu	1.28	Ti	1.46	Pb	1.75

1.6 Coordination

The immediate environment of the atom is characterised by the coordination number and the coordination polyhedron. The coordination number is the number of nearest neighbours of an atom. In simple structures, all the immediate neighbours of an atom are much closer to it than the second furthest neighbours and it is easy to state the coordination number; the situation is more ambiguous for more complex structures where the environment of the atom is heterogenous (through the nature and/or the distances of the atoms).

Unless otherwise stated, we shall assume that the central ion is a cation M surrounded by anions X. The coordination polyhedron is obtained by joining the centres of the anions. The simplest polyhedra are shown on figure 16.1; here, coordinated atoms are represented in 'compact' mode, with atomic radii to the scale of the diagram, and in 'ball-and-stick' mode, with atomic radii reduced so as to show the bonds. On the projections of the polyhedra the heights of the central cations are given in italics.

These polyhedra are useful in two ways: they can be used to depict chemical entities such as $(SiO_4)^{4-}$ tetrahedra and $(MF_6)^{4-}$ octahedra, and they enable a structure to be described by an assemblage of polyhedra which can be linked by the corners, the edges or the faces.

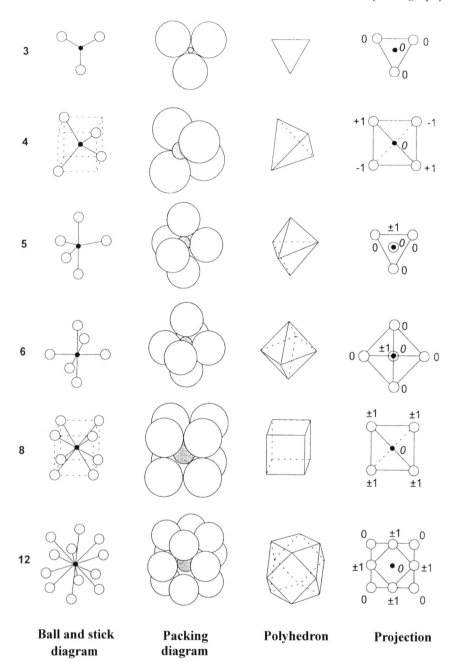

| Ball and stick diagram | Packing diagram | Polyhedron | Projection |

Figure 16.1

2 IONIC CRYSTALS

2.1 Conditions for Stability

◻ THE RELATION BETWEEN RADIUS AND COORDINATION
NUMBER

Consider a structure in which a small cation is surrounded by a number of anions. $Y =$ the maximum possible coordination will depend on the relative dimensions of the two ions (the radius ratio), and is obtained when the ions are touching each other and the central atom. If the anion radius increases any further, the anions will push each other apart and will no longer be touching the central ion; the potential energy will increase and the system become unstable. A new assemblage, with a different coordination number, is produced.

◆ 4-coordination
The four anions are located at the corners of a tetrahedron of side $2a$ at the centre of which is the cation. The limiting case with the anions touching each other occurs when:

$$2R^- = 2a; \ 2.(R^+ + R^-) = a\sqrt{6} \text{ (the diagonal of the tetrahedron)}; \ R^+ = a(\sqrt{3/2} - 1).$$
$$R^+ R^- = \sqrt{3/2} - 1 = 0.2247.$$

◆ 6-coordination (NaCl)
The six anions are located at the corners of an octahedron at the centre of which is the cation. The limiting case occurs when:

$$2R^- = a\sqrt{2}/2 \text{ i.e. } R^- = a\sqrt{2}/4; \text{ now, } R^+ + R^- = a/2. \ R^+ = a(2 - \sqrt{2})/4.$$
$$R^+/R^- = \sqrt{2} - 1 = 0.414.$$

◆ 8-coordination (CsCl)
The eight anions are located at the corners of a cube and the cations are at the centres of the cubes in octahedral interstices. The limiting case occurs when:

$$R^- = a/2; \ R^+ + R^- = a\sqrt{3}/2 \text{ hence: } R^+ = a(\sqrt{3} - 1)/2$$
$$R^+/R^- = \sqrt{3} - 1 = 0.732.$$

◆ 12-coordination (BaTiO$_3$)
The Ba^{2+} cation is at the centre of a cage of twelve O^{2-} ions constituting a cubeoctahedron (see figure 16.1). $2R^- = a\sqrt{2}/2; \ R^+ + R^- = a\sqrt{2}/2; \ R^+ = a\sqrt{2}/4$ hence: $R^+/R^- = 1$

If the ratio is lower than this limit, the cation will no longer be touching the anions and only a lowering of coordination can reduce the energy of the system. The stability conditions deduced from these purely geometrical considerations are given in table 4.

Table 16.4. Limiting values of radius ratios

Coordination	$a = R^+/R^-$	Examples	
12	$\geqslant 1$	$BaO_{12} : 1.08$	Perovskite
8	$0.732 \leqslant \text{ to } \leqslant 1.000$	$CaF_8 : 0.80$	Fluorite
6	$0.414 \leqslant \text{ to } \leqslant 0.732$	$TiO_6 : 0.50$	Rutile
4	$0.224 \leqslant \text{ to } \leqslant 0.414$	$SiO_4 : 0.30$	Quartz
3	$0.155 \leqslant \text{ to } \leqslant 0.224$		

The purely geometrical constraints of the rigid sphere model are not in themselves sufficient to specify the type of structure.

❒ PAULING'S RULES

Pauling stated a number of rules for ionic crystals, based on energy considerations; the three most important are as follows:

◆ A coordination polyhedron is formed around each ion. The anion-cation distance is determined by the sum of the ionic radii, and the coordination number by the radius ratio.
◆ In a stable ionic structure, the valency (or ionic charge) of each anion is equal or very close to, and opposite in sign from, the sum of the electrostatic valencies of the adjacent cations. The electrostatic valency v of a cation is its charge divided by its coordination number.

EXAMPLE: *The perovskite structure.*
–$LaAlO_3$ (Each La^{3+} is surrounded by twelve O^{2-} and each Al^{3+} is surrounded by six O^{2-})

$$v \text{ La–O} = 3/12 = 1/4; \quad v \text{ Al–O} = 3/6 = 1/2.$$

The oxygen is surrounded by two Al and four La, hence the sum of the electrostatic valencies of the adjacent cations $= 2 \times \frac{1}{2} + 4 \times \frac{1}{4} = 2$.

$KNbO_3$ (Each K^+ is surrounded by twelve O^{2-} and each Nb^{5+} is surrounded by six O^{2-})

$$v \text{ K} = 1/12; \quad v \text{ Nb–O} = 5/6.$$

The oxygen is surrounded by two Nb and four K, hence the sum of the electrostatic valencies of the adjacent cations $= 2 \times 5/6 + 4 \times 1/12 = 2$.

◆ The stability of the structure decreases if the coordination polyhedra are linked by edges and still more if they are linked by faces. The effect is greater the higher the charge on the cation and the lower its coordination.

2.2 Examples of Binary Structures

❒ THE CsCl STRUCTURE (CESIUM CHLORIDE)

This structure occurs when: $0.732 < R^+/R^- < 1$. There is a single formula unit per unit-cell (Cl: 0, 0, 0; Cs: 1/2, 1/2, 1/2). The lattice is cubic simple ($a = 4.123$ Å).
EXAMPLES: CsCl, CsBr, CsI, RbF, TlCl, NH_4Cl, AgI.

❒ THE NaCl STRUCTURE (SODIUM CHLORIDE)

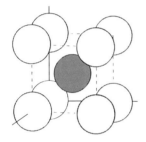 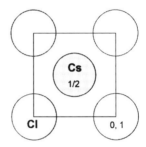

Figure 16.2a. CsCl   **Figure 16.2b**

This structure occurs when: $0.414 < R^+/R^- < 0.732$.
The space group is Fm3m. There are four formula units per unit-cell ($a = 5.64$ Å).
This is the structure of numerous halides and oxides such as SrO, MgO, BaO, CaO.

Table 16.5. The ratio R^+/R^- for the alkali halides

R^+/R^-	Li	Na	K	Rb	Cs
F	0.57	0.77	1.022	1.13	1.25
Cl	0.42	0.56	0.76	0.84	0.92
Br	0.39	0.52	0.70	0.78	0.85
I	0.35	0.46	0.63	0.69	0.76

Light shaded: NaCl type structure. Dark shaded: CsCl type.

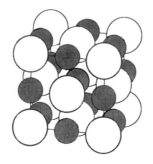
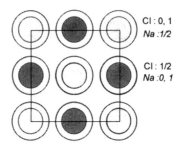

Figure 16.3a. NaCl   Figure 16.3b

◻ THE CAF₂ STRUCTURE (FLUORITE)

The space group is Fm3m ($a = 5.463$ Å). The reduced coordinates are:

Ca: 0, 0, 0 + cubic face-centred.

F: 1/4, 1/4, 1/4, 1/4, 1/4, 3/4 + cubic face-centred.

The coordination of the anion is four, and of the cation eight.

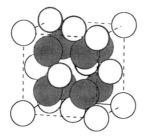

Figure 16.4a. Fluorite   Figure 16.4b

This structure occurs when: $0.732 < R^+/R^- < 1$. Examples include:

Table 16.6

Compound	CdF₂	CaF₂	HgF₂	SrF₂	PbF₂	BaF₂
a/Å	5.39	5.45	5.54	5.81	5.94	6.18
R A⁺/Å	0.97	0.99	1.10	1.13	1.21	1.35
RC/RA	0.73	0.744	0.827	0.849	0.909	1.015

Certain oxides such as ThO_2, UO_2, ZnO_2 also have this structure.

❒ THE TiO₂ STRUCTURE (RUTILE)

The space group is P4₂/mnm ($a = 4.954$ Å, $c = 2.958$ Å). There are two formula units per unit-cell.

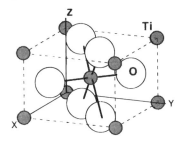

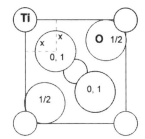

Figure 16.5a. Rutile   Figure 16.5b

The reduced coordinates are:

Ti: 0, 0, 0; 1/2, 1/2, 1/2;   O: $\pm(x, x, 0; 1/2 + x, 1/2 - x, 1/2)$.   ($x = 0.305$).

The titanium is six-coordinate and the oxygen three-coordinate.

This structure occurs when $0.41 < R^+/R^- < 0.73$ and is found in many oxides such as SnO_2, PbO_2, MnO_2, MoO_2, and also in fluorides such as FeF_2, CoF_2, ZnF_2, NiF_2, MnF_2.

❒ THE SiO₂ STRUCTURE (CRISTOBALITE)

This form of silica is stable only at high temperatures ($T > 1470°C$). Its space group is Fd3m ($a = 7.06$ Å) and there are eight formula units per unit-cell. Figure 16.6 shows the structure and its projection on the plane (001) where heights are given in units of 1/8 of the cell parameter. Each silicon is at the centre of a tetrahedron of oxygens, and the structure is built up with chains of SiO_4 tetrahedra linked through a corner. This compound has a very pronounced covalent character.

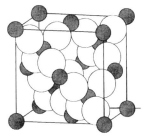

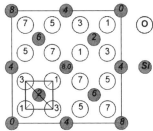

Figure 16.6a. SiO₂   Figure 16.6b

2.3 Ternary Structures

❑ THE SrTiO₃ STRUCTURE (PEROVSKITE)

The space group is Pm3m. There is one strontium at the centre of the unit-cell, one titanium at each corner and three oxygens at face centres (figure 16.7).

This type of ABX_3 structure is found in II–IV associations ($BaTiO_3$), III–III ($LaAlO_3$) and I–IV ($KNbO_3$). It is formed of a three-dimensional network of TiO_6 octahedra linked by their corners. The Sr^{2+} ions are at the centre of the octahedra of oxygens; if the cavity is too small, the structure is distorted. If we assume that the ions B (Ti) and X (O) are touching, the cell parameter

a is such that: $a/2 = R_B + R_X$. If the ions

A (Sr) and X are also touching, we have a $\sqrt{2}/2 = R_A + R_X$.
A coefficient g called the Goldsmidt coefficient can be defined such that:

$$R_A + R_X = g.\sqrt{2}.(R_B + R_X).$$

For an ideal perovskite structure, this coefficient is equal to one. If $0.8 \leqslant g \leqslant 1$ the perovskite is distorted, and if $g > 1$ there will be a different structure.

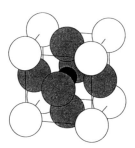

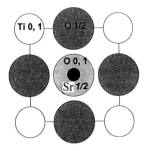

Figure 16.7a. $SrTiO_3$ Figure 16.7b

❑ THE MgAl₂O₄ STRUCTURE (SPINEL)

The spinels are double oxides with the general formula $MO.M'_2O_3$ or MM'_2O_4 (M = Mg, Fe, Mn, Zn, Ni; M′ = Al, Fe, Cr). The $MgAl_2O_4$ structure is cubic (space group Fd3m, $a = 8.08\ \text{Å}$) with eight formula units per unit-cell. Figure 16.8 shows projections on the plane (001) of the various types of atoms with their heights given in units of 1/8 of the cell parameter.

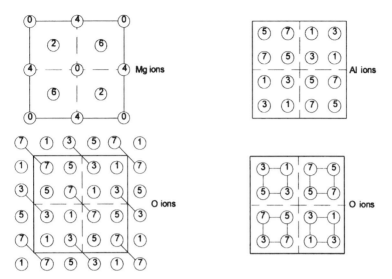

Mg ions

Al ions

O ions

O ions

Figure 16.8

The anions, which are larger than the metallic cations, are in a cubic close-packed arrangement; the cations are in tetrahedral (*A* sites) or octahedral (*B* sites) interstices. To every 32 oxygens there correspond 64 *A* sites and 32 *B* sites. In *normal spinels* eight *A* sites are occupied by the divalent cation and 16 *B* sites by the trivalent cation (e.g. $FeAl_2O_4$, $NiAl_2O_4$). In *inverse spinels*, the *A* sites are occupied by half of the trivalent cations and the *B* sites by the divalent and the rest of the trivalent cations (e.g. $FeNiFeO_4$, $ZnSnZnO_4$). From the point of view of magnetism, each type of site is considered to correspond to a sub-lattice whose occupying species have parallel spins, the two sub-lattices themselves being anti-parallel (spinel ferrimagnetism). Well-defined compounds are rare in nature, but there do exist many solid solutions.

2.4 Assemblages of Complex Ions: Calcite

In many cases, structures which have complex ions can be deduced from ones possessing only simple ions.

As an example, the structure of calcite, $CaCO_3$, may be derived from that of NaCl by replacing Na^+ ions with Ca^{++} ions and Cl^- ions with $(CO_3)^{2-}$ ions, and then stretching along the threefold axis.

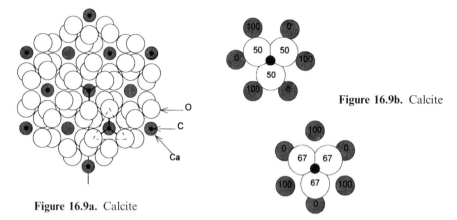

Figure 16.9b. Calcite

Figure 16.9a. Calcite

Figure 16.9c. Aragonite

The unit-cell is trigonal ($a = 6.31$ Å, $a = 46°6'$) with two formula units per unit-cell. The space groups is $R\bar{3}c$. Figure 16.9a is a projection in the pseudo-compact mode of the structure on the plane (001) of the multiple hexagonal unit-cell. Figures 16.9b and 16.9c show the arrangement of carbonate and calcium ions in calcite and in aragonite, which is another variety of calcium carbonate, belonging to the orthorhombic system with space group Pbnm. In both structures the carbonate ion is planar, with the carbon at the centre of an equilateral triangle of oxygens. In calcite each calcium ion is coordinated with six oxygens and each oxygen with two calcium ions. The (001) planes of the hexagonal unit-cell (the trigonal (111) planes) are alternate layers of calcium ions and carbonate ions.

This type of structure also occurs for nitrates (e.g. $AgNO_3$, KNO_3, $NaNO_3$), borates (e.g. $InBO_3$, $AlBO_3$) and carbonates (e.g. $FeCO_3$, $MgCO_3$, $NiCO_3$, $ZnCO_3$).

3 CLOSE PACKED STRUCTURES

Most metals crystallise in one of the following systems: face centred cubic, hexagonal close packed, body centred cubic.

The first two systems correspond to two ways of packing identical spheres in space so as to occupy a minimum volume. This applies to all the alkali metals, the alkaline earth elements, the transition metals, beryllium, magnesium, copper, gold and silver. The high coordination which characterises metallic structures is related to the physical properties of metals and especially their isotropy. Compounds which crystallise in the cubic close packed are more malleable than those which crystallise in the other two systems; this is due to slipping between dense lattice planes. In alloys, the presence of different atoms affects the regularity of the lattice and hinders slipping.

3.1 Close Packed Layers

We imagine that we wish to fill space as compactly as possible with identical spheres touching each other. If we set out to construct lattice layers with a maximum density, we shall obtain a maximum compactness in the layer by placing the centres of the spheres on the nodes of a hexagonal lattice; each sphere in the layer is then touching six other spheres (figure 16.10a). when we come to position the next layer, we note that there can be one, two or three points of contact (figure 16.10b). The arrangement is obviously most compact when each sphere **B** in the upper layer is touching three spheres **A** in the first layer. According to the position of the third layer, two arrangements are possible, leading to the highly compact packings 'cubic close packed' and 'hexagonal close packed'.

3.2 Cubic Close Packing

The third layer is projected onto the sites C (figure 16.10a) and the fourth is directly above the first. We thus obtain the sequence: ABCABC...(figure 16.10c).

❏ LATTICE SYMMETRY

The unit-cell, shown as a hatched diamond shape in figure 16.10a, is hexagonal. The fractional coordinates of the atoms are: A = 0, 0, 0; B = 1/3, 2/3, 1/3; C = 2/3, 1/3, 1/3. If D is the diameter of the spheres, the unit-cell parameters are: $a = D$, $b = D$ and $c = 3.D\sqrt{2/3}$ (c is equal to three times the height of the tetrahedron of side D). The ratio c/a is $3\sqrt{2/3} = 2.4495$.

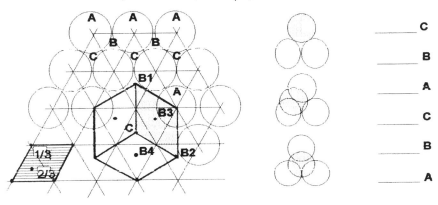

| Figure 16.10a | Figure 16.10b | Figure 16.10c |

This lattice can be described by a unit-cell of higher symmetry: the grey quadrilateral AB_1CB_2 with orthogonal diagonals equal to $2D$ is a square of side $2.R\sqrt{2}$. This packing actually corresponds to a cubic face-centred lattice with the unit-cell in bold lines in figure 16.10a.

❐ STRUCTURE PACKING

Examination of figure 16.10a will show that there are:

—12 nearest neighbours at a distance D (6 of type A in the layer, 3 of type B3 in the lower layer and 3 of type B_3 in the upper layer);
—6 second nearest neighbours at a distance $D.\sqrt{2}$ (type B_1);
—24 third nearest neighbours at a distance $D.\sqrt{3}$ (6 of type A. 6 of type C; 6 of type B_4 in the upper layer and 6 of type B_4 in the lower layer).

There exist four directions of maximum compactness (the threefold axes of the cube).

❐ INTERSTICES

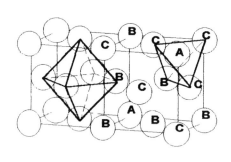

Figure 16.11

When spheres are packed together, there remain empty spaces called **interstices**. In the cubic close packed structure there are two types of interstice, tetrahedral and octahedral. In a cubic coordinate frame, the tetrahedral interstices are centred at 1/4, 1/4, 1/4 . . .; the octahedral interstices are located at 1/2, 1/2, 1/2 . . . The spheres labelled A, B, C in figure 16.11 correspond to the different packing sites.

3.3 Hexagonal Close Packing

The third layer is directly above the first, and we obtain the sequence ABAB . . . (figure 16.12b). This sequence is identical to the sequence ACAC . . .
The unit-cell is hexagonal (figure 16.12c) and contains two atoms: A at 0, 0, 0 and B at 1/3, 2/3, 1/2. The space group is $P6_3/mmc$ (atoms at the sites 2c).

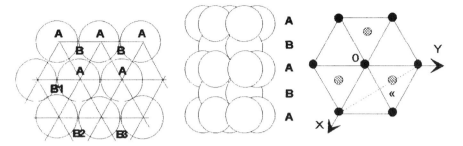

Figure 16.12

Examination of figure 16.12a will show that there are:

—12 *nearest neighbours* at a distance D (6 of type A in the layer, 3 of type B_1 in the lower layer and 3 of type B_1 in the upper layer).

—6 *second nearest* neighbours at a distance $D.\sqrt{2}$ (type B_2).

—18 *third nearest* neighbours at a distance $D.\sqrt{3}$ (6 of type A and 12 of type B_3).

This structure is thus slightly less compact than cubic close packing and has a single direction of maximum compactness (the [001] axis) instead of four.

Assuming the spheres are undeformable we have: $c/a = 2\sqrt{2/3} = 1.63299\ldots$ Elements with this structure actually have slightly different values because of distortion by electron clouds. The values of the ratio c/a for various metals are:

$$Zn \rightarrow 1.86; \quad Co \rightarrow 1.633; \quad Mg \rightarrow 1.6235; \quad Zr \rightarrow 1.59$$

NOTE:

◆ Other sequences of layers are also possible, e.g. for the lanthanides La, Nd, Pm and Pr the sequence is ... ABAC ... and for Sm the sequence is ... ABACACBCB ...

◆ Whatever the compact packing arrangement of touching spheres, the space filling ratio is the same and equal to $\pi/(3\sqrt{2})$ i.e. 74%.

3.4 Body Centred Cubic

In body centred cubic packing of identical spheres, the space filling ratio is $\frac{\pi}{8}\sqrt{3}$ i.e. 68%.

Figure 16.13

The difference is slight compared with close packing, but there is a change of coordination from 12 to 8. Figure 16.13 shows the eight nearest neighbours and six next nearest neighbours. A number of elements occur with this structure: Li, Na, K, Rb, Cs, Ta and W.

3.5 Structures Derived from Close Packing

Close packing also occurs with compounds of neighbouring elements such as certain metal alloys. Solid solutions are obtained by quenching a liquid mixture and in the disordered alloys obtained the atoms are distributed randomly. When the mixture is cooled slowly, however, a partial or total segregation can occur, or there may be crystallisation of an ordered alloy of specific composition. Thus gold and silver are miscible in all proportions and form solid solutions with a random distribution of atoms.

❐ COPPER–GOLD ALLOYS

Gold and copper are miscible in all proportions, but for the compositions AuCu and AuCu₃, ordered phases may be obtained. The alloy AuCu is tetragonal and in the ordered state there is a succession of layers of gold and copper along the tetragonal axis. AuCU₃ is face centred cubic in the disordered state; in the ordered state the fractional coordinates of the atoms are:

Au: 0, 0, 0;   Cu: 1/2, 1/2, 0;   1/2, 0, 1/2;   0, 1/2, 1/2

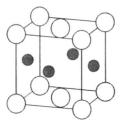

Figure 16.14. AuCu

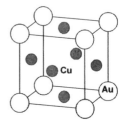

Figure 16.15. AuCu₃

❐ CsCl Type Packing

This is a structure often found in binary 1:1 compounds such as MgAg, AlFe and CuZn.

There exist also 'superstructures' of the CsCl type with a unit-cell which is a multiple of the CsCl unit-cell and sites occupied by different types of atoms. For example, consider the super-cell composed of eight unit-cells with four types of site A, B, C and D and having projections with heights as shown in figure 16.16.

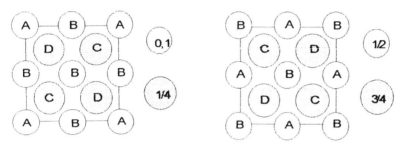

Figure 16.16a   Figure 16.16b

According to the nature of the occupied sites, the following structures occur:

Table 16.7

A	B	C	D	Type	Examples
Al	Fe	Fe	Fe	Fe_3Al	Li_3Bi, Fe_3Si
Al	Mn	Cu	Cu	$MnCu_2Al$	
Tl	Na	Tl	Na	NaTl	LiA, LiZn
As		Mg	Ag	MgAgAs	LiMgAs
Ca		F	F	CaF_2	CuF_2, $BaCl_2$, Li_2O
Zn		S		ZnS (zinc blende)	SiC, GaAs, CuCl
C		C		Diamond	Si
Na	Cl			NaCl	LiH, AgF, MgO

4 COVALENT STRUCTURES

4.1 The Diamond Structure

Diamond, silicon and geranium all have the same structure; the space group is Fd3m and the unit-cell contains 8 atoms (the formula unit is formed of atoms at $(0, 0, 0)$, $(1/4, 1/4, 1/4)$ + cubic face centred). The unit-cell parameter is

$a = 3.5668$ Å. Each carbon atom is at the centre of a regular tetrahedron of carbon atoms. The length of the C–C bond is 1.54 Å and the bond angles are 109°28'.

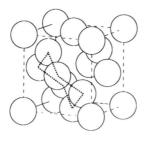

Figure 17a. Diamond   **Figure 17b**

If the structure is viewed along the threefold axes, a lattice of distorted hexagons in the 'chair' configuration can be seen.

4.2 The Zinc Blende or Sphalerite Structure

If the carbon atoms in diamond are alternately replaced with sulphur and zinc atoms, the sphalerite or zinc blende structure is obtained. In this type of structure, the total number of valence electrons is four times the number of atoms.

Other examples include IV–IV compounds such as βSiC, III–V compounds such as BP, GaAs and InSb, II–VI compounds such as BeS, CdS, HgS and ZnSe and I–VI compounds such as CuCl and AgI. The space group of zinc blende is F$\bar{4}$3m, the unit-cell parameter is 5.409 Å and the ZnS distance is 2.43 Å.

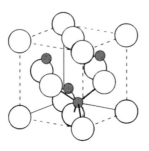

Figure 18a. Zinc blende   **Figure 18b**

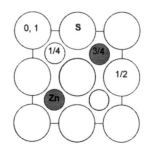

For each type of atom the coordination number is four (figure 16.18), with a regular coordination tetrahedron. If the structure is viewed along a threefold axis it can be seen to consist of a cubic close packed stacking (ABCABC . . .) with alternate layers of sulphur and zinc.

4.3 The ZnS or Wurtzite Structure

In wurtzite the layers are hexagonal close packed (ABAB . . .) with alternate layers of sulphur and zinc. The space group is P6$_3$mc; the fractional coordinates of the atoms are:

$$\text{Zn: } 0,0,0 \qquad \tfrac{1}{3},\tfrac{2}{3},\tfrac{1}{2}$$
$$\text{S: } 0,0,0.375 \qquad \tfrac{1}{3},\tfrac{1}{3},0.875$$

The unit-cell parameters are: $a = 3.81$ Å and $c = 6.23$ Å ($c/a = 1.635$). Zn–S bond lengths parallel to [001] (2.336 Å) are slightly greater than those parallel to the plane (001). Again the coordination is four for both types of atom, but the coordination tetrahedron is no longer regular. Figure 16.19b shows the projection of the hexagonal unit-cell on the plane (001) and figure 16.19a shows a semi-compact model assembled from three unit-cells.

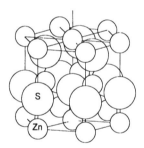

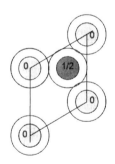

| Figure 16.19a. Wurtzite | Figure 16.19b |

4.4 The Graphite Structure

This is a hexagonal structure with the space group P63mc; the parameters are: $a = 2.456$ Å and $c = 6.696$ Å. The fractional coordinates of the atoms are:

$$0, 0, 0; \quad 0, 0, \tfrac{1}{2}; \quad \tfrac{2}{3}, \tfrac{1}{3}, 0; \quad \tfrac{1}{3}, \tfrac{2}{3}, \tfrac{1}{2}$$

This structure is obtained by a stacking of layers. Layers of integral heights are type AB and those of half-integral heights are type AC. Within the planes of the layers there are strong covalent bonds, each carbon atom being linked to three others (C–C = 1.42 Å), whereas the layers themselves are linked by Van der Waals forces (C–C = 3.35 Å). The conductivity is high within the layers, but very low in the perpendicular direction.

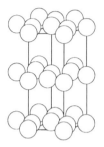

Figure 16.20a. Graphite

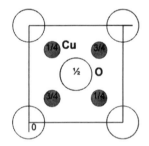

Figure 16.20b

4.5 The Cu₂O Structure (Cuprite)

The oxygen lattice is body centred cubic while the copper lattice is face centred cubic; the unit-cell parameter is 4.27 Å. Each oxygen is at the centre of a tetrahedron of copper and the Cu–O distance is 1.849 Å. This structure might be considered as an interstitial solution of oxygen atoms in a lattice of metallic copper, though the oxygen atoms are too large to fit the cavities in the copper lattice. The bonds are not really covalent and the compound has a pronounced metallic character as evidenced by the relatively high conductivity of cuprite.

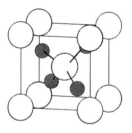

Figure 16.21a. Cuprite

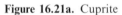

Figure 16.21b

5 STRUCTURE BUILDING WITH POLYHEDRA

Crystal chemists now work on more and more complex compounds for which a description based simply on the positions of the atoms is no longer adequate. However, a number of structures can be described in terms of assemblages of polyhedra linked in various ways. This kind of approach simplifies structure descriptions and brings to the fore certain properties of the materials studied such as cages, channels and preferred orientations. This introduction to crystal chemistry will be limited to a brief presentation of a few types of assemblages of Mx_n octahedra. For an account of the silicates, where numerous and complex arrangements of SiO_4 tetrahedra are found, the reader is referred to specialised works.

5.1 Octahedra Linked by Corners

◻ THE PEROVSKITE STRUCTURE (ABX_3)

The structure of normal cubic perovskites ($CaTiO_3$, $KZnF_3$ etc.) can be described as a framework of BX_6 octahedra linked by their corners, each X being shared by two octahedra; the ions A occupy the cavities between the octahedra. In this way we obtain a three-dimensional assemblage of octahedra whose tetragonal axes coincide with those of the unit-cell. If the relative dimensions of the octahedra of ions A are incompatible with this configuration, the basic structure becomes distorted (the octahedra may rotate about one, two or three axes, become distorted etc.), causing a lowering of the symmetry.

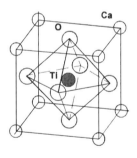
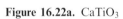
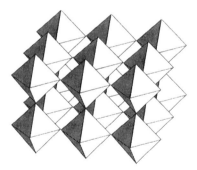

Figure 16.22a. $CaTiO_3$ **Figure 16.22b**

◻ THE TETRAFLUOROALUMINATE STRUCTURE (ABF_4)

The typical structure ($TlAlF_4$, figure 16.23) has layers of octahedra AlF_6 linked by corners and separated by layers of thallium ions. The octahedra have tetragonal symmetry and their symmetry axes are parallel to those of the unit-cell.

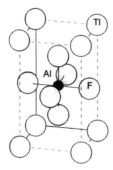

Figure 16.23a. TlAlF$_4$ **Figure 16.23b**

The bonding is very strong within the layers but weaker in the direction of the fourfold axis (giving a laminar structure which cleaves readily in the plane of the layers). There exist numerous variations of this structure such as those with rotation of the octahedra about one, two or three axes and staggering of the layers.

❐   THE RUTILE STRUCTURE (RX$_2$)

This structure, which has already been described above, can be considered either as a compact assemblage of oxygens with the titanium ions occupying half the octahedral cavities, or as an assemblage of TiO$_6$ octahedra linked by corners. In either case the oxygen is shared between three octahedra.

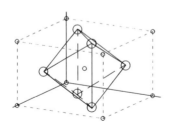

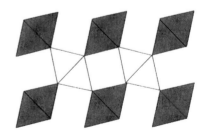

Figure 16.24a. Rutile   **Figure 16.24b**

5.2 Octahedra Linked by an Edge

☐ MX$_4$ TYPE CHAINS

Two octahedra linked by an edge correspond to a composition (MX$_5$)$_2$, that of the pentahalides. A whole chain of octahedra linked by an edge will correspond to a composition MX$_4$.

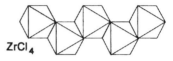

NbCl$_4$ ZrCl$_4$

Figure 16.25

In this way there occur either linear chains such as NbCl$_4$ and NbI$_4$ or chains with more complex conformations such as the zigzag configuration of ZnCl$_4$.

☐ MX$_3$ TYPE LAYERS

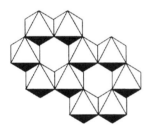

Figure 16.26. AlCl$_3$

Assemblages of octahedra linked by an edge can also produce layers of composition MX$_3$. The layers are stacked in such a way that the atoms X form a close packed arrangement. This type of structure occurs in many trihalides such as AlCl$_3$ and CrCl$_3$.

☐ MX$_2$ TYPE LAYERS

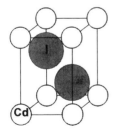
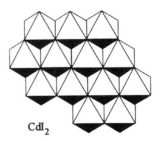

Figure 16.27a. CdI$_2$ **Figure 16.27b**

In this type of arrangement (figure 16.27) each octahedron in the layer is linked to six neighbours. Each halogen is shared between three octahedra in the layer and the assemblage of the halogens is either hexagonal close packed as for example in CdI_2, PbI_2, $MgBr_2$, $CrBr_2$, $Mg(OH)_2$, $Fe(OH)_2$, SnS_2 and AgF_2, or cubic close packed as in $CdCl_2$, $MgCl_2$ and $FeCl_2$.

5.3 Polyhedra Linked by Faces (NiAs)

The compound nickel arsenide is hexagonal ($P6_3/mmc$) with two formula units per unit-cell:

$$Ni: 0, 0, 0; \quad 0, 0, 1/2$$

$$As: 1/3, 1/3, u; \quad 2/3, 1/3, u + 1/2; \quad u = 1/4$$

This structure occurs in many compounds RX such as AuSn, CrS, CrSb, FeS, FeSb, MnAs, MnBi and NiSb.

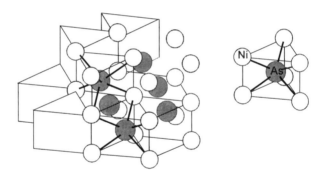

Figure 16.28. NiAs

Each atom X is at the centre of a right triangular prism of atoms R (an example of non-octahedral six-coordination). Each atom R has eight nearest neighbours (six X and two R). The structure can be considered as an assemblage of prisms stuck together by a face.

Chapter 17
Special Techniques

In part two of the book we gave a detailed account of the classical techniques in X-ray crystallography. Here we give a brief description of techniques which require special apparatus and techniques for dealing with substances which are not crystalline in the strict sense of the word.

1 SCATTERING BY NON-CRYSTALLINE SUBSTANCES

We shall simply outline the principles involved in studying scattering by non-crystalline substances; for a more detailed study, which is outside the scope of this work, the reader should consult, e.g. 'X-ray Diffraction' by A. Guinier.

❏ SCATTERING CAPACITY

The diffracted amplitude in a direction characterised by the vector

$$\mathbf{S} = \frac{\mathbf{s} - \mathbf{s}_0}{\lambda}$$

is the Fourier transform ($Tr\ F$) of the electron density $\rho(\mathbf{r})$:

$$A(\mathbf{S}) = \int \rho(\mathbf{r}).e^{-2j\pi.\mathbf{S}.\mathbf{r}}.dv_{\mathbf{r}} = \sum_{n=1}^{N} f_n.e^{-2j\pi.\mathbf{S}.\mathbf{r}_n}$$

(the integral is over all the object space)

The observed property is still the intensity, which is the square of the modulus of the amplitude: $I_N(\mathbf{S}) = |A(\mathbf{S})|^2$. For a diffracting object composed of N identical entities (atoms, unit-cells), we define a unit scattering capacity: $I(\mathbf{S}) = I_N(\mathbf{S})/N$.

If F is the structure factor of the elementary entities (atomic scattering factor or structure factor), we can then introduce the interference function:

$$\mathcal{J}(S) = \frac{I(S)}{F^2} = \frac{I_N(S)}{N.F^2}$$

☐ SCATTERED INTENSITY

The total scattered intensity is the product of the unit scattering capacity, the number of entities (after correcting for absorption) and the scattered intensity of an isolated electron. It can be expressed in the following two ways:

◆ *As a function of atomic scattering factors:*

$$I_N(S) = A(S).A^*(S) = \sum_1^N f_n.e^{-2j\pi S.r_n} . \sum_1^N f_{n'}.e^{-2j\pi.S.r_{n'}}$$

$$I_N(S) = \sum \sum f_n.f_{n'}.e^{-2j\pi S.(r_n - r_{n'})}$$

For $n = n'$ there are N terms whose sum is: Σf_n^2.
For $n \neq n'$ there are $N(N-1)/2$ pairs of conjugate terms equal to:

$$f_n.f_{n'}.[\cos 2\pi.S(r_n - r_{n'}) + \cos 2\pi.S(r_{n'} - r_n)]$$

Writing $r_{nn'} = r_n - r_{n'}$, we obtain:

$$I_N(S) = \sum_1^N f_n^2 + \sum_{n \neq n'} \sum f_n.f_{n'}.\cos 2\pi.S.r_{nn'}$$

◆ *In terms of the electron density*

$$I_N(S) = A(S).A^*(S) = \int \int \rho(u).\rho(v).e^{-2j\pi S.(v-u)}.dv_u.dv_v$$

Writing $r = v - u$ we obtain:

$$I_N(S) = \int \int \rho(u).\rho(u+r).e^{-2j\pi.S.r}.dv_u.dv_r$$

Using the generalised Patterson function $P(r) = \int \rho(u).\rho(u+r).dv_u$ the expression for the intensity becomes:

$$I_N(S) = \int P(r).e^{-2j\pi.S.r}.dv_r \Leftrightarrow P(r) = \int I_N(S).e^{-2j\pi.S.r}.dv_S$$

The intensity in reciprocal space is the Fourier transform of the Patterson function in direct space. The observable property is actually an average intensity which is a function of the statistics of the distribution of diffracting objects in direct space.

$$\overline{I_N}(S) = Tr\, F(\overline{P(r)})$$

❐ SCATTERED INTENSITY FROM AN INFINITE HOMOGENEOUS OBJECT

Consider a volume V containing on average $\overline{N}$ objects (N is large); the average volume available to an object is $v_0 = V/\overline{N}$ and the probability of finding an atom in the volume dv at a distance $\mathbf{r}$ from the object chosen as origin is:

$$dp(\mathbf{r}) = p(\mathbf{r}).dv/v_0.$$

$p(\mathbf{r})$ is the distribution function for the atoms. In a crystal, this function is null if $\mathbf{r}$ is not a lattice vector. In disordered substances an atom's radius of influence extends no further than a few atomic distances and so at large distances the positions of the atoms are not correlated and for large $\mathbf{r}$, $p(\mathbf{r}) = 1$; fluctuations of $p(\mathbf{r})$ around 1 correspond to short-range order. To take account of this, a Dirac peak $\pi(\mathbf{r}) = \delta(\mathbf{r}) + p(\mathbf{r})/v_0$ can be introduced into the function for the probability of finding an atom at a distance $\mathbf{r}$ from the atom at the origin.

To show the short-range variable part we can write this in the form:

$$\pi(\mathbf{r}) = 1/v_0 + \delta(\mathbf{r}) + (p(\mathbf{r}) - 1)/v_0.$$

This probability is related to the average value $Pa(\mathbf{r})$ of the Patterson function applied to the volume being considered. In calculating $Pa(\mathbf{r})$, the terms $\rho(\mathbf{u}).dv_{\mathbf{u}}$, which contain one atom, are equal to 1 and the others are null. The integral is equivalent to the sum of $1/v_0$ terms $\rho(\mathbf{u} + \mathbf{r})$ whose average value is $\pi(\mathbf{r})$:

$$Pa(\mathbf{r}) = \pi(\mathbf{r})/v_0$$

The Fourier transform of $\pi(\mathbf{r})$ is:

$$\Pi(\mathbf{S}) = 1 + \delta(\mathbf{S})/v_0 + 1/v_0.Tr\,F(p(\mathbf{r}) - 1)$$

For a perfect crystal made up of N unit-cells of volume Vc the distribution function is a series of Dirac peaks centred on the lattice nodes. The Fourier transform is the written:

$$\Pi(\mathbf{S}) = \frac{1}{V_c}\sum_{hkl}\delta(\mathbf{S} - \mathbf{N}^*_{hkl})$$

❐ SCATTERED INTENSITY FROM A FINITE HOMOGENEOUS OBJECT

Let $\varphi(\mathbf{r})$ be a function (the form factor) equal to 1 inside the specimen and null elsewhere. If $\rho(\mathbf{r})$ is the electron density of the infinite object, that of the finite object becomes $\rho'(\mathbf{r}) = \rho(\mathbf{r}).\varphi(\mathbf{r})$. The Fourier transform of $\varphi(\mathbf{r})$ is:

$$\Phi(\mathbf{S}) = \int \varphi(\mathbf{r}).e^{-2j\pi.\mathbf{S}.\mathbf{r}}.dv_{\mathbf{r}}$$

The Fourier transform of $\rho'(\mathbf{r})$, which is the amplitude scattered by the finite object, is the product of the convolution of the transforms of $\rho(\mathbf{r})$ and $\varphi(\mathbf{r})$, i.e.:

$$A'(\mathbf{S}) = \int A(\mathbf{S}).\Phi(\mathbf{S} - \mathbf{u}).dv_{\mathbf{u}}$$

It can be shown that the interference function can be written:

$$\mathcal{J}(\mathbf{S}) = \frac{1}{V} \int \Pi(\mathbf{u}).|\Phi(\mathbf{S} - \mathbf{u})|^2.dv_{\mathbf{u}} = \frac{1}{V}\Pi(\mathbf{u})*|\Phi(\mathbf{S})|^2$$

$$\mathcal{J}(\mathbf{S}) = \frac{1}{V}\left[\left(1 + \frac{1}{V_0}\int (p(\mathbf{r}) - 1)e^{2j\pi.\mathbf{S}.\mathbf{r}}.dv\right)*|\Phi(\mathbf{S})|^2 + \frac{1}{v_0}\delta(\mathbf{S})*|\Phi(\mathbf{S})|^2\right]$$

The second term of the sum is equal to:

$$\frac{1}{v_0}\delta(\mathbf{S})*|\Phi(\mathbf{S})|^2 = \frac{1}{v_0}\int \delta(\mathbf{u}).|\Phi(\mathbf{S} - \mathbf{u})|^2.dv_{\mathbf{u}} = \frac{|\Phi(\mathbf{S})|^2}{V.v_0}$$

While the first factor of the first product only varies slowly with $\mathbf{S}$ in an object displaying long range order, the function $\Phi(\mathbf{S})$ has a very sharp maximum for $\mathbf{S} = 0$ for all but very small objects. The convolution product can be assimilated to the product of the first factor and the integral of $|\Phi(\mathbf{S})|^2$ in the neighbourhood of the origin, i.e. V.

$$\mathcal{J}(\mathbf{S}) = 1 + \frac{1}{v_0}\int (p(\mathbf{r}) - 1)e^{-2j\pi.\mathbf{S}.\mathbf{r}}.dv + \frac{|\Phi(\mathbf{S})|^2}{v_0.V}$$

In this general expression for the interference function, the second term represents the peak in the form function $\Phi(\mathbf{S})$. Physically, it corresponds to *very* small diffraction angles and it is only detectable for diffracting objects of very small size for which there is negligible spreading of the form function in reciprocal space. The first term is a function only of the statistical distribution of objects in the specimen.

For a perfect crystal, the interference function

$$\mathcal{J}(\mathbf{S}) = \frac{1}{V}\Pi(\mathbf{u})*|\Phi(\mathbf{S})|^2$$

is equal to:

$$\mathcal{J}(\mathbf{S}) = \frac{1}{V.V_c}\sum_{hkl}\delta(\mathbf{S} - \mathbf{N}_{hkl}^*)^*|\Phi(\mathbf{S})|^2 + \frac{1}{V.V_c}\sum_{hkl}|\Phi(\mathbf{S} - \mathbf{N}_{hkl}^*)|^2$$

The domain of reflection around the actual nodes in the reciprocal lattice depends on the shape and size of the specimen.

On the reciprocal lattice nodes ($\mathbf{S} = \mathbf{N}^*_{hkl}$) we have $\mathcal{J}(\mathbf{S}) = V^2/V.V_c = N$ i.e.:

$$I_N = N^2 f^2.$$

❒ THE DEBYE FORMULA

An object moving in such a way that all orientations relative to the incident beam are equally probable, is equivalent to a perfect powder (or a gas or a dilute solution). For a given orientation of the scattering object, we have:

$$I_N(\mathbf{S}) = \sum_1^N f_n^2 + \sum_{n\neq n'}\sum f_n.f_{n'}.\cos 2\pi.\mathbf{S}.\mathbf{r}_{nn'}$$

The average scattering capacity is the sum of the means:

$$\overline{I_N(\mathbf{S})} = \sum_1^N f_n^2 + \sum_{n\neq n'}\sum f_n.f_{n'}.\overline{\cos 2\pi.\mathbf{S}.\mathbf{r}_{nn'}}$$

To calculate the average value, we may write $\alpha = \{\mathbf{S},\mathbf{r}_{nn'}\}$ and β is the angle between the normals to an arbitrary reference plane and the plane $\mathbf{S}.\mathbf{r}_{nn'}$.

$$\overline{\cos 2\pi.\mathbf{S}.\mathbf{r}_{nn'}} = \int_0^{2\pi}\frac{\mathrm{d}\beta}{2\pi}\int_0^\pi\cos(2\pi.\mathbf{S}.\mathbf{r}_{nn'}.\cos\alpha).2\pi.\sin\alpha.\mathrm{d}\alpha = \frac{\sin(2\pi.\mathbf{S}.\mathbf{r}_{nn'})}{2\pi.\mathbf{S}.\mathbf{r}_{nn'}}$$

$$\overline{I_N(\mathbf{S})} = \sum_1^N\sum_1^N f_n.f_{n'}\frac{\sin(2\pi.\mathbf{S}.\mathbf{r}_{nn'})}{2\pi.\mathbf{S}.\mathbf{r}_{nn'}}$$

In this relation, only the *lengths* of interatomic vectors are involved. The maximum occurs at $\mathbf{S}^* = 0$ and secondary maxima occur for $\mathbf{S}.\mathbf{r}_{nn'} = 1.2295$ 2.2387; 3.242... (zeros of $\tan(u) - u$).

This series when calculated gives the intensity of the lines in a powder diagram but it is by no means a trivial exercise to show that it is formally equivalent to the classical expression for the scattered intensity.

❒ X-RAY SCATTERING BY AMORPHOUS SUBSTANCES

◆ **Perfect gases**

−*Monatomic gases*: the distribution function $p(\mathbf{r})$ is always equal to 1 since the probability of finding an atom at a given point is constant by hypothesis: $\pi(\mathbf{r}) = 1/v_0 + \delta(\mathbf{r})$. The interference function is always equal to 1 and the scattering capacity is equal to the square of the atomic scattering factor; no interference occurs. Incoherent or Compton scattering must also be considered.

For a gas, the two effects are of the same order of magnitude, but for coherent scattering the intensity decreases with θ, whereas the opposite is true for incoherent scattering.

−*Polyatomic gases*: For diatomic gases with interatomic distance a and atomic scattering factor f, the Debye formula gives:

$$\overline{I_N(\mathbf{S})} = 2.f^2.\left(1 + \frac{\sin(2\pi.S.a)}{2\pi.S.a}\right)$$

At small angles, the intensity is equal to $4f^2$ (the interference effect), tending to $2f^2$ at large angles, corresponding to dissociated molecules. Oscillations should be observed on the $I(\theta)$ curve, with a first maximum for an angle θ_0 such that $2a.\sin\theta \approx 1.23\lambda$; however, the oscillations are damped and displaced by the effect of f decreasing with θ and by incoherent scattering. For polyatomic gases the curve of scattered intensity is obtained by summing the contributions of each type of atom pair. The maxima overlap and the scattering curve is difficult to interpret.

X-ray diffraction is thus unsuitable for the study of gases. Intensities are not negligible, however, and in diffraction experiments on solids, scattering by air is a considerable source of interference.

◆ Condensed amorphous states

The condensed amorphous states of matter—compressed gases, liquids and glasses—are intermediate between perfect gases without any order and the crystalline state where the order is perfect. Diffraction patterns obtained depend on the distribution $p(\mathbf{r})$ of the atoms. In the case of a liquid, $p(\mathbf{r})$ can be calculated using as a model hard impenetrable spheres of diameter a. The curve of intensities deduced from this model shows very weak intensities at small angles, rising to a first maximum at $S = 1/a$; there then follow damped oscillations with an interference function tending to 1 at large values of S. In the absence of long-range order, fluctuations in interatomic distance completely average the phase term $2\pi Sr$ in the Debye relation.

The diffraction pattern of an amorphous substance is characterised by the appearance of *one or more diffuse rings*; if there exists a large number of pairs of atoms separated by a distance x_0, then corresponding to them in the Debye relation there will be the same term $\sin(2\pi Sx_0)/2\pi Sx_0$, which has a first maximum at $S_0 = 2\sin\theta_0/\lambda \approx 1.23/x_0$.

The order of magnitude of the average distance between nearest neighbours can be determined from the diameter of the first ring, but for a rigorous analysis of the patterns, the Fourier transform of the diffraction curve (corrected for interfering phenomena) must be calculated, and the radial distribution of atoms in direct space is then obtained.

Using a suitable thermal process, it is sometimes possible to crystallise glass. It then becomes evident that the diffuse rings of the amorphous material are in fact the envelope of the diffraction lines of the crystallised material.

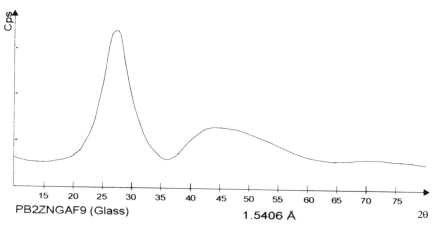

DS spectrum of a fluoride glass

2 EXAFS

☐ PRINCIPLE

An electron can be ejected from an atom by X-ray photons provided their energy hv is higher than the binding energy E_i of the electron. The curve of absorption as a function of energy shows edges K, L ... corresponding to the excitation of $1s$, $2s$, $2p$ ... electrons. Beyond the edge corresponding to the element, absorption decreases uniformly for a gas, but oscillates in a solid. These oscillations can extend from 600 to 1000 eV with periods of the order of 50–100 eV depending on the environment of the photo-excited atom.

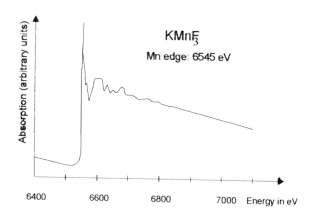

The reason for these oscillations is that the photo-electron ejected by the atom is propagated in the form of a spherical wave which is then back-scattered by neighbouring atoms. The phenomenon is known as EXAFS (Extended X-ray absorption fine structure).

The effect provides information on the symmetry of the local environment; although it has been known since 1930, use has only been made of it since the development of synchrotron radiation generators giving 'white' sources of sufficient intensity.

❐ STERN'S FORMULA

Spectral oscillations are characterised by: $\chi(k) = \dfrac{\mu(k) - \mu_0(k)}{\mu(k)}$

$\mu_0(k)$ is the absorption by an isolated atom.

Stern, Lytle and Sayers established the following formula which is now used in interpreting spectra:

$$\chi(k) = -\frac{1}{k} \sum_j 3\cos^2(\mathbf{R}_j, \mathbf{E}) \frac{|f_j(\pi, k)|}{R_j^2} . \sin(2k.R_j + 2\delta + \vartheta_j).e^{-2\sigma^2 k^2}.e^{2R_j/\Lambda(k)}$$

j:	the label for any neighbour located at R_j
$|f_j(\pi,k)|$:	the back-scattering amplitude for the atom
ϑ_j:	the back-scattering phase for the atom j
$3\cos^2(\mathbf{R}_j, \mathbf{E})$:	a polarisation term related to the angle between the electric field $\mathbf{E}$ and $\mathbf{R}_j$
δ:	a phase term characterising the excited atom

In the Debye–Waller term, which takes into account the variations in the values of R_j, σ is the standard deviation of the R_j distribution. The final exponential term corresponds to damping from inelastic scattering; the mean free path of the electrons $\Lambda(k)$ is close to $k/4$ (in eV). The wave-number (in $\mathring{A}^{-1}$) is written:

$$k = \frac{1}{0.529} \left[\frac{E - E_0}{13.605} \right]^{1/2} \text{ (energies in eV)}$$

The edge energy E_0 is slightly dependent on the environment of the excited atom, and is adjustable. The phase shifts ϑ_j and δ together with back-scattering amplitudes have been calculated for most atoms. It is often preferred to carry out phase determination in a compound with known structure as similar as possible to the one being studied; phase shifts obtained in this way, when applied to the compound of unknown structure, usually give good results.

❐ EXPERIMENTAL APPARATUS

The incident beam passes through a two-crystal monochromator which keeps the direction of the beam constant. The intensity is measured before and after passing through the specimen, using ion chambers. Solid specimens (sieved powders) are stuck in a thin layer onto adhesive tape and placed in the beam.

A data collection system controls the rotation of the monochromator, enabling the curve $\mu(k)$ to be calculated and recorded.

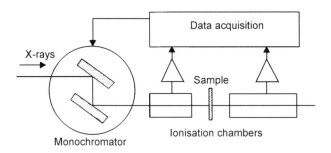

❐ ANALYSIS OF EXAFS SPECTRA

First the modulation $\chi(k)$ of the curve $\mu(k)$ is extracted by smoothing the spectrum to give the curve $\mu_0(k)$. If possible a comparison is made with a reference specimen having chemical and crystallographic properties as similar as possible to the compound being studied, so as to ensure a reliable transfer of the phases. Stern's formula shows that $k\chi(k) \propto \sin(2kR + \varphi(k))$ and hence a Fourier transform must be performed which will give, in real space, the peaks of a function for the distributions of distances between pairs of atoms.

Assuming that $\varphi(k) = \alpha k + \beta$, the peaks of the transform give the distances $Ra = R + \alpha/2$.

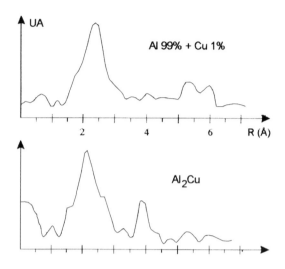

Fourier transforms of χ(k) *(From Fontaine et al. LURE document, 1979)*

The upper curve is that of a solid solution of 1% copper in an aluminium matrix; the reference compound is the alloy Al_2Cu. In the Fourier transforms obtained, the main peaks correspond to layers of first and second nearest neighbours. In the alloy, the Al–Cu peak is at 2.13 Å while the value obtained by diffraction is 2.487 Å. For the solid solution, the Al–Cu peak is at 2.37 Å = 2.13 Å + 0.24 Å; in the solid solution, the Al–Cu peak is thus close to 2.49 Å + 0.24 Å i.e. 2.73 Å.

In more complex cases (e.g. where there are several different atoms in the first layer), or if spectral amplitudes are required (for the measurement of N and σ), a different strategy must be used: an attempt is made to reconstruct the spectrum from the theoretical formula. In order to limit the number of variables, the peak to be studied in the Fourier transform is filtered and inverted to obtain the specific EXAFS signal; the reconstruction is then attempted by adjusting the parameters relating to the layer in question.

❏ APPLICATIONS

◆ This technique is a local one since only nearest neighbours can be distinguished.

◆ EXAFS is highly selective since the edges of the various elements in the compound are excited separately. In a binary compound AB, the pairs BB are not involved when the edge of A is studied. EXAFS can be used to analyse dilute systems and in particular to determine chemical impurities.

◆ EXAFS can be used provided there exists a *local radial order*, but there need be no long-range order; the technique is thus usable with liquids, amorphous substances and glasses. It should be remembered however that fluctuations in the radial order strongly dampen the signal and generally in disordered media only the first layer is visible.

A well-resolved EXAFS spectrum will yield:

−The distances between the excited atom and its neighbours ($\Delta r \approx 0.01$ Å).
−The number N of nearest-neighbour scatterers ($\Delta N \approx 0.2$ to 0.5).
−An estimate of the fluctuations in the values of Ri ($\Delta \sigma^2 \approx 0.01$ Å^2).

The technique is now widely used to study amorphous substances, glasses, saline solutions and crystal defects.

3 X-RAY FLUORESCENCE SPECTROMETRY

◻ PRINCIPLE AND APPARATUS

A suitably excited element will emit characteristic radiation. The excitation can be by impact with accelerated particles or by high energy photons from an anode or a radioactive source. For analytical purposes, electrons (the spectrometer is linked to a scanning electron microscope) or X-rays are generally used. Here we shall only give a brief description of X-ray fluorescence and the problems involved in its use.[1]

In X-ray fluorescence, energy analysis is performed on the emission spectrum of a specimen bombarded with primary photons. The photons ionise the target atoms, which then return to their ground state by emission of a line spectrum with characteristic wavelengths. There are few lines in the spectra, which are therefore simpler to interpret than those of classical emission spectrometry. The intensity of a emission line depends on:

−the probability of ionisation in the initial energy level;
−the probability of the hole being filled by an electron at the final energy level;
−the probability of the photon leaving the atom before being re-absorbed.

This latter probability gives rise to fluorescence defined by: $\eta = n_f/n$, where n is the number of primary photons causing ionisation of a given level, and n_f is the number of secondary photons emitted by the atom; $n - n_f$ is the number of re-absorbed photons (the Auger effect). The amount of fluorescence depends on the initial ionised layer and on the element; it is very small when Z is small (0.018 for carbon) and tends to 1 for large values of Z (0.859 for tin).

---

[1]For a detailed study, consult, e.g., R. Tertian and F. Claisse, *Principles of Quantitative X-ray Fluorescence Analysis*, Heyden, London (1982).

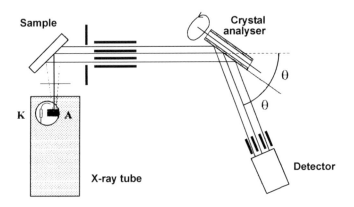

In this type of spectrometer, high efficiency frontal anodes, often of rhodium, are used. The electrons are emitted by an annular cathode and focused onto the anode by electronic optics. The emitted photons pass through a beryllium window followed by a primary filter. The secondary photons emitted by the specimen pass through a diaphragm and a collimator into the analysing crystal where the different wavelengths are separated according to the Bragg equation $n\lambda = 2d\sin\theta$. The detector is geared to the analysing crystal with a 2:1 speed ratio, and measures the intensity $I(\lambda)$ of the beam.

Several analysing crystals are mounted on a turret, and the crystal is selected to cover the required wavelength range. The most widely used crystals are (100) or (110)-cut lithium fluoride, in first or second order, and PET (pentaerythritol, $C(CH_2OH)_4$). The output of the detector, which can be a scintillation counter or, for light elements, a gas flow counter, is fed to wave shaping circuitry and energy discriminators. To avoid air fluorescence the whole system is under vacuum. Powder specimens are compressed into pellets or sintered. Binding agents used contain only light elements which are invisible under fluorescence (e.g. borax, lithium tetraborate). The bead technique is also used: the sample is fused with a flux (borax), giving a highly homogeneous solid solution.

For *quantitative analysis*, the analyser and detector are placed in the reflection of a line of the element. Intensities measured on a series of samples can be converted into concentrations of the element.

For *qualitative analysis*, the analyser and detector rotate at uniform speed and the scattered intensities are measured; it is then only necessary to identify the characteristic lines to identify the elements present in the sample.

❏ Primary and Secondary Fluorescence

— Primary fluorescence

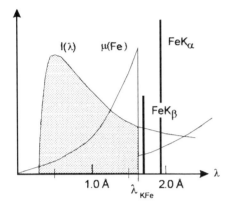

The figure shows the absorption curve μ of iron, in which there is a discontinuity at λ_{KFe} and lines $K_{\alpha Fe}$ and $K_{\beta Fe}$, together with the emission curve of the generator. Only those photons with a wavelength shorter than λ_{KFe} can ionise the K level of iron; this is primary fluorescence, which is the type which occurs for pure elements.

— Secondary and tertiary fluorescence

This occurs when the element is associated with another element of higher atomic number. An example is a stainless steel (Fe–Ni–Cr) in which the nickel fluorescence is also excited, with emission of the line $K_{\alpha Ni}$ (whose wavelength is shorter than for $K_{\alpha Fe}$) which then induces secondary fluorescence of the iron and which in turn can cause tertiary fluorescence of the chromium ($K_{\alpha Ni} < K_{\alpha Fe} < K_{\alpha Cr}$).

— The intensity of fluorescence radiation

Let φ_1 and φ_2 be the incident and emergent beam angles for a sample, $A = \sin\varphi_1 / \sin\varphi_2$, I_0^{λ} the incident intensity and E_i the excitation factor (equal to the product of the ionization probabilities of the initial level, the probability of emission of the line in question and the fluorescence efficiency).

C_i is the concentration of element i in the sample, μ_i^{λ} its absorption coefficient and

$$\mu_t^{\lambda} = \sum_i C_i . \mu_i^{\lambda}$$

the absorption coefficient of the sample.

In calculating the intensity of fluorescence, absorption by the sample of incident and emergent radiation according to the Beer law must be taken into account.

For a thick sample, it can be shown that the intensity of primary fluorescence radiation is:

$$I_i \propto E_i.C_i. \int_{\lambda_0}^{\lambda_i^K} \frac{\mu_i^\lambda.I_0^\lambda.\mathrm{d}\lambda}{\mu_t^\lambda + A.\mu_t^\lambda}.$$

For monochromatic excitation of intensity J^λ, this formula becomes:

$$I_i \propto E_i.C_i \frac{\mu_j^\lambda.J^\lambda}{\mu_t^\lambda + A.\mu_t^\lambda} = B_i \frac{C_i}{\mu_t^*} \quad \text{where: } \mu_t^* = \mu_t^\lambda = \mu_t^\lambda + A.\mu_t^\lambda$$

Thus the intensity of fluorescence is proportional both to the concentration of the element *and* to the reciprocal of the effective attenuation coefficient μ^* of the compound, which itself depends on C_i. Hence the *intensity of fluorescence is not a linear function of concentration*. The other elements in the compound increase or decrease the contribution of the element in complicated processes through matrix effects. The situation becomes even more complex if secondary and tertiary radiations are taken into account and if a polychromatic excitation source is used.

❐ QUANTITATIVE ANALYSIS

The previous paragraph gives an idea of the complexity of the problem and underlines the need to use external calibrations to take into account inter-element effects. Once the experimental conditions have been optimised for the element being analysed, sample and reference data are collected and subjected to one of the many methods of correction which have been developed to take into account the inter-element effects. The overall quality of an X-ray fluorescence spectrometer depends as much on the quality of the hardware as on data processing software.

This method of non-destructive testing is a powerful tool for metallurgists, enabling rapid (a quarter of an hour) and accurate (a few percent) determination of the composition of an alloy. Although it can be applied to liquids, it is used above all for finely divided solids (powders and dusts). It is a highly sensitive method and traces (tens of parts per million) can be detected.

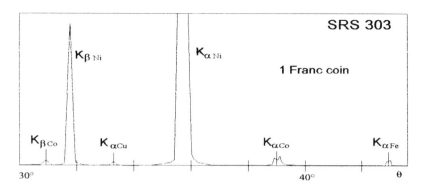

As an example, the figure above shows part of a spectrum ($30° < \theta < 45°$) of a coin of one French franc; semi-quantitative analysis gives the following composition: nickel 97%, magnesium 0.4%, silicon 0.4%, iron 0.13%, cobalt 0.5%, copper 0.1%, fluorine 0.5%, bromine 0.1%, rubidium 0.2%, tantalum 0.2% and thallium 0.2%.

4 DIFFRACTION OF X-RAYS BY SURFACES

◻ X-RAYS AND SURFACES

The refractive index of a pure element to X-rays is given by:

$$n = 1 - r_e.Z.\frac{\lambda^2}{2\pi V} \quad (V = \text{volume of unit-cell})$$

r_e is the 'classical' radius of the atom (2.82×10^{-15} m).

For a crystallised compound with m atoms per unit-cell, the index is:

$$n = 1 - \varepsilon - j\delta$$

$$\text{where } \varepsilon = r_e \frac{\lambda^2}{2\pi V} \sum_{k=1}^{m}(Z_k + g_k) \text{ and } \delta = r_e \frac{\lambda^2}{2\pi V} \sum_{k=1}^{m} h_k$$

Z_k, g_k and h_k are respectively the atomic number, the real part and the imaginary part of the dispersion correction for the atom k. Below a critical angle of incidence $\alpha_c \approx \sqrt{2\varepsilon}$ there is total reflection; the incident wave is evanescent in the crystal and the penetration depth becomes extremely small (a few tens of angstroms) compared with several thousand angstroms by the time the angle of incidence is three or four times the critical angle. As ε is of the order of 10^{-5}, α_c is of the order of milliradians.

Furthermore transmission and reflection are strongly dependent on the angle of incidence; the Fresnel formulae gives the transmission as:

$$T^2(\alpha_i) = \left| \frac{2 \sin \alpha_i}{\sin \alpha_i + \sqrt{n^2 - \cos^2 \alpha_i}} \right|^2$$

which has a pronounced maximum at $\alpha_i = \alpha_c$: a very small change in the angle of incidence near the critical angle has a large effect on the transmitted and reflected intensities. The shallow depth of penetration for glancing incidence means that diffraction at very small angles can be used to study surfaces.

When elastic scattering of X-rays occurs in a periodic structure, with a form function L and a structure factor $F_{(S)}$ of the formula unit, the scattered intensity is of the form $I(S) \propto \| F_{(S)} \|^2 .L^2$.

S is a reciprocal vector equal to $h\mathbf{A}^* + k\mathbf{B}^* + l\mathbf{C}^*$, and in a three-dimensional lattice formed from $N_1.N_2.N_3$ cells, the form function becomes:

$$L = \frac{\sin \pi.N_1.\mathbf{a}.\mathbf{S}}{\sin \pi.\mathbf{a}.\mathbf{S}} \frac{\sin \pi.N_2.\mathbf{b}.\mathbf{S}}{\sin \pi.\mathbf{b}.\mathbf{S}} \frac{\sin \pi.N_3.\mathbf{c}.\mathbf{S}}{\sin \pi.\mathbf{c}.\mathbf{S}}$$

The scattered intensity is concentrated around the reciprocal nodes. For a two-dimensional lattice with base vectors $\mathbf{a}$ and $\mathbf{b}$, the third term L disappears owing to the loss of periodicity in the $\mathbf{c}$ direction and there is no longer any restriction on the values of the index l which now appears as a continuous variable.

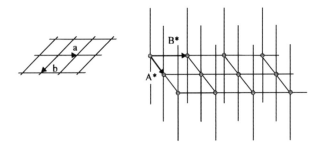

The reciprocal lattice of a plane lattice is composed of 'rods' along which the intensity decreases continuously with distance away from the nodes in the plane defined by $\mathbf{A}^*$ and $\mathbf{B}^*$. Any modulation of the intensity along these rods would be due to a finite thickness of the lattice. Note that during diffraction at glancing angles the diffraction pattern of the interior of the crystal will be superimposed on that of the surface.

❒ EXPERIMENTAL EQUIPMENT AND DATA TREATMENT

The angular divergence and energy of the incident beam are critical and the angles of incidence and observation must be perfectly controlled.

Because of adsorption by surfaces, it is generally necessary to work in an ultra-high vacuum, which sets additional constraints on the geometry of the diffractometer. Measurements of intensity profiles along normals to the plane of the reciprocal lattice can be facilitated by using position detectors. Surface atoms are few in number and diffracted intensities are weak, and so high fluxes are required. The numerous geometrical constraints make it necessary to use apparatus having many degrees of freedom, such as four-circle devices.

In the treatment of the data, the same problems arise as for three-dimensional structures; if a model of the surface can be obtained (e.g. by tunnel effect microscopy or electron diffraction), the structure is refined by comparison of measured and calculated intensities. Otherwise the Patterson method must be used to determine the inter-atomic vectors.

❒ THE STRUCTURES OF SURFACES

The real surface of a crystal cannot be treated as its simple section by a plane parallel to the surface; surface atoms have dangling bonds and a reorganisation process occurs to minimise the energy. This surface reorganisation results in movements of atoms (both normal and parallel to the surface) leading to base vectors $\mathbf{a}_i^s$ for the surface cell which are different from the vectors $\mathbf{f}_i^V$ defining the parallel internal planes. This effect is particularly marked in the case of covalent crystals, which have highly directional bonds. Even when the plane under investigation is a lattice plane it is difficult to obtain perfectly flat surfaces; real surfaces are stepped and in addition are contaminated by pollutants, making it necessary to work in an ultra-high vacuum as mentioned above to eliminated adsorbed atoms.

❒ EXAMPLES OF SURFACE STUDIES

An understanding of the surfaces and interfaces in semiconductor materials is vital in electronics technology, which is why much of the work on surfaces has been carried out on these materials. Surface diffraction on semiconductors has shown the existence of surface cells made up of adventitious atoms positioned at the nodes of a plane lattice whose base vectors are related to those of the crystal lattice by relations of the form: $\mathbf{a}^s = m\mathbf{f}^v$ and $\mathbf{b}^s = n\mathbf{g}^v$ where m and n

depend on the type of surface. Thus for the (001) surfaces of gallium arsenide, GaAs, arrangements of the following type have been suggested:

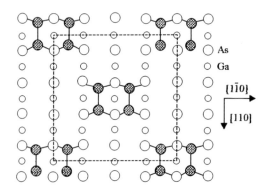

The surface plane is composed of pairs of arsenic dimers bound to arsenics in the first internal layer.
The surface cell here is of the 4 × 4 type.
(From Sauvage-Simkin *et al.*, *Phys. Rev.* **83** 1989).

For silicon and germanium (111) surfaces, it has been found that the surface atoms are situated above the internal atoms and form 2 × 2 cells. In the case of the interface Si (111)–metal (Au, Al), the metal atoms form a hexagonal lattice with a cell having parameters $\sqrt{3}$ times those of the surface.

This method has also been used to study the transition zone in heteroepitaxy and to follow the state of surface roughness during epitaxial growth in the vapour phase or by bombardment. By choosing a suitable angle of incidence it can be arranged that two successive stepped surfaces diffract in opposite phase to each other, and it is then possible to follow the growth of the specimen layer by layer.

Present day techniques such as tunnel effect microscopy enable models to be proposed for surfaces, but for materials with an ordered structure these models must be refined by diffraction to obtain precise information on the positions of the atoms. Surface diffraction is, however, a technique requiring the skills and equipment only found in larger laboratories.

Chapter 18

Crystallographic Computing

Progress in crystallography is closely linked to the growth of computing power, and crystallographers today have at their disposal powerful computing tools that have been developed over the last forty years.

Personal computers can now be used to display simple structures and to calculate the positions of diffraction spots for a crystal of known structure; however, to determine a structure, the long and complex calculations on a large amount of data require the use of complex software on high-power computers.

Manufacturers of diffraction apparatus supply a suite of programs with their equipment which will handle all the tasks involved:

◆ Controlling the instrument (accurately positioning the specimen and detector).
◆ Entering and storing the data.
◆ Processing the data.
◆ Displaying the results (files, 2D and 3D structure projections, stereoscopic views, thermal agitation ellipsoids etc.)

Processing and display software has been developed by pluridisciplinary teams of mathematicians, computer scientists and crystallographers in research centres served by major laboratories.

After first explaining briefly the basic notions[1] we shall examine structure refinement by the least squares technique and then give a few simple examples of programs.

---

[1]For a study in depth of programming methods for crystallography, see, e.g., the summary articles featured in volumes B and C of the International Tables.

1   THE BASIC NOTIONS

1.1   Crystallographic Frames

❏   THE DIRECT LATTICE

The 'natural' frame is that of the base vectors in which lattice translations are represented by triplet of integers or semi-integers. For low-symmetry unit-cells, there exists more than one choice of base vectors. There is, however, a cell, called the Niggli reduced cell, which enables the lattice to be described unequivocally; this is a primitive cell constructed on the three shortest non-coplanar translations. There are various programs for determining this cell.

The defining characteristics of the direct lattice are the base vector lengths a, b and c, the angles between the vectors α, β and γ and the unit-cell volume:

$$V = abc[1 - \cos^2\alpha - \cos^2\beta - \cos^2\gamma + 2\cos\alpha.\cos\beta.\cos\gamma]^{1/2}$$

In this frame, a direct row $[uvw]$ is represented by the vector:

$$\mathbf{n}_{uvw} = \mathbf{r} = u.\mathbf{a} + u.\mathbf{b} + u.\mathbf{c}$$

When programming, matrices are often use to represent the vectors:

$$\mathbf{r} = u.\mathbf{a} + v.\mathbf{b} + w.\mathbf{c} = (u,\ v,\ w).\begin{pmatrix} \mathbf{a} \\ \mathbf{b} \\ \mathbf{c} \end{pmatrix} = (\mathbf{a},\ \mathbf{b},\ \mathbf{c}).\begin{pmatrix} u \\ v \\ w \end{pmatrix}$$

The matrix formulation of the scalar product is:

$$\mathbf{r}_1.\mathbf{r}_2 = (u_1,\ v_1,\ w_1).\begin{pmatrix} \mathbf{a}^2 & \mathbf{a.b} & \mathbf{a.c} \\ \mathbf{a.b} & \mathbf{b}^2 & \mathbf{b.c} \\ \mathbf{a.c} & \mathbf{b.c} & \mathbf{c}^2 \end{pmatrix}.\begin{pmatrix} u_2 \\ v_2 \\ w_2 \end{pmatrix} = \mathbf{u}_1^{\mathrm{T}}.\mathbf{M}.\mathbf{u}_2$$

❏   THE RECIPROCAL LATTICE

The parameters defining the reciprocal lattice are expressed in terms of the direct values by the following relations (cf. § II.6):

$$A^* =\| A^* \|= b.c.\sin\alpha.V^{-1};\ B^* = a.c.\sin\beta.V^{-1};\ C^* = a.b.\sin\gamma.V^{-1}$$

$$\cos\alpha^* = \frac{\cos\gamma.\cos\beta - \cos\alpha}{\sin\gamma.\sin\beta} \quad \cos\beta^* = \frac{\cos\alpha.\cos\gamma - \cos\beta}{\sin\alpha.\sin\gamma} \quad \cos\gamma^* \frac{\cos\alpha.\cos\beta - \cos\gamma}{\sin\alpha.\sin\beta}$$

◻ LATTICE CALCULATIONS

From the scalar product, the modulus of the rows and the angle between them can be calculated. In the direct lattice, the modulus of a row is the square root of the scalar product $\mathbf{r} \cdot \mathbf{r}$; the reciprocal of this modulus is equal to the interplanar spacing D^*_{uvw} between the planes $(u\ v\ w)^*$ of the reciprocal lattice to which the row $[u\ v\ w]$ is normal.

The modulus of a reciprocal row is the square root of the scalar product $\mathbf{N}^*_{hkl} \cdot \mathbf{N}^*_{hkl}$; the reciprocal of this modulus is equal to the interplanar spacing d_{hkl} between the planes of the family $(h\ k\ l)$ in the direct lattice. To determine the angles between these rows, the scalar product is again used. Interatomic distances and bond angles are calculated in the same way.

◻ LATTICE TRANSFORMATIONS

A structure described in a frame chosen according to standard crystallographic practice is not always the most convenient; thus a physicist studying the transition between a monoclinic 2/m phase and a tetragonal 4/m phase will prefer to describe the monoclinic phase with the two-fold axis oriented along Oz rather than the normal choice, Oy.

Lattice transformations are frequent in calculations since at least two reference frames are involved: that of the diffraction pattern, in relation to the laboratory, and that of the crystal, in relation to the positions of the atoms.

In general, the base vector frame related to the crystal is not orthonormal and is therefore unsuitable for numerical calculations. For example, in this frame the distance between two atoms of coordinates (x_1, y_1, z_1) and (x_2, y_2, z_2) is:

$$D^2 = X^2 + Y^2 + Z^2 + 2XY\cos\gamma + 2YZ\cos\alpha + 2XZ\cos\beta$$

where: $X = a(x_1 - x_2)$; $Y = b(y_1 - y_2)$; $Z = c(z_1 - z_2)$

The calculations are therefore performed in an orthonormal frame related to the crystal. There are an infinite number of ways of orthogonalizing the crystal frame but in most cases the 'international' frame, IF, is used.

◻ THE INTERNATIONAL FRAME

This right-handed orthonormal frame $(\mathbf{O}, \mathbf{i}, \mathbf{j}, \mathbf{k})$ is defined by[2]

$$A = \begin{pmatrix} \mathbf{a} \\ \mathbf{b} \\ \mathbf{c} \end{pmatrix} \quad I = \begin{pmatrix} \mathbf{i} \\ \mathbf{j} \\ \mathbf{k} \end{pmatrix} \quad \text{where } \mathbf{i} = \frac{\mathbf{a}}{a}; \ \mathbf{j} = \frac{\mathbf{a} \wedge \mathbf{C}^*}{a.C^*.\sin\{\mathbf{a},\mathbf{C}^*\}}; \ \mathbf{k} = \frac{\mathbf{C}^*}{C^*}$$

---

[2] There exists another definition of IF in which $\mathbf{k}$ is parallel to $\mathbf{c}$.

The frame-changing matrix M in the transformation $I = (M).A$ and its inverse are respectively (cf. § II.6):

$$(M) = \begin{pmatrix} 1/a & 0 & 0 \\ -\cos\gamma/(a.\sin\gamma) & 1/b.\sin\gamma & 0 \\ A^*.\cos\beta^* & B^*.\cos\alpha^* & C^* \end{pmatrix}$$

$$A = \begin{pmatrix} a & 0 & 0 \\ b.\cos\gamma & b.\sin\gamma & 0 \\ c.\cos\beta & -c.\sin\beta.\cos\alpha^* & 1/C^* \end{pmatrix} .I = (M^{-1}).I$$

☞ | *When using this transformation, the* covariance *of the Miller indices of the lattice planes and the* contravariance *of the row indices must be taken into account.*

Consider a row in the direct lattice $\mathbf{OD} = u.\mathbf{a} + v.\mathbf{b} + w.\mathbf{c}$; the coordinates of the point D in IF are:

$$x = u.a + v.b.\cos\gamma + w.c.\cos\beta$$
$$y = v.b\,\sin\gamma - w.c\,\sin\beta.\cos\alpha^*$$
$$z = w.c\,\sin\beta.\sin\alpha^*$$

Consider the reciprocal row OE: $\mathbf{N}^*_{hkl} = h.\mathbf{A}^* + k.\mathbf{B}^* + l.\mathbf{C}^*$. The co-ordinates of E in IF are:

$$x = h.A^*.\sin\beta^*.\sin\gamma$$
$$y = h.A^*.\sin\beta^*.\cos\gamma + k.B^*.\sin\alpha^*$$
$$z = h.A^*.\cos\beta^* + k.B^*.\cos\alpha^* + l.C^*$$

In IF, the distance between two atoms of coordinates (x_1, y_1, z_1) and (x_2, y_2, z_2) is simply (in a cartesian frame):

$$D^2 = X^2 + Y^2 + Z^2$$

where: $X = (x_1 - x_2)$; $Y = (y_1 - y_2)$; $Z = (z_1 - z_2)$

❐ THE TRIGONAL LATTICE

It is best with this lattice to transform to trigonal hexagonal before going into IF. The base vectors of the trigonal cell are converted to those of the hexagonal cell by:

$$\begin{pmatrix} A \\ B \\ C \end{pmatrix} = \begin{pmatrix} 1 & \bar{1} & 0 \\ 0 & 1 & \bar{1} \\ 1 & 1 & 1 \end{pmatrix} \cdot \begin{pmatrix} a \\ b \\ c \end{pmatrix}$$

If the results are required in the trigonal frame, the inverse transformation is simply applied.

1.2 Representation of Rotations

NOTE: Rotation matrices give the coordinates of the image of the object which has undergone a rotation in terms of the coordinates of the object. The base vectors are unchanged.

❐ ROTATION IN AN ORTHONORMAL FRAME.

In an orthonormal frame $\mathbf{i}$, $\mathbf{j}$, $\mathbf{k}$, rotations through an angle φ of a vector $\mathbf{r}$ about axes directed along $\mathbf{i}$, $\mathbf{j}$, $\mathbf{k}$ generate vectors $\mathbf{t}^i = \mathbf{R}^i_\varphi.\mathbf{r}$, $\mathbf{t}^j = \mathbf{R}^j_\varphi.\mathbf{r}$, $\mathbf{R}^k_\varphi.\mathbf{r}$ where (anticlockwise rotations are positive):

$$\mathbf{R}^i_\varphi = \begin{pmatrix} 1 & 0 & 0 \\ 0 & \cos\varphi & -\sin\varphi \\ 0 & \sin\varphi & \cos\varphi \end{pmatrix} \quad \mathbf{R}^j_\varphi = \begin{pmatrix} \cos\varphi & 0 & \sin\varphi \\ 0 & 1 & 0 \\ -\sin\varphi & 0 & \cos\varphi \end{pmatrix}$$

$$\mathbf{R}^k_\varphi = \begin{pmatrix} \cos\varphi & -\sin\varphi & 0 \\ \sin\varphi & \cos\varphi & 0 \\ 0 & 0 & 1 \end{pmatrix}$$

For rotoinversions, all the $+1$ must be replaced by -1.

☞ *These relations are only valid if the frame is orthonormal and if the rotation axis is a base vector.*

❐ ROTATION IN THE FRAME OF THE CRYSTAL

The simplest way of representing a rotation through an angle φ about a direction $\mathbf{OS}$ in the crystal (frame $A = 0$, $\mathbf{a}$, $\mathbf{b}$, $\mathbf{c}$) is to choose an orthnormal frame $(I = 0, \mathbf{i}, \mathbf{j}, \mathbf{k})$ whose base vector is parallel to OS: $I = (M).A$.
   Rotation about $\mathbf{i}$ is represented by the matrix R^i_φ. In frame A this rotation will be: $R = (M^{\mathrm{T}}).R^i_\varphi.(M^{\mathrm{T}})^{-1}$.

◻ COMPOUND ROTATIONS

◆ About a single axis:
We simply take the product of the two matrices representing the rotations (respecting the order!)
◆ About different axes:
If the axes are orthogonal, we can take the product of the matrices R^i_φ which represent the rotations. Otherwise we have to use, for example, Eulerian angles.

◻ EULERIAN ANGLES

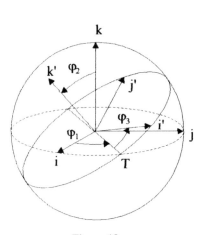

Figure 18.a

Consider two orthonormal frames:
I (0, **i, j, k**) and I^1 (0, **i′, j′, k′**)
These can be made to coincide by three successive rotations:
1—Rotation through φ_1 about **k** (**i** ≡ OT)
2—Rotation through φ_2 about **OT** (**k** ≡ **k′**)
3—Rotation through φ_3 about **k′** (**i** ≡ **i′**)
The coordinates X (x, y, z) and $X′$ ($x′$, $y′$, $z′$) in the two frames are related by:

$$X′ = R_E . X$$

where R_E is the Eulerian matrix:

$$R_E = R^k_{\varphi 3} . R^i_{\varphi 2} . R^k_{\varphi 1}$$

Any rotation is decomposed into a series of three rotations.

1.3   Generating Equivalent Positions

To obtain the coordinates of all the atoms in the unit-cell it is sufficient to know the positions of the atoms in the formula unit and to apply the symmetry operations of the group generators. The group symmetry operations can be represented by affine transformations of the type:

$$\begin{pmatrix} x'_1 \\ x'_2 \\ x'_3 \end{pmatrix} = \begin{pmatrix} r_{11} & r_{12} & r_{13} \\ r_{21} & r_{22} & r_{23} \\ r_{31} & r_{32} & r_{33} \end{pmatrix} . \begin{pmatrix} x_1 \\ x_2 \\ x_3 \end{pmatrix} + \begin{pmatrix} t_1 \\ t_2 \\ t_3 \end{pmatrix}$$

The elements r_{ij} of the matrix represent a proper or improper rotation and t_i a translation. Homogeneous or Seitz matrices are generally used; with these 4×4 matrices the new coordinates can be calculated from the old ones by the relation:

$$(x'_1 \quad x'_2 \quad x'_3 \quad 1) = (x_1 \quad x_2 \quad x_3 \quad 1) . \begin{pmatrix} r_{11} & r_{12} & r_{13} & 0 \\ r_{21} & r_{22} & r_{23} & 0 \\ r_{31} & r_{32} & r_{33} & 0 \\ t_1 & t_2 & t_3 & 1 \end{pmatrix}$$

An n-fold axis will give rise to n homogeneous matrices (including the identity matrix).

Examples of homogeneous matrices:

2_1 axis $\parallel [001]$ at $(1/4, 0, z)$ 21 axis $\parallel [010]$ at $(0, y, 0)$ Mirror $n \parallel (010) y = 1/4$

$$\begin{pmatrix} -1 & 0 & 0 & 0 \\ 0 & -1 & 0 & 0 \\ 0 & 0 & 1 & 0 \\ 1/2 & 0 & 1/2 & 1 \end{pmatrix} \qquad \begin{pmatrix} -1 & 0 & 0 & 0 \\ 0 & 1 & 0 & 0 \\ 0 & 0 & -1 & 0 \\ 0 & 1/2 & 0 & 1 \end{pmatrix} \qquad \begin{pmatrix} 1 & 0 & 0 & 0 \\ 0 & -1 & 0 & 0 \\ 0 & 0 & 1 & 0 \\ 1/2 & 1/2 & 1/2 & 1 \end{pmatrix}$$

1.4 Calculation of the Structure Factor

The algebraic expression for the structure factor, which is:

$$F_{\mathbf{S}} = F_{hkl} = \sum_{i=1}^{n} (f_i)_t . e^{2j\pi . \mathbf{r}_i . \mathbf{S}} = \sum_{i=1}^{n} (f_i)_t . e^{2j\pi(h.x_i + k.y_i + l.z_i)}$$

will be transformed into $F_{\mathbf{s}} = A_{\mathbf{s}} + i.B_{\mathbf{s}}$ where:

$$A_{\mathbf{s}} = \sum_{i=1}^{n} (f_i)_t (\cos 2\pi(h.x_i + k.y_i + l.z_i)); \quad B_{\mathbf{s}} = \sum_{i=1}^{n} (f_i)_t . \sin 2\pi(h.x_i + k.y_i + l.z_i)$$

If g is the order of the crystal space group and m the number of atoms in the formula unit, the total number of atoms in the unit-cell will be $N = mg$, provided that all the atoms are in general positions. To include atoms in special positions we let p_j be the number of equivalent positions for the atom j and $d_j = p_j/g$.

If (β_{ij}^k) is the 3×3 matrix for anisotropic thermal agitation of the atom k we write $\lambda_k = (\mathbf{S}^T).(\beta_{ij}^k).(\mathbf{S})$ where: $(\mathbf{S}^T) = h\mathbf{A}^* + k\mathbf{B}^* + l\mathbf{C}^*$.

Following these conventions we obtain the general relation:

$$A_{\mathbf{S}} = \sum_{j=1}^{m} d_j . f_j^{\mathbf{S}} . \sum_{i=1}^{g} e^{-\lambda_k} . \cos 2\pi(h.x_i + k.y_i + l.z_i)$$

Atomic scattering factors are tabulated for all the elements in terms of $\sin\theta/\lambda$; to find the value of f_j for a given value of the vector $\mathbf{S}$ we use interpolation. Since these tables take up considerable memory space, the variation of f with $\sin\theta/\lambda$ is often represented as a sum of Gaussian functions $f(\sin\theta) = \sum_{i=1}^{n} \mathbf{a}_i . \mathrm{e}^{-\mathbf{b}_i \sin^2\theta/\lambda^2} + c_i$, and f_s is calculated from this. For very precise calculations we take $n = 4$; twelve values are sufficient for each type of atom.

Structure determination by numerical methods requires fairly powerful calculation facilities; for example, consider a cubic structure with a unit-cell of 4 Å where it is required to know the positions of the atoms to 0.05 Å. The electron density must be determined at $(4/0.05)^3 = 512\,000$ points; if 750 reflections are used, a Fourier series to 750 terms will have to be calculated for each point, giving a total of around 4×10^8 values.

2 STRUCTURE REFINEMENT

In refinement methods, the parameters of each atom (coordinates, thermal agitation factors) are varied so as to minimise the reliability factor:

$$R = \sum_{\mathbf{S}} \omega_{\mathbf{S}}(|F_{\mathbf{S}}^{Obs}| - k|F_{\mathbf{S}}^{Cal}|)^2 = \sum_{\mathbf{S}} \omega_{\mathbf{S}} . \Delta F_{\mathbf{S}}^2$$

$\omega_{\mathbf{s}}$ is the weighting of each refraction spot of the reciprocal vector $\mathbf{S}$ and k is a scaling factor relating calculated and observed factors.

□ LEAST SQUARES METHOD

R is a function of the N parameters x_i (nine parameters per atom in the anisotropic thermal agitation model and four in the isotropic model):

$$R = R(x_1, \ldots, x_i, \ldots, x_N)$$

When the minimum value of R is reached, all the derivatives $\partial R/\partial x_i$ are zero and:

$$\sum_{\mathbf{S}} \omega_{\mathbf{S}} . \Delta F_{\mathbf{S}} . \frac{\partial \Delta F_{\mathbf{S}}}{\partial x_i} = 0 \Rightarrow \sum_{\mathbf{S}} \omega_{\mathbf{S}} . \Delta F_{\mathbf{S}} . \partial \frac{k.F_{\mathbf{S}}^{Cal}|}{\partial x_i} = 0$$

Initial determination of the structure gives the approximate values x_i' of the parameters x_i. We now seek the set of best values $\Delta x_i = x_i - x_i'$.

For the initial values we have:

$$\frac{\partial R}{\partial x_i} = -\sum_{\mathrm{S}} \omega_{\mathrm{S}}.\Delta F_{\mathrm{S}}.\partial| \frac{k.F_{\mathrm{S}}^{Cal}|}{\partial x_i} \neq 0$$

To the first order we can write:

$$\Delta\left(\frac{\partial R}{\partial x_i}\right) = \sum_i \left[\frac{\partial}{\partial x_i}\left(-\sum_{\mathrm{S}} \omega_{\mathrm{S}}.\Delta F_{\mathrm{S}}.\frac{\partial|k.F_{\mathrm{S}}^{Cal}|}{\partial x_j}\right)\right].\Delta x_i$$

The best values of x_i are solutions of the system of N linear equations (normal equations):

$$\sum_i \Delta x_i.\sum_{\mathrm{S}} \omega_{\mathrm{S}} \frac{\partial|k.F_{\mathrm{S}}^{Cal}|}{\partial u_i}.\frac{\partial|k.F_{\mathrm{S}}^{Cal}|}{\partial u_j} = \sum_{\mathrm{S}} \omega_{\mathrm{S}}.\frac{\partial|k.F_{\mathrm{S}}^{Cal}|}{\partial u_j}$$

Writing:

$$\alpha_{ij} = \sum_{\mathrm{S}} \omega_{\mathrm{S}} \frac{\partial|k.F_{\mathrm{S}}^{Cal}|}{\partial u_i}.\frac{\partial|k.F_{\mathrm{S}}^{Cal}|}{\partial u_j} ; \quad y_j = \sum_{\mathrm{S}} \omega_{\mathrm{S}}.\frac{\partial|k.F_{\mathrm{S}}^{Cal}|}{\partial u_j},$$

the system of normal equations is then: $\alpha_{ij}.\Delta x_i = y_i$.

Solving this system requires the inverse matrix of α_{ij} which is symmetrical:

$$\Delta x_i = (\alpha_{ij})^{-1}.y_j$$

The operation is by no means a trivial one. Consider for example a structure with thirty atoms in the formula unit; in an anisotropic thermal agitation model there will be 30×9 parameters plus one more (the value of k). The matrix of rank 271 contains $271 \times 35 = 36\,585$ different terms, and each term is the sum of several hundred elements!

To simplify the problem, we note that each element in the matrix α_{ij} is the sum of the product of partial derivatives whose signs are random. If i is different from j, the sum will *a priori* be small. On the other hand, if $i = j$, all the products will be positive and it will then be possible to ignore all but the terms on the main diagonal of the matrix; this will very substantially simplify the calculations. This method becomes invalid if there are interactions between the elements of the matrix; there is, necessarily, coupling between the values of the coordinates and the thermal agitation parameters of a given atom and hence a middle path is usually adopted between calculation of the full matrix and use of the diagonal least mean squares approximation.

With terms relating to a single atom, 9×9 or 4×4 blocks are constructed (depending on the thermal agitation model used). These blocks are spread along the main diagonal; the other elements are assumed to be zero in this block diagonal approximation. To take into account interactions between atoms (in the case of molecules), larger blocks can be used. In this way, there are far fewer terms to calculate; furthermore, specific methods for the inversion of this type of matrix are available.

Recently new algorithms using fast Fourier transform (FFT) methods have been developed and successfully used to reduce the calculation time.

❏ STRUCTURE DETERMINATION PROGRAMS

There are available today complete systems for structure determination. With this modular software, the initial structure is obtained and then refined. A survey of articles in *Acta Crystallographica* shows that the most widely used among them are SHELX (for one third of all structures), OAK RIDGE Program (one third) and XRAY (one sixth). The first versions of these programs were developed in the late sixties and they are now constantly updated to keep pace with theoretical and technological progress.

❏ THE PROGRAM SHELX

This was developed by the Sheldrick team and can be used in computers of modest size, which explains its popularity.

It has two modules: SHELXS for structure solving and SHELXL for refinement. Each module has about 6 000 lines of Fortran code.

The SHELLXS module calculates normalised structure factors E_s, performs the Patterson interpretation and uses direct methods of structure calculation. The structure factors of all the elements (but not ions) are stored internally. From the name of the group the program checks the symmetry data entered by the user. Because of the presence of symmetry elements, the reciprocal domain being studied will have equivalent spots; the experimental data are surveyed to enable a set of independent reflections to be selected.

The module SHELXL automatically takes account of the constraints on the values of the parameters U_{ij} imposed by the symmetry; it offers the choice between the 'full matrix' method and the 'block cascade' method; it also takes account of the fact that the crystal used for recording the intensities may be twinned. There is now a PC version of the program.

As an illustration, here is an example of a data file used by the program. The crystal was NaCaCrF6, and was twinned.

The commentaries are in italics; the sign = indicates line continuation.

```
TITL NACACRF6 STOE
CELL   0.70926 9.103 9.103 5.120 90. 90. 120. λ, a, b, c, α, β, γ
ZERR   3 0.002 0.002 0.0015 0 0 0      standard deviations
LATT   -1                             Centrosymmetric
SYMM   -Y,X-Y,Z                       Symmetry elements
SYMM   Y-X,-X,Z
```

```
SYMM    Y,X,-Z
SYMM    -X,Y-X,-Z
SYMM    X-Y,-Y,-Z                        Calculating Fhkl values (ions)
SFAC    NA 3.2565,2.6671,3.9362,6.1153,1.3998,0.2001,1.0032,14.039 =
        0.4040,0.030,0.025,112.2,1.02 22.9898
SFAC    CA 15.6348 -0.00740 7.9518 0.6089 8.4372 10.3116 0.8537 =
        25.9905 -14.875 0.203 0.306 1264 1.00 40.08
SFAC    CR 9.6809,5.59463,7.81136,0.334393,2.87603,12.8288 =
        0.113575,32.8761,0.518275,0.284,0.624,2526,0.615 51.996
SFAC    F 3.632,5.27756,3.51057,14.7353,1.26064,0.442258,0.940706 =
        47.3437,0.653396,0.014,0.010,49.99,1.30 18.9984
UNIT    3 3 3 18
L.S.    5
ACTA
TWIN    1 1 0 -1 0 0 0 0 1 2             Twin (type and number)
WGHT    0.010300 0.0190
EXTI    0.106090
BASF    0.43138                          Twin coefficient
FVAR    0.68620                          Initial data
CA      2 0.62946 10.00000 10.00000 10.50000 0.01004 0.01103 =
        0.00860 0.00009 0.00005 0.00552
NA      1 0.28859 10.00000 10.50000 10.50000 0.01411 0.01311 =
        0.01390 0.00234 0.00117 0.00656
CR1     3 10.00000 10.00000 10.00000 10.16667 0.00833 0.00833 =
        0.01082 0.00000 0.00000 0.00417
CR2     3 10.33333 10.66667 0.50089 10.33333 0.00820 0.00820 =
        0.00858 0.00000 0.00000 0.00410
F1      4 0.53017 0.77857 0.28599 11.00000 0.01302 0.01893 =
        0.01698 0.00886 0.00501 0.00752
F2      4 0.79857 0.91159 0.79549 11.00000 0.01293 0.01914 =
        0.01748 -0.00898 -0.00471 0.00870
F3      4
        0.13689 0.55037 0.71597 11.00000 0.01223 0.01524 =
        0.01684 -0.00214 0.00509 0.00470
HKLF    4                                Read intensity file
END
```

After refinement (4709 reflections analysed of which 1778 were independent), the following results were obtained (the program also calculates all the interatomic distances and bond angles):

NACACRF6

ATOM	x	y	z	U11	U22	U33	U23	U13	U12
Ca	0.62947	0.00000	0.00000	0.01004	0.01107	0.00862	0.00013	0.00007	0.00553
	0.00002	0.00000	0.00000	0.00005	0.00007	0.00006	0.00010	0.00005	0.00004
Na	0.28863	0.00000	0.50000	0.01420	0.01327	0.01385	0.00237	0.00119	0.00664
	0.00006	0.00000	0.00000	0.00014	0.00020	0.00019	0.00041	0.00020	0.00010

```
Cr1   0.00000 0.00000 0.00000 0.00835 0.00835 0.01084 0.00000 0.00000 0.00418
      0.00000 0.00000 0.00000 0.00006 0.00006 0.00009 0.00000 0.00000 0.00003

Cr2   0.33333 0.66667 0.50093 0.00821 0.00821 0.00862 0.00000 0.00000 0.00410
      0.00000 0.00000 0.00007 0.00004 0.00004 0.00006 0.00000 0.00000 0.00002

F1    0.53017 0.77859 0.28619 0.01295 0.01877 0.01719 0.00892 0.00499 0.00745
      0.00009 0.00010 0.00013 0.00024 0.00027 0.00020 0.00024 0.00022 0.00023

F2    0.79854 0.91157 0.79552 0.01287 0.01906 0.01748 -.00891 -.00471 0.00860
      0.00009 0.00009 0.00011 0.00023 0.00032 0.00019 0.00026 0.00022 0.00023

F3    0.13687 0.55036 0.71599 0.01229 0.01524 0.01693 -.00219 0.00513 0.00471
      0.00008 0.00008 0.00012 0.00023 0.00030 0.00021 0.00021 0.00019 0.00021
```

```
R1 = 0.0148 for 1767 Fo > 4.sigma(Fo) and 0.0150 for all 1778 data
wR2 = 0.0348, GooF = S = 1.109, Restrained GooF = 1.109 for all data
```

In spite of these powerful programs, structure determination is still an art requiring experience and a critical mind.

3   EXAMPLES OF SIMPLE PROGRAMS

3.1   Diffraction Pattern Synthesis

It is relatively simple to determine the position of diffraction spots for a *known* crystal structure; the calculation of intensities on the other hand requires many more calculations: structure factors, and thermal agitation, Lorentz and adsorption corrections.

These pattern synthesis programs are mainly for teaching purposes, but there do exist commercial programs which, from a single Laue photograph, enable calculation of the angle through which the crystal must be turned for a given direction to be parallel to the beam; such programs are extremely time-saving when setting up the crystal.

Diffraction pattern synthesis for rotating crystal methods is also possible. As an example we give the flow diagram for calculating the spectrum of a rotating crystal in a Bragg camera with complete rotations. For partial rotations or Weissenberg photographs, the programming is rather more tricky since the moment must be determined at which the reciprocal nodes penetrate the Ewald sphere.

Cylindrical film rotation method

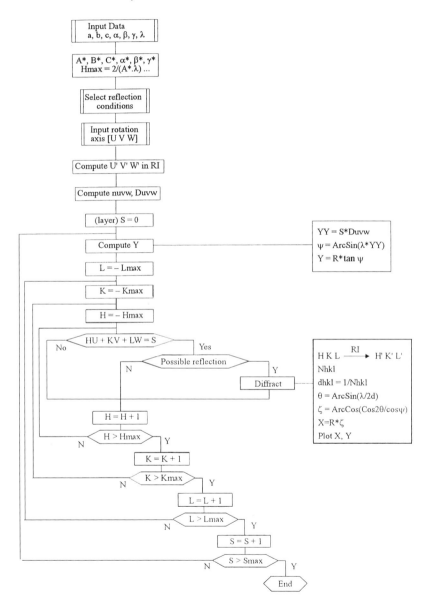

Figure 18.1

3.2   Stereographic Projections

Plotting a stereographic projection by classical drawing methods is a long operation requiring care and method even for an experienced operator. These projections can be calculated and either drawn on paper or displayed on a monitor.

In what follows, we assume that the crystal is oriented such that the base vectors **i** *and* **j** *in the international frame are contained in the plane of the stereographic projection. The projections will then be such that the crystal symmetry elements are in agreement with the conventions of the International Tables.*

❒   PLOTTING A POLE WHEN φ AND ρ ARE KNOWN

We wish to determine the coordinates x and y, in the plane of the projection, of the point p, the projection of the pole of the face, in terms of the angles φ and ρ of the pole. The axis Ox is at the origin of the azimuths.

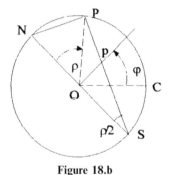

Figure 18.b

Given that SOp and SPN are similar triangles, we have:

$$Op = \frac{SO.PN}{SP}$$

$PN = 2R \sin \rho/2$, $SP = 2R \cos \rho/2$.
$Op = R \tan \rho/2$. Hence :
$x = R \cos \varphi.\tan \rho/2$
$y = R \sin \varphi.\tan \rho/2$

❒   PLOTTING A POLE OF INDICES h, k AND l

Except in the case of cubic crystals, the parameters of the direct and reciprocal lattices must be known. The coordinates x_0, y_0 and z_0 of the pole $(h\ k\ l)$ in the international frame are then sought. Let P be the pole of the face $(h\ k\ l)$; this will be the intersection of the normal to the face, OP, with the projection sphere. **OP** is collinear with the vector $\mathbf{N}^*_{hkl}$ of the reciprocal lattice:

$$\mathbf{OP} = \lambda.(h.\mathbf{A}^* + k.\mathbf{B}^* + l.\mathbf{C}^*)$$

$$\| \mathbf{OP} \| = R = \lambda. \| \mathbf{N}_{hkl} \|$$

$$\lambda = \frac{R}{\sqrt{\mathbf{N}.\mathbf{N}}}$$

Hence in the international frame (refer back to the section) the coordinates are:

$$x_0 = \lambda(hA^*\sin \beta^* \sin \gamma)$$
$$y_0 = \lambda(-hA^*\sin \beta^* \cos \gamma + k.B \sin \alpha^*)$$
$$z_0 = \lambda(hA^*\cos \beta^* + kB^* \cos \alpha^* + lC^*)$$

from which we deduce φ and ρ where: $\tan \varphi = \dfrac{y_0}{x_0}$ $\qquad \cos \rho = \dfrac{z_0}{R}$

❐ PLOTTING A ZONE CIRCLE

Let $(h_1 \ k_1 \ l_1)$ and $(h_2 \ k_2 \ l_2)$ be the two planes of the poles p_1 and p_2 through which we wish to pass a zone circle. The stages in the construction are as follows:
a) Determine the zone axis $[u \ v \ w]$ of $(h_1 \ k_1 \ l_1) - (h_2 \ k_2 \ l_2)$.
This is a direct row where:

$$u = k_1.l_2 - l_1.k_2$$
$$v = l_1.h_2 - h_1.l_2$$
$$w = h_1.k_2 - h_2.k_1$$

The row is collinear with the direct lattice vector **OQ**.
The coordinates of Q in the international frame are x', y' and z'.

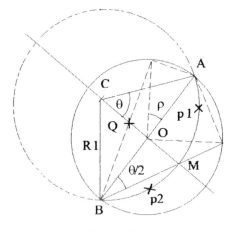

Figure 18.c

b) Calculate the angles φ and ρ of the zone axis where:

$$\varphi = \arctan{(y'/x')}$$
$$\rho = \arccos{(z'/R)}.$$

c) Calculate the angles φ_M and ρ_M of the point M which is the intersection of the zone circle with OQ.

d) Calculate the coordinates x_M and y_M of the point M.

e) Calculate the radius of the zone circle $R_1 = R/\sin \rho_M$.

f) Calculate the coordinates X and Y of the centre C of the circle.

$$X = x_M - R_1 \sin \varphi_M \qquad Y = y_M - R_1 \cos \varphi_M$$

g) Draw the arc AB of the great zone circle, inside the projection circle. This arc, with centre at (X, Y) has a radius R_1, starts at the point B defined by the angle $(\varphi_M - \rho_M)$ and subtends an angle $2\rho_M$.

3.3   Representation of a Structure

❐   GENERATING THE ATOMS

We take the reduced coordinates of the atoms of the formula unit and the homogeneous matrices of the space group to which the structure belongs, and by successive application of these we generate all the equivalent positions. Any non-integral translations related to the lattice mode are added to these. Next, whole lattice translations are applied to generate all of the atoms contained in the volume in question. This method leads to 'doubletons' which must be identified and eliminated. The coordinates of the atoms are then calculated in an orthonormal frame.

To show the threefold and sixfold symmetries of trigonal and hexagonal structures, two extra unit-cells can be generated from the initial unit-cells, by using the following relations:

$$x_1 = y - x \quad y_1 = -x \quad z_1 = z$$
$$x_2 = -y \quad y_2 = x - y \quad z_2 = z$$

❏ PRINCIPLE OF THE PROJECTION METHOD

The positions of the atoms are calculated in an orthonormal frame (O, x, y, z).
To observe the structure from different angles, the observer is moved.

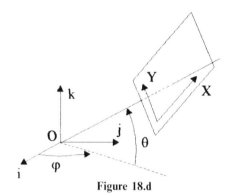

Figure 18.d

The viewing direction is defined by the angles θ and φ.
The structure is projected onto a normal plane (a window screen) to the right of the viewing direction.
The distance of the plane from the origin is determined so as to fill as much of the screen as possible.

If the direction of observation is a row [{u v w}], the angles φ and θ are deduced from the scalar products $(u\mathbf{a} + v\mathbf{b} + w\mathbf{c}).(u\mathbf{a} + v\mathbf{b})$ and $(u\mathbf{a} + v\mathbf{b}).(u\mathbf{a})$.

In the frame of the plane of projection, the coordinates X, Y and Z of an atom situated at x, y, z are:

$$X = -x \sin\theta + y \cos\theta$$
$$Y = -x \cos\theta.\sin\varphi - y \sin\theta.\sin\varphi + z \cos\varphi$$
$$Z = x \cos\theta.\cos\varphi + y \sin\theta.\cos\varphi + z \sin\varphi$$

The coordinate Z is inverted. Once the calculations have been made, the atoms are sorted according to the value of Z. To draw the projection, we begin with the atoms furthest from the observer; these will be progressively hidden by nearer atoms, giving a simple and effective method (the Z-buffer method) of handling the hidden parts.

Hidden zone elimination of coordination polyhedra and bonds is much more difficult, requiring complex algorithms which need considerable computing power.

Part 3
EXERCISES AND PROBLEMS

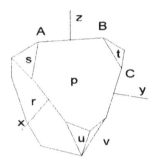

Exercises

LATTICES AND MILLER INDICES

1—A small crystal of barytine (orthorhombic) was measured with a two-circle goniometer. The angles between the normals to the faces were as follows:

(110)–$(010) = 48°47'$	(001)–$(o) = 36°32'$
(001)–$(111) = 63°26'$	(001)–$(p) = 20°51'$
(001)–$(m) = 56°30'$	(001)–$(q) = 52°44'$

Calculate the ratios a/b, b/c and c/a.
State the indices of the faces p, o, m and q.

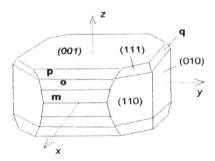

2—Construct the stereographic projection of a rhombododecahedron (cubic system), given that the angles of azimuth and inclination are as follows:

Faces	a	b	c	d	e
φ	$0°$	$45°$	$90°$	$315°$	$0°$
ρ	$45°$	$90°$	$45°$	$90°$	$135°$

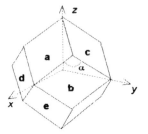

—Indicate all the symmetry elements on this projection.
—Calculate the angle between faces a and b and between b and d.
—Determine the indices of the edges between faces a and b and between faces b and c. Deduce from these the value of α.

3—Consider a crystal of lead chloride, $PbCl_2$. It is in the mmm class. The indices of faces b and c are respectively (010) and (001).

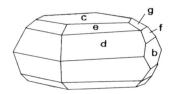

The angles of azimuth and inclination are:

Faces	b	c	d	e	f	g
φ	90°	–	30.72°	30.72°	90°	90°
ρ	90°	0°	66.76°	49.34°	67.21°	30.75°

Given that face d is a (111) face, determine the values of the ratios a/b, c/a and b/c and determine the indices of the faces e, f and g.

4—In a crystal lattice, can there be a row $[uvw]$ normal to a plane (hkl)? Make a general survey and then examine the following particular cases:

—cubic lattice,
—tetragonal lattice,
—monoclinic lattice.

5—Express the base vectors $\mathbf{A}^*$, $\mathbf{B}^*$ and $\mathbf{C}^*$ of a reciprocal monoclinic lattice in terms of the values of a, b, c and β in the direct lattice.

6—For a monoclinic crystal with angle $\beta = 94°12'$, the following row parameters were determined by X-ray diffraction:

$$[100] = 5.81 \text{ Å} \quad [010] = 8.23 \text{ Å} \quad [001] = 6.11 \text{ Å}.$$
$$[110] = 5.04 \text{ Å} \quad [011] = 10.3 \text{ Å} \quad [101] = 8.76 \text{ Å}.$$

What type of lattice has the crystal?

7—Show that in a cubic crystal the row [110] is normal to the plane (110). Show that this plane contains the rows [001], and [1$\bar{1}$0] and [$\bar{1}$11]. Calculate the angle between [001] and [$\bar{1}$11] and that between [$\bar{1}$11] and [1$\bar{1}$0]. Determine the angle between an A_4 axis and an A_2 axis, and then the angle between two A_2 axes.

8—In a hexagonal crystal the parameters of the rows [101]=6.16 Å and [1$\bar{1}$0]=6.22 Å were measured by diffraction. Calculate the ratio c/a.

9—Show that the rows [21$\bar{1}$], [120], [$\bar{1}$42] are coplanar. Does the lattice type have to be taken into account? What are the Miller indices of the plane containing these rows?

10—Calculate the spacing between the planes (321) and (123) of a tetragonal lattice, then of a hexagonal lattice, when $a = 4$ Å and $c = 6$ Å.

11—Exercise on transforming Miller indices with change of frame:

The primitive unit-cell (C) of a face-centred cubic lattice (F) is actually a rhombohedral cell (R) with $\alpha = 60°$. Any rhombohedral unit-cell can be represented by a multiple hexagonal unit-cell (H).

$$\mathbf{a}_H = \mathbf{b}_R - \mathbf{c}_R$$
$$\mathbf{b}_H = \mathbf{c}_R - \mathbf{a}_R$$
$$\mathbf{c}_H = \mathbf{a}_R + \mathbf{b}_R + \mathbf{c}_R$$

—Determine the transformation matrices $\mathbf{R} = (C \rightarrow R)$, $\mathbf{H} = (R \rightarrow H)$ and $\mathbf{K} = (C \rightarrow H)$.
—Determine the multiplicity of the cells R, C and H.
—Give the Miller indices in the lattice R of a plane indexed (111) in C, and then those of a plane with indices (345).
—Repeat the previous exercise for the lattice H.
—Give the indices in C of a row indexed [001] in H, then those of a row indexed [135].

12—Calcite, $CaCO_3$, crystallises in the trigonal (rhombohedral) system. The primitive unit cell has the parameters $a = 6.36$ Å and $\alpha = 46°10'$. The crystals cleave into rhombohedra whose edges define a multiple unit-cell with parameters a' and α' where:

$$\mathbf{a}' = 3\mathbf{a} - \mathbf{b} - \mathbf{c}$$

—Determine the multiplicity of the cleavage cell.
—Calculate $a' = f(a, \alpha)$, and then $\alpha' = g(a,\alpha)$.
—Give the Miller indices of the faces of the cleavage rhombohedron in the initial frame.

13—Using spherical trigonometry, calculate the volume of the cell constructed on the base vectors **a**, **b** and **c**.

14—The pentagontrioctahedron

Consider the general form {321} of the cubic class 432; the corresponding polyhedron is a pentagontrioctahedron; two enantiomorphs (right- and left-handed) exist.

—Complete the indexing of the stereographic projection.
—Show that one of the edges of face a is parallel to a threefold axis.
—Determine the indices of the edges of face a and calculate the angles between the faces.
—Assuming all the faces have the same development, show that there exists a simple geometrical relation between certain edges.

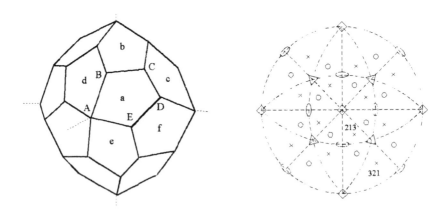

15—**The characteristic angles of the icosahedral point groups**

The regular icosahedron, in which the twenty faces are equilateral triangles, belongs to the $\overline{5}\,\overline{3}\,\mathrm{m}$ point group. The symmetry elements of this non-crystallographic group are: 6 A5, 10 A3, 15 A2, 15 M and 1 C.
Build a model of the solid and use it to identify the symmetry elements.

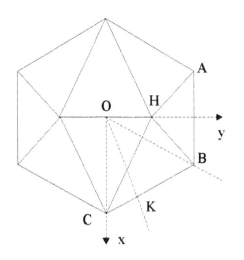

The following method can be used to determine the characteristic angles of this point group:

A projection is constructed of the icosahedron on the xOy plane (Ox, Oy and Oz are the twofold axes). Given that the side of a pentagon is related to the radius R of its circumscribed circle by the equation:

$$a = \frac{R}{2}\sqrt{10 - 2\sqrt{5}}$$

calculate the angles between the twofold and fivefold axes, the threefold and fivefold axes and the twofold and threefold axes.

Using a Wulff net, construct the stereographic projection of the point group and $\overline{5}\,\overline{3}\,m$ of the icosahedron.

STEREOGRAPHIC PROJECTIONS

1—A crystal of iodic acid, HIO_3, was measured using a two-circle goniometer. The following results were obtained:

Faces	a	b	c	d	e	f	g	h	i
φ	0°	93°22′	180°	273°22′	226°41′	273°22′	316°41′	316°41′	46°41′
ρ	90°	90°	90°	90°	52°48′	43°51′	54°25′	34°57′	52°48′

Faces	j	k	l	m	n	p	q	r	s
φ	93°22′	136°41′	136°41′	46°41′	226°41′	136°41′	316°41′	180°	0°
ρ	43°50′	54°25′	34°57′	127°12′	127°12′	125°35′	125°35′	136°10′	136°10′

Construct the stereogram of the crystal. Indicate the symmetry elements and determine the class for iodic acid. X-ray diffraction gives the indices (110), ($1\overline{1}0$) and (101) respectively to the faces a, d and g.

Determine the values of the ratios b/a, c/b and c/a and then index all the faces.

2—Measurements on a crystal of barium bromate with a two-circle goniometer gave the following results:

Faces	a	b	c	d	e	f	g	h
φ	0°	48°50′	73°45′	106°15′	131°10′	180°	228°50′	253°45′
ρ	90°	90°	90°	90°	90°	90°	90°	90°

Faces	i	j	k	l	m	n	o	p
φ	286°15′	311°10′	180°	0°	30°21′	329°39′	149°39′	210°21′
ρ	90°	90°	3°30′	45°12′	67°43′	67°43′	125°35′	136°10′

Construct the stereogram of the crystal. Indicate the symmetry elements and determine the class for barium bromate (faces with $\rho > 120°$ are invisible). The faces f and k are given the indices (100) and (001); place the hypothetical face (010). Determine the angles α, β and γ. If the indices of l and m are ($\overline{1}01$) and ($\overline{2}\overline{1}1$), determine the ratio c/a and then index all the faces.

3—The unit-cell parameters for topaz (class mmm) are:

$$a = 4.65\,\text{Å} \quad b = 8.80\,\text{Å} \quad c = 8.40\,\text{Å}; \quad \alpha = \beta = \gamma = \pi/2.$$

Consider a crystal having the associated forms {001}, {101}, {111}, {110}, {120}, {011}, {021} and {112}. For each form, calculate the angles φ and ρ and construct the stereographic projection of the crystal.

4—Show that if two faces (pqr) and (xyz) are in a zone, the face ($h\,k\,l$), where $h = p + x$, $k = q + y$ and $l = r + z$, belongs to the same zone.

APPLICATION: Using this relation together with the zones drawn on the stereogram, complete the indexing of the following trigonal stereographic projection in Miller coordinates.

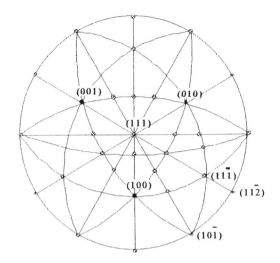

THE RECIPROCAL LATTICE

A crystal in the class $\frac{4}{m}$ mm (tetragonal holohedron) exhibits a set of associated forms. From measurements with a two-circle goniometer the following angles φ (azimuth) and ρ (inclination) of the five faces were made:

Faces	*hkl* indices	φ	ρ
p	(001)	. . .	0°
q	?	0°	68°18′
r	?	0°	39°57′
s	?	45°	74°17′
t	?	45°	60°38′

1—Construct the stereographic projection of the crystal.

2—From a number of possibilities for the Miller indices of the face q, two have been selected:

hypothesis *a*: q = (111).

hypothesis *b*: q = (011).

For each hypothesis, calculate the ratio c/a and determine the indices of the faces r, s and t.

3—Assuming that the crystal lattice is type P, that q is the face (011) and that the parameter a is equal to 3.777 Å, determine the indices and the parameter of the row [uvw] parallel to the edge between faces q and t.

4—We wish to construct the reciprocal lattice planes (hkl)* normal to the row [uvw] above.

—Establish a relation between the indices of the nodes in the plane (hkl)* which passes through the origin, and the indices of nodes in the next higher (hkl)* plane.

—Construct the plane (hkl)* which passes through the origin ($\sigma^2 = 5 \times 10^{-8}$ cm²).

— Position the projection of the next higher plane (hkl)* on this plane.

SYMMETRY AND SPACE GROUPS

1—The product of symmetry elements.

Perform the following multiplications:

a: Two intersecting twofold axes at an angle α.
b: Two intersecting mirror planes at an angle α.
c: A twofold axis and a mirror plane whose normal makes an angle α with the axis.
d: We assume that the intersections of the symmetry in the three cases above are also centres of inversion.

Determine the resultant overall symmetry for $\alpha = 90°$, $60°$, $45°$ and $30°$, and construct the sixteen corresponding stereographic projections.

2—Multiply the one-, two-, three-, four-and sixfold axes by the inversion element and indicate the resulting symmetries.

3—Working in an orthonormal frame Ox, Oy, Oz:

a: Show that the rotation matrix:

$$\begin{pmatrix} 1 & 0 & 0 \\ 0 & 0 & \bar{1} \\ 0 & 1 & 0 \end{pmatrix}$$

represents a rotation through $\pi/2$ about the Ox axis.

b: Using analogous matrices, show that the existence of two fourfold axes implies the existence of four threefold axes, and state their orientation.

c: Show that the inverse of the above implication is false. What does the presence of four threefold axes oriented as in question b imply?

d: Obtain the product of a fourfold axes oriented along [001] and a threefold axis along [111].

4—Multiply a rotation C_n ($n = 2, 3, 4, 6$) by a translation **t** normal to the axis.

5—Multiplication of symmetry operators

The following notation is used for symmetry operators:

Pure translation $\Rightarrow$ (E|t).   Pure rotation $\Rightarrow$ (C_n|0).

Rotation followed by translation $\Rightarrow$ (C_n|t).

A mirror plane (010) normal to Ox is denoted σ_x, a rotation through π about an axis $\parallel$ to Ox is denoted $C2_x$, and so on.

In a tetragonal frame, a $(1\bar{1}0)$ mirror plane is denoted σ_{xy} or σ_{45}.

Perform the following multiplications:

$(C2_y|\mathbf{0}).(C2_x|\mathbf{0})$ $(\sigma_x|\mathbf{0}).(\sigma_y|\mathbf{0})$

$(\sigma_z|\mathbf{0}).(I|\mathbf{0})$ $(C2_z|\mathbf{0}).(I|\mathbf{0})$

$(C4_z|\mathbf{0}).(C2_x|\mathbf{0})$ $(C2_x|\mathbf{0}).(C4_z|\mathbf{0})$

$(C6_z|\mathbf{0}).(C2_x|\mathbf{0})$ $(C2_x|\mathbf{0}).(C6_z|\mathbf{0})$

$(\sigma_{60}|\mathbf{0}).(\sigma_x|\mathbf{0})$ $(\sigma_{30}|0).(\sigma_x|0)$

Perform the following multiplications in an orthorhombic lattice by seeking those displacements from the origin which reduce to a minimum the translation parts:

$(C2_x|\mathbf{0}).(E|1/2(\mathbf{a}+\mathbf{b}))$ $(\sigma_x|\mathbf{0}).(E|1/2(\mathbf{a}+\mathbf{b}))$

$(\sigma_y|\mathbf{0}).(E|1/2(\mathbf{a}+\mathbf{b}))$ $(\sigma_x|\mathbf{0}).(\sigma_y|1/2(\mathbf{b}+\mathbf{c}))$

$(\sigma_x|1/2(\mathbf{a}+\mathbf{b}).(\sigma_y|1/2)(\mathbf{b}+\mathbf{c}))$.

Repeat the above exercise for a tetragonal lattice:

$(C4_z|\mathbf{0}).(E|\mathbf{a})$ $(C2_z|\mathbf{0}).(E|\mathbf{a}$

$(C2_{xy}|\mathbf{0}).(E|\mathbf{a})$.

$(C4_z^!|\mathbf{0}).(E|1/2(\mathbf{a}+\mathbf{b}+\mathbf{c}))$ $(C4_z^2|\mathbf{0}).(E|1/2(\mathbf{a}+\mathbf{b}+\mathbf{c}))$

$(C2_x|\mathbf{0}).(E|1/2)\mathbf{a}+\mathbf{b}+\mathbf{c}))$ $(C2_{xy}|\mathbf{0}).(E|1/2(\mathbf{a}+\mathbf{b}+\mathbf{c}))$

$(\sigma_x|\mathbf{0}).(E|1/2(\mathbf{a}+\mathbf{b}+\mathbf{c}))$ $(\sigma_{xy}|\mathbf{0}).(E|1/2(\mathbf{a}+\mathbf{b}+\mathbf{c}))$

$(\sigma_z|\mathbf{0}).(E|1/2(\mathbf{a}+\mathbf{b}+\mathbf{c}))$.

6—Determine all the space groups derived from the class m.

7—Does the group Pmbm exist?

8—Explain the apparent contradiction between the two diagrams below:

2_1 axis   Mirror plane b

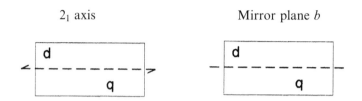

9—Group Cmc$_{21}$
State to which point group this group belongs and draw the stereographic projection of the symmetry elements on (001).

Deduce from this the number of general positions for the group.

Complete the projection of the given group below.

($m \Rightarrow$ mirror plane $m \perp$ to Ox; c $\Rightarrow$ mirror plane c $\perp$ to Oy; $2_1 \Rightarrow 2_1$ axis $//$ to Oz at O)

Determine the coordinates of the general positions equivalent to x, y, z.

Indicate the positions of a few special coordinates.

Determine the conditions for systematic absences in the group.

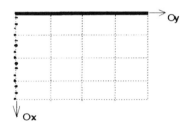

10—Group Amm2
Do the same exercises as for the group Cmc2$_1$.

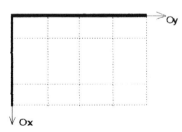

11—Consider a p-fold axis, normal to O in the plane of the figure. Show that a rotation $\mathcal{R}(O,\theta)$ through $2\pi/p$ about O, followed by a translation **t** perpendicular to the rotation axis is equivalent to a rotation $\mathcal{R}(I,\theta)$.

$$A \xrightarrow{\mathcal{R}(O,\theta)} A' \xrightarrow{\mathbf{t}} A''$$

$$A \xrightarrow{\mathcal{R}(I,\theta)} A''$$

HINT: Take the axis Ox parallel to **t** and use the rotation matrix:

$$\begin{pmatrix} \cos\theta & -\sin\theta \\ \sin\theta & \cos\theta \end{pmatrix}$$

and then determine the coordinates of the point I in the plane xOy.

APPLICATION: Study of group I4.
With which three lattice translations can this group be generated?
   Using the theorem above, state the symmetry elements obtained when a fourfold rotation is combined with the lattice translations.

NOTE: fourfold axis $= \{\mathcal{R}(O, \pi/2); \ \mathcal{R}(O, \pi); \ R(O, 3\pi/2); \ \text{Identity}\}$.
   Draw the projection of the group symmetry elements on (001) and indicate the equivalent positions.

12—Construct the group I4$_1$/a.

13—Show that a tetragonal C lattice is equivalent to a tetragonal P lattice.
   —Show that a monoclinic F lattice is equivalent to a monoclinic C lattice.
   —Show that a hypothetical hexagonal F lattice is actually an orthorhombic I lattice.

14 Write the matrix for rotation about Oz in a hexagonal lattice.

STRUCTURE FACTORS

1—Diamond crystallises in the cubic F (face centred) system with eight atoms per unit-cell (0, 0, 0) and (1/4, 1/4, 1/4) + face centred translations.
   State the fractional coordinates of the eight atoms in the cell.
   Determine the values of h for which the reflections (hhh) have a systematic absence.

2—The alloy **Fe₃Al** exists in three forms:

A phase; Cubic cell of side a. Occupied sites: $(0, 0, 0)$ and $(1/2, 1/2, 1/2)$. The iron and aluminium atoms are distributed randomly between the two sites.

B phase: Cubic cell of side a. The $(0, 0, 0)$ sites are all occupied by iron; the $(1/2, 1/2, 1/2)$ sites are occupied randomly by iron or aluminium.

C phase: Cubic cell of side $2a$. The sites occupied by iron atoms are:

(0, 0, 0)	(1/2, 1/2, 0)	(1/2, 0, 0)	(0, 1/2, 0)	(1/4, 3/4, 1/4)	(3/4, 1/4, 1/4)
(0, 0, 1/2)	(1/2, 1/2, 1/2)	(1/2, 0, 1/2)	(0, 1/2, 1/2)	(1/4, 1/4, 3/4)	(3/4, 3/4, 3/4)

Those occupied by aluminium atoms are:

(1/4, 1/4, 1/4)	(3/4, 3/4, 1/4)	(1/4, 1/4, 3/4)	(1/4, 3/4, 3/4).

—What are the Bravais lattices for each phase?

—If the atomic scattering factors for iron and aluminium are f_{Fe} and f_{Al}, what atomic scattering factor must be attributed to the sites occupied randomly in phases **A** and **B**?

—Diffraction experiments are carried out on the three phases using the same apparatus. Give the theoretical interpretation of the spectra and present the results in the form of a table in the rows of which are indicated the line indices and the structure factors for each phase. In a given row, enter the lines corresponding to a given diffraction angle. Complete the table up to line (111) of phase **A**. Draw conclusions.

POWDER DIFFRACTION PATTERNS

1—The interplanar spacing d_{hkl} obtained from a Debye–Scherrer photograph are as follows (values in Å):

| 3.24 | 3.13 | 2.81 | 2.21 | 1.985 | 1.81 | 1.69 | 1.62 | 1.56 | 1.40 |

Show that this pattern corresponds to a mixture of two face centred cubic species. One of the species has extra systematic absences compared with the F lattice. The series of spacings associated with this species could be considered as those of a P lattice; however, index them in an F lattice. Given these assumptions, determine the parameter of each species and index all the lines.

We assume that the chemical formulae of the two species are AB and AC. By making a simple assumption on the values of the atomic scattering factors f_A and f_C, explain the extra systematic absences shown by compound AC.

2—Barium titanate, $BaTiO_3$, is cubic above 120°C with a parameter $a_\alpha = 4.01$ Å. At room temperature it is tetragonal with $a_\beta = 3.99$ Å and

$c_\beta = 4.03$ Å. Show that the powder photograph of the β phase has lines at about the same angles as for the α phase but that certain lines are doublets or triplets. Under what conditions will an hkl line remain single?

3—From a JCPDS file have been extracted the intensities for four cubic compounds. For each compound deduce the unit-cell parameter, the indices of the lines and the type of lattice.

Barium (Ba)			CsCl			Diamond			Copper		
$D/$Å	I	hkl	$D/$Å	I	hkl	$D/$Å	I	hkl	$D/$Å	I	hkl
3.55	100		4.12	45		2.06	100		2.088	100	
2.513	20		2.917	100		1.261	25		1.808	46	
2.051	40		2.380	13		1.0754	16		1.278	20	
1.776	18		2.062	17		0.8916	8		1.090	17	
1.590	12		1.844	14		0.8182	16		1.0436	5	
1.451	6		1.683	25					0.9038	3	
1.343	14		1.457	6					0.8293	9	
1.1852	6		1.374	5					0.8083	8	
1.1236	4		1.304	8							

4—From a JCPDS file have been extracted the lists of d_{hkl} values and intensities for four cubic compounds. For each compound deduce the unit-cell parameter, the indices of the lines and the type of lattice.

❐ For NaCl and KCl explain the low intensity of the lines whose three indices are all odd. Why is the phenomenon more pronounced with KCl?

❐ For α iron, state why there are not enough data to be able to draw unambiguous conclusions.

❐ For silicon, explain why there are absences in addition to those induced by the lattice mode.

α iron			NaCl			KCl			Si		
$D/$Å	I	hkl	$D/$Å	I	hkl	$D/$Å	I	hkl	$D/$Å	I	hkl
2.0268	100		3.26	13		3.633	1		3.1355	100	
1.4332	20		2.821	100		3.146	100		1.9201	55	
1.1702	30		1.994	55		2.2251	37		1.6375	30	
1.0134	10		1.701	2		1.8972	<1		1.3577	6	
0.9064	12		1.628	15		1.8169	10		1.2459	11	
0.8275	6		1.410	6		1.5730	5		1.1086	12	
			1.294	1		1.4071	9		1.0452	6	
			1.261	11		1.2839	5		0.9600	3	
			1.1515	7		1.1121	1		0.9180	7	

5—Time-of-flight method

Packets of neutrons are produced by periodic bombardment ($f = 24\,Hz$) of a heavy element target. This bombardment, of typical duration $0.4\,\mu s$, produces a large number of neutrons (25 n for 1 p) by spallation of the target. The neutrons produced have too high an energy to use directly in diffraction experiments and they are slowed down by a moderator.

The resulting neutrons, produced at the same original time, have different energies and hence speeds (a continuous range of speeds is observed) and can be used for fixed angle diffraction. Using the de Bröglie relation, show that if the neutrons have different speeds, they will have different wavelengths λ. Using the Bragg equation with a fixed angle θ_0, show that there exists a relation between the time of flight t of the neutrons between the source and the detector through the specimen (distance L) and the interplanar spacing d_{hkl} of the diffracting layers. State the relation between t (ms), L (m) and d_{hkl} (Å).

Given that $h = 6.62 \times 10^{-34}\,Js$ and $m_n = 1.675 \times 10^{-27}\,kg$, which layers are the first to diffract?

The spectrum of a cubic substance was recorded at the 'Argonne National Laboratory' with $L\sin\theta_0 = 13.94\,m$. The last six diffraction peaks emerged at the following times t:

Peak	t/ms	$d_{hkl}/Å$	hkl
6	13.503		
5	11.515		
4	9.549		
3	8.760		
2	7.796		
1	7.351		

Complete the table and deduce the unit-cell parameter of the substance.

Structure Analysis: Rutile, TiO_2

The space group is $P4_2/mnm$ ($a = 4.594\,Å$, $c = 2.958\,Å$ and $z = 2$). The fractional coordinates are:

Ti: 0, 0, 0   1/2, 1/2, 1/2

O: $\pm(x, x, 0$; $1/2 + x, 1/2 - x, 1/2$) ($x = 0.305$).

Draw a projection of the structure on (001) and determine the coordination of the atoms. Show that there are two types of Ti–O bond and three types of

O–O bond and calculate their lengths. Draw a projection of the structure on (110), taking several unit-cells along Oz. Deduce from this a model for the structure.

Calcium Titanate (CaTiO$_3$)

CaTiO$_3$ occurs naturally as the mineral perovskite; it crystallises in the group Pcmn (standard name: Pnma), with:

$$a = 5.37\,\text{Å},\ b = 7.63\,\text{Å},\ c = 5.44\,\text{Å},\ z = 4.$$

The coordinates of the atoms are:

Ti (site 4a): 1/2 , 0, 0.   Ca (site 4c): 0, 1/4 , 0.03.

O1 (site 4c): 1/2 − 0.037, 1/4, −0.018.

O2 (site 8d): 1/4 −0.018, −0.026, 1/4 −0.018.

Draw a projection on (010) and show that the structure can be described by a pseudo-cubic unit-cell with $a = c \approx 3.82\,\text{Å}$, $b \approx 3.82\,\text{Å}$, $\beta \approx 90°$.

Determine the transformation matrix.

State how the structure differs from that of an ideal perovskite.

Problems

Cuprite, Cu_2O, crystallises in the cubic m3m class. Draw the stereographic projection of the point group, indicating the positions of the poles of the faces for the forms {100} (cube) and {111} (octahedron).

On a powder photograph the following lattice spacings were measured (values in Å):

$$3.020; \ 2.465; \ 2.135; \ 1.743; \ 1.510; \ 1.287; \ 1.233;$$

— From these measurements, deduce the cell parameter, the indices of the lines and the lattice mode.

The fractional coordinates of the atoms are:

$$O: \quad \tfrac{1}{4}, \tfrac{1}{4}, \tfrac{1}{4}; \quad \tfrac{3}{4}, \tfrac{3}{4}, \tfrac{3}{4};$$

$$Cu: \quad 0, 0, 0; \quad 0, \tfrac{1}{2}, \tfrac{1}{2}; \quad \tfrac{1}{2}, 0, \tfrac{1}{2}; \quad \tfrac{1}{2}, \tfrac{1}{2}, 0.$$

— Draw the stereographic projection of the structure on (001) and indicate the positions of the symmetry elements; deduce the space group from this.
— Using the structure factor, indicate the h, k and l values for which there are systematic extinctions.
— The diffraction pattern of a crystal rotating about [110] is obtained using a camera with a circumference of 360 mm and a copper target ($\lambda_{K\alpha 1} = 1.5402$ Å). Calculate the spacing on the film between the layers $K = 2$ and $K = -2$. What relation is there between the indices of the spots in layer K and those of the rotation layer?

Using this relation, construct the reciprocal plane passing through the origin; state the scale used.

THE CUBIC-TRIGONAL TRANSITION

1 Express the base vectors **A***, **B*** and **C*** of the reciprocal lattice in terms of the base vectors **a**, **b** and **c** of the direct lattice. The value of each reciprocal vector should be expressed in the form of the quotient of a vector product and a scalar triple product. *We let* $\sigma^2 = 1$.

2 The rhombohedral cell is characterised by:

$$\|\mathbf{a}\| = \|\mathbf{b}\| = \|\mathbf{c}\| \qquad \alpha = \beta = \gamma \neq \frac{\pi}{2}$$

Show that the volume of this cell is equal to:

$$V = a^3 \sqrt{1 - 3\cos^2\alpha + 2\cos^3\alpha}$$

A method of calculation is suggested in the following diagram.

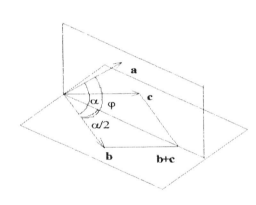

$$\varphi = \{\mathbf{a}, \mathbf{b} + \mathbf{c}\} \quad \alpha = \{\mathbf{a}, \mathbf{b}\}$$

$$\frac{\alpha}{2} = \{\mathbf{b}, \mathbf{b} + \mathbf{c}\}$$

3 Knowing that:

$$(\mathbf{a} \wedge \mathbf{b}).(\mathbf{c} \wedge \mathbf{d}) =$$

$$(\mathbf{a}.\mathbf{c}).(\mathbf{b}.\mathbf{d}) - (\mathbf{a}.\mathbf{d}).(\mathbf{b}.\mathbf{c})$$

and using the results of questions 1 and 2, show that in a trigonal lattice:

$$\frac{1}{d_{hkl}^2} = \frac{(h^2 + k^2 + l^2)\sin^2\alpha + 2(hk + kl + hl).(\cos^2\alpha - \cos\alpha)}{a^2(1 + 2\cos^3\alpha - 3\cos^2\alpha)}$$

4 Since calculations in a trigonal cell are complex (see above), we generally work in the triple hexagonal cell containing the nodes whose fractional coordinates are:

$$0, 0, 0 \quad 2/3, 1/3, 1/3 \quad 1/3, 2/3, 2/3.$$

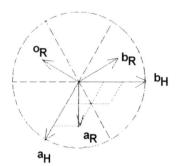

Using the projection of the two cells on (001), express the base vectors of the trigonal cell in terms of those of the hexagonal cell and vice versa.

5 Certain compounds exhibit a cubic ⇔ trigonal transition involving a stretching (or compressing) of the cubic cell along a threefold axis of the cube.

The angle α of the rhombohedron obtained is close to $\pi/2$ ($\alpha = \pi/2 - \varepsilon$). By comparing the expressions for d_{hkl} values for the cubic and trigonal systems (see 3 above), show how the phase transition lifts the h, k and l degeneracy in the diffraction lines of a Debye–Scherrer photograph. In the expression for the d_{hkl} values in the trigonal phase, the approximations arising from the smallness of ε should be made. State how the cubic lines (111), (200) and (110) change during the cubic ⇔ trigonal transition.

6 Application: Consider the compound PbLi, which is cubic (CsCl structure) above 214°C and trigonal at lower temperatures. At room temperature, a powder photograph is produced using a 360 mm circumference camera and a copper anode ($\lambda_{K\alpha} = 1.540$ Å). The diameter of the first diffraction ring is 50.22 mm. Complete the table below, using the simplified expression for d_{hkl} in the calculations. ($a \approx \pi/2$).

$d_{hkl}/$Å	I_{hkl}	$h\ k\ l$
$a = ?$	$\alpha = ?$	
?	100	1 0 0
2.515	70	1 1 0
2.493	70	$\bar{1}$ 1 0
?	10	1 1 1
2.040	?	1 1 $\bar{1}$
1.770	40	2 0 0
1.590	50	2 1 0
1.580	?	$\bar{2}$ 1 0
?	?	2 $\bar{1}$ 0

DETERMINATION OF THE SPACE GROUP OF NiO

We wish to determine to which structural family (NaCl or ZnS) the compound NiO belongs. The diffraction angles and line intensities are given for the powder photograph of NiO obtained with a copper anode (wavelength $\lambda_{K\alpha Cu} = 1.5402$ Å). Determine the values of d_{hkl} and index the lines.

2θ	Intensities	d_{hkl}(Å)	hkl
37.283	57		
43.279	100		
62.729	45		
75.265	14		
79.310	10		
94.887	4.5		
106.592	6.3		
110.636	14.9		
128.423	13.2		

Calculate the intensities of the reflections and compare the values with the experimental intensities; limit the exercise to the reflections (111), (200), (220), (311) and (222).

The following expression for the atomic scattering factor of the atom k (λ in Å) should be used:

$$f_k(\theta, \lambda) = A_k . \exp(-a_k . \sin^2 \theta / \lambda^2) + B_k . \exp(-b . \sin^2 \theta / \lambda^2) + C_k$$

	A_k	a_k	B_k	b_k	C_k
Ni^{2+}	12.76	2.637	8.638	19.88	5.65
O^{2-}	4.758	7.831	3.637	30.05	1.594

The diffracted intensity is given by: $I_{hkl} = m.L.P.F^2_{hkl}$ where:
   m = the multiplicity of the reflection,
   $L = 1/\sin^2 \theta . \cos \theta$, the Lorentz factor.
   $P = (1 + \cos^2 2\theta)$, the X-ray polarisation factor,
   F_{hkl} the structure factor for the reflection (hkl) given by:

$$F_{hkl} = \sum_k f_k \cdot \exp(-2j\pi.\mathbf{S}.\mathbf{r}_k)$$

The atom positions are given in the following table:

	NaCl type			ZnS type		
O^{2-}	0	0	0	0	0	0
	$\frac{1}{2}$	$\frac{1}{2}$	0	$\frac{1}{2}$	$\frac{1}{2}$	0
	0	$\frac{1}{2}$	$\frac{1}{2}$	$\frac{1}{2}$	0	$\frac{1}{2}$
	$\frac{1}{2}$	0	$\frac{1}{2}$	0	$\frac{1}{2}$	$\frac{1}{2}$
Ni^{2+}	$\frac{1}{2}$	0	0	$\frac{1}{4}$	$\frac{1}{4}$	$\frac{1}{4}$
	0	$\frac{1}{2}$	0	$\frac{3}{4}$	$\frac{3}{4}$	$\frac{1}{4}$
	0	0	$\frac{1}{2}$	$\frac{3}{4}$	$\frac{1}{4}$	$\frac{3}{4}$
	$\frac{1}{2}$	$\frac{1}{2}$	$\frac{1}{2}$	$\frac{1}{4}$	$\frac{3}{4}$	$\frac{3}{4}$

After adjusting the strongest reflection to 100 (multiply I_{hkl} by a coefficient of proportionality k), fill in the table for the two types of structure proposed.

							NaCl			ZnS		
hkl	*Int*	θ	*LP*	*m*	fO^{2-}	fNi^{2+}	F_{hkl}	I_{hkl}	kI_{hkl}	F_{hkl}	I_{hkl}	kI_{hkl}
1 1 1												
2 0 0												
2 2 0												
3 1 1												
2 2 2												

Draw conclusions from the exercise.

OXIDES OF IRON

Consider the three crystallised compounds A, B and C which are all oxides of iron.

Compound A is cubic with a type F lattice, the unit-cell parameter a is 4.31 Å and the density is 5.97 g/cm³.

Compound B is also cubic with an F type lattice, the unit-cell parameter a is 8.37 Å and the density is 5.20 g/cm³.

Compound C is trigonal, the triple hexagonal unit-cell has the parameters a = 5.03 Å and c = 13.74 Å. Its density is 5.26 g/cm³.

The atomic mass of iron is 55.85; that of oxygen is 16.

— Calculate the volume of the unit-cell volume for each compound.
— Deduce from this the chemical formulae of each compound and the number of formula units per cell.
— Give the indices of the first three diffraction lines that would normally be observed on a powder photograph of compound A. Calculate the diffraction angles that would be obtained using a copper anode ($\lambda_{K\alpha} = 1.54$ Å).
— A rotating crystal photograph of compound A is obtained using a 360 mm diameter camera with an iron anode ($\lambda_{K\alpha} = 1.98$ Å). The rotation row is [001]. Give the x and y coordinates of the spot (111) using the point of impact of the incident beam with the film as origin.
— Diffraction spots with all indices even are seen to be stronger than the other spots on the photograph. Propose a structure for compound A which takes account of this phenomenon.

THE STRUCTURE OF KIO$_2$F$_2$

The appearance of the crystals and measurements with a two-circle goniometer suggest the possibility of a unit-cell with orthogonal axes. Using a rotating crystal camera of diameter 180 mm and a cobalt anode ($\lambda_{K\alpha} = 1.7889$ Å), the following exposures are made:

— a rotation about [001]; the distance between layers 2 and −2 is found to be 27.05 ± 0.09 mm.
— a rotation about [001]; the distance between layers 2 and −2 is found to be 26.90 ± 0.09 mm.

The parameter b is 5.97 ± 0.02 Å.

Determine a and c and indicate which crystal systems are possible.

When examined under a polarising microscope, the crystal is found to be biaxial. To which system does it belong?

Its density is $3.8\,\mathrm{g\,cm^{-3}}$; deduce the number of formula units per cell ($K = 39$, $I = 127$, $O = 16$, $F = 19$).

After indexing, a survey of systematic absences showed that the possible groups are Pbcm and Pca2_1.

— Give the number of general equivalent positions for the group Pbcm. What conclusions can be drawn if the compound belongs to this group? Draw the group on a (001) plane. Take as origin a centre of symmetry in the mirror plane b (the mirror plane is at a height 1/4). Indicate the coordinates of the general equivalent positions.

— Draw a projection of the group Pca2_1 on (001). Indicate the coordinates of the general equivalent positions. State the conditions limiting the possible reflections for the spots hkl, $0kl$, $hk0$, $h00$, $0k0$, $00l$, $h0l$. Take as origin a 2_1 axis in the mirror plane a.

When examined for the piezoelectric effect, the crystal gave doubtful results, but it was found to be pyroelectric. Indicate the space group and point group of the compound.

PSEUDOSYMMETRY

A monoclinic holohedral crystal (class 2/m) has the following unit-cell parameters:

$a = 10.07\,\text{Å}$ $b = 14.28\,\text{Å}$ $c = 8.64\,\text{Å}$ $\beta = 125°40'$

— Two rotating crystal photographs were obtained using a 240 mm diameter camera with a useful height of 80 mm. The incident beam entered the camera normal to its axis at half-height ($\lambda_{K\alpha} = 1.54\,\text{Å}$).

How many layers are obtained if the crystal rotates about the row [001] and then about the row [010]?

— Calculate the angle between the planes ($\bar{2}01$) and (001).

— A lattice has a unit-cell bounded by the planes ($\bar{2}01$), (010) and (001) and has a symmetry other than monoclinic; what is this lattice, and how could its actual symmetry be revealed?

— After indexing, the only systematic absences were found to be for $0k0$ spots where $k = 2n + 1$. Explain this and give the space group of the compound.

— Give the general equivalent positions together with a few special positions.

THE AuCu PHASE TRANSITION

A powder photograph of a sample of AuCu quenched at high temperature (phase A) is obtained. In this cubic phase the atoms are distributed completely randomly. The measured values of d_{hkl} (in Å) are:

$$2.293 \quad 1.982 \quad 1.405 \quad 1.195 \quad 1.146 \quad 0.992 \quad 0.912$$

— Calculate the unit-cell parameter, the lattice mode and the theoretical density (Cu = 64, Au = 197).

After suitably annealing, the structure of the alloy becomes completely ordered (phase B); the (001) planes are alternately entirely copper or entirely gold. In the initial cubic frame, the fractional coordinates are then:

$$\text{Cu: } 0, 0, 0; \quad 1/2, 1/2, 0 \quad \text{Au: } 1/2, 0, 1/2; \quad 0, 1/2, 1/2$$

— Show that the lattice is tetragonal and construct a primitive unit-cell. In order to compare the diffraction spectra of the two phases, phase A is represented in this new frame; show that phase A is tetragonal I with $c/a = \sqrt{2}$. Index the lines in phase A in this frame.

A powder photograph of a sample of phase B is obtained. This photograph has many more lines than that of phase A; the increase in the number of lines arises from two phenomena: tetragonalisation and the appearance of superstructure rings.

Tetragonalisation. Using the expression for interplanar spacing, show that in the transformation from a cubic P unit-cell to a tetragonal P unit-cell, there is a tripling or a doubling of certain lines (set $c/a = \sqrt{2} - \varepsilon$). How do the first eight lines in a cubic P lattice change?

Superstructure. Using the structure factor (calculated in a unit-cell containing four atoms), show that the phase B photograph has the same lines as for phase A, with additional weak lines.

Reorganise these results and give the appearance of the powder spectrum of phase B.

CESIUM BROMATE ($CsBrO_3$)

This compound crystallises in the tetragonal system. Measurements with a two-circle goniometer gave the following results:

Faces	a	b	c	d	f
ρ	$0°$	$54°27'$	$54°27'$	$54°28'$	$90°$
φ	—	$105°20'$	$225°18'$	$345°21'$	$135°19'$

Only those faces used in the following part of this problem are featured in the table. In particular, there exist other faces in zone with face f. Face a is assumed to be a (111) face and b a (100) face.

Determine the ratio c/a of the corresponding hexagonal unit-cell.

Using a copper anode and a camera with a 360° circumference, a rotating crystal photograph is made with the axis of rotation normal to face a. The distance between the zero layer and layer 3 is 38.9 mm.

When the rotation axis is a row normal to face f, the distance between the zero layer and row 3 is 53.5 mm; deduce the ratio c/a.

TETRAHEDRON AND OCTAHEDRON

A crystal has tetrahedral and octahedral forms associated.

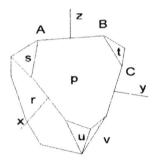

Five faces were measured with a goniometer:

Faces	p	r	t	u	v
ρ	54°20′	125°40′	63°05′	116°55′	116°55
φ	45°	315°	90°	0°	90°

Construct the stereogram and deduce from it the crystal class.

The crystal habit suggests that the unit-cell axes are in the directions Ox, Oy and Oz in the figure, and that face p has the indices (111).

Deduce the ratios a/b and c/a and the indices of face t.

An X-ray study confirms the choice of axis direction and shows that spots such as $h + k + l = 2n + 1$ are absent. With rotating crystal photographs, the rotation row parameters were determined with:

— the crystal rotating about the edge AB;
— the crystal rotating about the edge BC between p and t.

The respective values $n_1 = 5.27$ Å and $n_2 = 6.33$ Å were found.
Show that these results are incompatible with the notation (111) for face p.
Index this face correctly and determine the unit-cell parameters.

SODIUM CHLORATE (NaClO₃)

A crystal of sodium chlorate (cubic system) exhibits the following association of forms: pentagon dodecahedron {021} and tetrahedron {11$\bar{1}$}. Which class corresponds to the simultaneous presence of these two forms?
A rotating crystal photograph is obtained with rotation about the row common to (210) and (2$\bar{1}$0). The parameter is 6.56 Å.

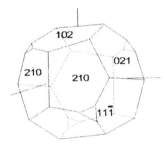

Starting from a glancing beam on (210), by how much must the crystal be rotated to obtain a reflection from this family of planes? ($\lambda = 1.54$ Å).
What is the parameter of the row common to (11$\bar{1}$) and (210)?

HEXAGONAL CLOSE-PACKED STRUCTURE

A powder photograph of magnesium, which crystallises with a hexagonal close-packed structure, has been obtained.

— Establish the relation giving the values of d_{hkl} in a hexagonal lattice.
— Complete the following table and determine the values of the parameters a and c.

d_{hkl}/Å	2.778	2.605	2.452	?	1.6047	?
hkl	?	002	101	102	110	200

— Calculate the ratio c/a for the hexagonal close-packed structure using the rigid-sphere model. Compare it with the value for magnesium.

The fractional coordinates of the unit-cell atoms are:

$$2/3, 1/3, 1/4 \qquad 1/3, 2/3, 3/4;$$

— List the tetrahedral and octahedral vacancies in the unit-cell.

What is the maximum radius of atoms that can be placed in the two types of vacancy without modifying the packing?

— What is the index n of the 6_n screw axis positioned at the origin?
— Complete the projection of the space group and name it.
— Determine the general reflection conditions of the group.

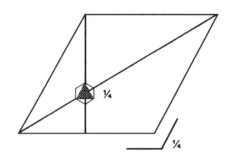

BORON PHOSPHATE (BPO$_4$)

This compound crystallises in the group $\bar{4}$.

The distance between layers 3 and 4 in a rotating crystal photograph with rotation about [001], a 360 mm diameter camera and a copper anode ($\lambda_{K\alpha} = 1.5406$ Å) is 55.5 mm.

The values of d_{hkl} (in Å) obtained from a powder photograph :

$$3.6351 \qquad 3.3207 \qquad 3.0699 \qquad 2.254 \qquad 1.9719$$

— Determine c, a and c/a.

The fractional coordinates of the atoms are:

$$\text{B: } 0, 1/2, 1/4 \quad \text{P: } 0, 0\ 0 \quad \text{O: } 0.14.\ 0.26.\ 0.13$$

— Draw a projection on (001), quoting heights.

A unit-cell with axes x' and y' at 45° to the initial axes is required.

— Write the transformation matrices to change the axes and coordinates.
— Use these to deduce the new coordinates of the boron and potassium atoms in the new unit-cell. Calculate c'/a'.
— Show that if this ratio is slightly modified, the boron and potassium atoms can be seen to be assembled in a classic binary structure.

DETERMINATION OF THE SPACE GROUP

A compound to be analysed is orthorhombic; its density is $6.05 \, \text{g cm}^{-3}$ and its molar mass is 375 g.

A rotating crystal photograph is obtained with the crystal rotating about the row [100], a camera of circumference 180 mm and a copper anode ($\lambda_{K\alpha} = 1.541 \, \text{Å}$). Determine the parameter of the rotation row given that the distance between layers 7 and -7 is 81 mm.

The figures below show the positions of the spots in the central area of a set of Buerger photographs obtained with a molybdenum anode (($\lambda_{K\alpha} = 0.7903 \, \text{Å}$). For each of the four photographs the geometry was such that $\mathbf{A.a} = \sigma^2 = \mathbf{R}_0 . \lambda$ where $\mathbf{R}_0 = 60 \, \text{mm}$. The direction of the incident beam is shown for each photograph. The notations layer 0 and layer 1 correspond respectively to the reciprocal planes containing the origin and the next higher parallel plane.

— Determine the unit-cell parameters for the compound.
— Calculate the number of formula-units in the unit-cell.
— Draw the projection on (001) of the space group Cmca. The generators to be used are a mirror plane m (100) at $x = 0$, a mirror plane c (010) at $y = 1/4$ and a centre of inversion at (0, 0, 0).
— Indicate the positions equivalent to the general positions x, y, z and the conditions for systematic absences of this group.
— Show from the Buerger photographs that the compound could belong to the group Cmca.

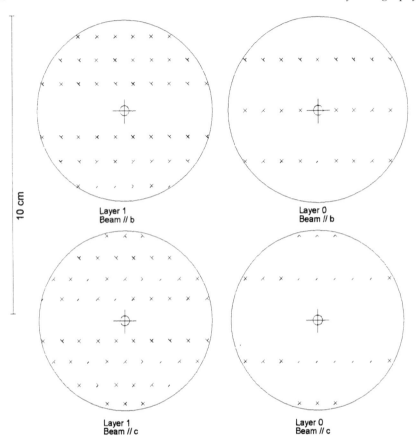

10 cm

Layer 1
Beam // b

Layer 0
Beam // b

Layer 1
Beam // c

Layer 0
Beam // c

Solutions to the Exercises

LATTICES AND MILLER INDICES

1—Barytine

> faces (110) and $(010) \Rightarrow a/b = 0.876$;
> faces (111) and $(110) \Rightarrow c/a = 1.504$; $b/c = 0.758$
> $p = (104)$ $o = (102)$ $m = (101)$ $q = (011)$.

2—Dodecahedron

> $a = (101)$ $b = (110)$ $c = (011)$ $e = (10\bar{1})$ $d = (1\bar{1}0)$
> One edge is a row: $\{a,b\} = \pi/3$ $\{b,d\} = \pi/2$. $\alpha = 109°28'$.

3—Lead chloride

> $a/b = 0.5942$ $c/a = 2.002$ $b/c = 0.8406$.
> $e = (112)$ $f = (021)$ $g = (012)$.

4—Rows normal to the lattice planes

$[uvw] \perp (hkl) \Rightarrow [uvw] // [hkl]^*$ hence: $h\mathbf{A}^* + k\mathbf{B}^* + l\mathbf{C}^* = \lambda(u\mathbf{a} + v\mathbf{b} + w\mathbf{c})$ (1)

The scalar products of (1) and $\mathbf{a}$, $\mathbf{b}$ and $\mathbf{c}$ give:

$$h = \lambda(ua^2 + v\mathbf{b.a} + w\mathbf{c.a})$$
$$k = \lambda(u\mathbf{a.b} + vb^2 + w\mathbf{c.b})$$
$$l = \lambda(u\mathbf{a.c} + v\mathbf{b.c} + wc^2)$$

Integers h, k, l must be found to obtain integers u, v, w.
Cubic: $h = Ku$ $k = Kv$ $l = Kw$
We can take $K = 1 \Rightarrow h = u$, $k = v$, $l = w \forall u$, v, w
For all h, k, l we have $(hkl) \perp [hkl]$.
Tetragonal: $h = Ku$ $k = Kv$ $l = Jw$
$J \neq K$ since a $\neq$ c. If we take $K = 1 \Rightarrow J \neq 1$: there is no general solution
where h k and l take any value.
The solutions are $(hk0) \perp [hk0]$ and $(00l) \perp [00l]$.
Monoclinic: $h = Ku + Jw$ $k = Mv$ $l = Ju + Nw$
A single solution exists: the twofold axis $(010) \perp [010]$.
The symmetry axes are normal to the lattice planes.

5—Monoclinic reciprocal lattice

Noting that $\mathbf{A^*}$ and $\mathbf{C^*}$ are linear combinations of $\mathbf{a}$ and $\mathbf{c}$ we have:
(The vectors $\mathbf{a}$, $\mathbf{c}$, $\mathbf{A^*}$ and $\mathbf{C^*}$ are coplanar)

$$\mathbf{A^*} = \frac{1}{a\sin^2\beta}\left(\frac{\mathbf{a}}{a} - \frac{\mathbf{c}}{c}\cos\beta\right); \quad \mathbf{B^*} = \frac{\mathbf{b}}{b}; \quad \mathbf{C^*} = \frac{1}{c\sin^2\beta}\left(\frac{\mathbf{c}}{c} - \frac{\mathbf{a}}{a}\cos\beta\right).$$

6—Monoclinic lattice

The lattice is type C since the row [110] contains a node at $a/2$, $b/2$, 0.

7—Cubic lattice

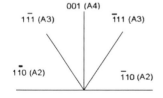

001 (A4)
1$\bar{1}$1 (A3) $\bar{1}$11 (A3)
1$\bar{1}$0 (A2) $\bar{1}$10 (A2)

The row $[uvw]$ is normal to the plane (hkl) if:

$$hu + kv + lw = 0.$$

The figure shows those rows contained within
the (110) plane which are symmetry axes.

8—Hexagonal crystal [101] $\Rightarrow$ a + c $c/a = 1.3937$.

9—Coplanar rows

$$\begin{vmatrix} 2 & 1 & \bar{1} \\ 1 & 2 & 0 \\ \bar{1} & 4 & 2 \end{vmatrix} = 0$$

The determinant of the indices is null and hence the 3 rows are coplanar [and contained within the plane $(2\bar{1}3)$]. This result is general and does not depend on the crystal system.

10—Interplanar spacings

Using the distance relations in chapter 6, we have:

Tetragonal: $d_{321} = 1.0909$ Å $\qquad d_{123} = 1.3333$ Å.

Hexagonal: $d_{321} = 0.7878$ Å $\qquad d_{123} = 1.0954$ Å.

11—Change of frame

$$R = \begin{pmatrix} 0 & \frac{1}{2} & \frac{1}{2} \\ \frac{1}{2} & 0 & \frac{1}{2} \\ \frac{1}{2} & \frac{1}{2} & 0 \end{pmatrix} \qquad H = \begin{pmatrix} 0 & 1 & -1 \\ -1 & 0 & 1 \\ 1 & 1 & 1 \end{pmatrix} \qquad K = \begin{pmatrix} 0 & -\frac{1}{2} & \frac{1}{2} \\ \frac{1}{2} & 0 & -\frac{1}{2} \\ 1 & 1 & 1 \end{pmatrix}$$

$$\text{Det} = 1/4 \qquad\qquad \text{Det} = 3 \qquad\qquad \text{Det} = 3/4$$

Multiplicities obtained either by calculating the volume or from direct counting of nodes:

$$\text{cell C} = 4, \quad \text{cell R} = 1, \text{ cell H} = 3.$$

The Miller indices are covariant with base vectors:

$$(111)_C \Rightarrow (111)_R \ (345)_C \Rightarrow (987)_R \ (111)_C \Rightarrow (003)_H \ (345)_C \Rightarrow (1\bar{2}24)_H$$

The row indices are contravariant with base vectors:

$$[001]_H \Rightarrow [111]_C \qquad\qquad [135]_H \Rightarrow [13\,9\,8]_C$$

12—Calcite

$$v = (\mathbf{a}, \mathbf{b}, \mathbf{c}) \quad v' = (\mathbf{a}', \mathbf{b}', \mathbf{c}') = 16.v \quad a' = a\sqrt{11 - 10\cos\alpha} = 12.83 \text{ Å}.$$
$$\mathbf{a}'.\mathbf{b}' = a^2(6.\cos\alpha - 5) = a^2(11 - 10.\cos\alpha).\cos\alpha': \ \alpha' = 101°58'.$$

Cleavage face $(100) \Rightarrow (112)$.

13—Volume of any general unit-cell

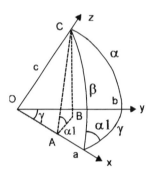

Let $\alpha 1$, $\beta 1$ and $\gamma 1$ be dihedral angles between the planes containing the base vectors.

The area of the parallelogram constructed on the vectors $\mathbf{a}$ and $\mathbf{b}$ is $ab.\sin\gamma$.

From $C(OC = c)$ we drop a perpendicular BC onto xOy and from B we draw BA normal to Ox.

$AC = c.\sin\beta \quad BC = AC.\sin\alpha 1 = c.\sin\beta.\sin\alpha 1$.

$V = abc.\sin\alpha 1.\sin\beta.\sin\gamma$.

In the spherical triangle we have: $\cos\alpha = \cos\beta.\cos\gamma + \sin\beta.\sin\gamma.\cos\alpha 1$. Hence:

$$\cos\alpha 1 = \frac{\cos\alpha - \cos\beta.\cos\gamma}{\sin\beta.\sin\gamma}$$

followed by the values of $\sin\alpha 1$ and finally that of

$$V = a.b.c.\sqrt{1 - \cos^2\alpha - \cos^2\beta - \cos^2\gamma + 2\cos\alpha.\cos\beta.\cos\gamma}.$$

14—Pentagontrioctahedron

Indices of the faces:

$$a = (321), \ b = (213), \ c = (132), \ d = (3\bar{1}2), \ e = (31\bar{2}), \ f = (23\bar{1})$$

Indices of the edges:

$$BC = [\bar{5}71], \ AB = [\bar{5}39], \ AE = [5\bar{9}3], \ ED = [\bar{1}11] \equiv A3, \ CD = [1\bar{5}7]$$

Angles between the edges:

{ABC} = 126°; {BAE} = 77°; {AED} = 126°;
{EDC} = 94°; {DCB} = 116°

The row OC is a threefold axis, hence: BC = CD.
The row OA is a fourfold axis, hence: AB = AE.

15—Icosahedron

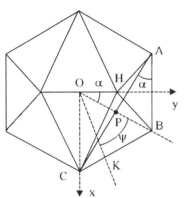

The points O, A, B, C, K are in the plane of the diagram. OB is a fivefold axis. OK is normal to the face whose projection is BC and it is a threefold axis.

Let α be the angle Oy (*A*2) and OB (*A*5) and P be the intersection between the two orthogonal lines OB and AC. The five triangles with a common vertex B are projected onto a plane normal to OB as a pentagon with centre P. We have:

$$\cos \alpha = \frac{AP}{AB} = \frac{R}{a} = \frac{2}{\sqrt{10 - 2\sqrt{5}}}$$

From this we deduce $\alpha = 31.717°$. From: $\sin \alpha = AB/2OB$, we have: $OB = a/2 \sin \alpha$.

We write $\Psi = \{OB, OK\}$. K is the centre of gravity of the triangle whose projection is BC hence:

$$\sin \Psi = \frac{KB}{OB} = \frac{a\sqrt{3}}{3.OB}$$

and $\Psi = 37.38°$. {OC,OK} = 20.90°.

Spherical trigonometry can also be used:

Consider a spherical triangle whose vertices are stereographic projections of three fivefold axes. The angles of triangle ($\hat{A}$) are $360°/5 = 72°$. The sides of triangle (a) are defined by the angle between two fivefold axes. Now, in a spherical triangle, we have:

$$\cos \hat{A} = -\cos \hat{B}. \cos \hat{C} + \sin \hat{B}. \sin \hat{C}. \cos a$$

From which we deduce $a = 63.4349°$ which is twice the angle between a twofold axis and a fivefold axis.

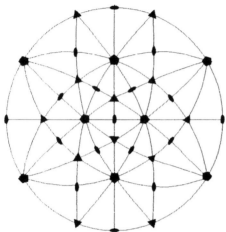

The above calculations enable us to position the symmetry axes within a plane xOy. Four of the threefold axes of the group occupy the same position as in the cubic group m3m. The projection shown opposite is obtained. The poles of the faces of the icosahedron coincide with the threefold axes. (The poles of a pentagonal dodecahedron coincide with the fivefold axes). The faces can be indexed with (hkl) notation but h, k and l are then irrational.

Thus the pentagonal dodecahedron can be denoted (01τ) where $\tau = \frac{1}{2}(\sqrt{5}+1)$.

THE STEREOGRAPHIC PROJECTION

1—Iodic acid

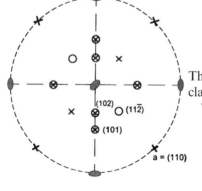

The only symmetry elements are twofold: the class is 222 (D_2) (orthorhombic lattice)
The parameter ratios are:

$$b/a = 1.0606$$
$$c/a = 1.3976$$
$$c/b = 1.3176$$

2—Barium bromate

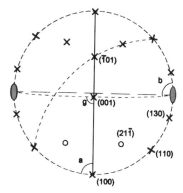

Class 2/m (monoclinic)

$$\alpha = \gamma = 90°$$
$$\beta = 93°30'$$
$$c/a = 1.072$$

Indexing is easier if zone circles are drawn. Faces for which ρ is 90° have [001] as a zone axis (their index l is zero).

b is in the plane of the projection and **c** is perpendicular to this plane.

3—Topaz

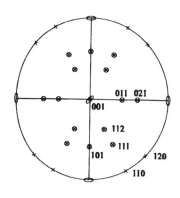

(001):	$\varphi = \cdots$	$\rho = 0°$
(101):	$\varphi = 0°$	$\rho = 61.03°$
(111):	$\varphi = 27.85°$	$\rho = 63.92°$
(110):	$\varphi = 27.85°$	$\rho = 90°$
(120):	$\varphi = 46.58°$	$\rho = 90°$
(011):	$\varphi = 90°$	$\rho = 43.67°$
(021):	$\varphi = 90°$	$\rho = 62.35°$
(112):	$\varphi = 27.85°$	$\rho = 45.61°$

4—Trigonal

Let $[uvw]$ be the zone axis; hence we can write:

$$pu + qv + rw = 0$$
$$xu + yv + zw = 0$$

Taking the sum we have: $(p + x)u + (q + y)v + (r + z)w = 0$

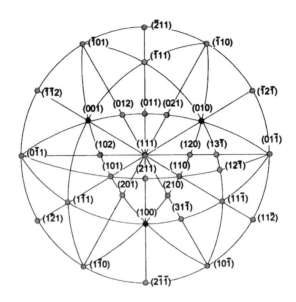

RECIPROCAL LATTICE

Hypothesis a: $c/a = 1.778$ r = (113) s = (021) t = (011).

Hypothesis b: $c/a = 2.515$ r = (013) s = ($\bar{1}$11) t = ($\bar{1}$12).

For the chosen hypothesis we have: $c = 9.4916$ Å.

The row common to q and t is [1$\bar{1}$1]. The parameter of this row is 10.9 Å.

A direct row (vector $u\mathbf{a} + v\mathbf{b} + w\mathbf{c}$) is normal to a reciprocal row (vector $h\mathbf{A}^* + k\mathbf{B}^* + l\mathbf{C}^*$) if: $h.u + k.v + l.w = 0$.

The reciprocal plane normal to [1$\bar{1}$1] and passing through the origin is such that its rows [hkl]* have indices which satisfy the relation: $h - k + l = 0$.

For the next higher plane we have: $h - k + l = 1$. To generate this plane we must find two non-collinear rows which define a primitive plane unit-cell.

We can take [110] and [10$\bar{1}$] with $N[110] = 1.875$ cm and $N[10\bar{1}] = 1.424$ cm.

The angle between these two rows is 48°54′. Note that rows [110] and [$\bar{1}$12] are orthogonal but do not define a primitive unit-cell. Another possible choice is [110] and [011]. The next higher plane is identical to the plane containing the origin (add 1 to the third index), but its origin (the node 001) does not project onto node 000. To construct the projection of node 001 on the plane containing the origin, we note that [001] ⊥ [110] and that [001] and [$\bar{1}$12] are coplanar and at an angle of 60.6°.

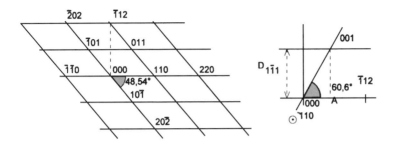

This exercise illustrates the method used for indexing the various layers of a rotating crystal photograph. (Rotate the reciprocal lattice about node 000 for the zero layer, about A for layer 1 and so on).

SYMMETRY AND SPACE GROUPS

1—Multiplication of symmetries

Using the composition laws for symmetry elements, we find the following point groups:

α	a) A2–A2	b) m–m	c) A2–m	d) +inversion
90°	222	mm2	mm2	mmm
60°	32	3m	$\bar{3}$m	$\bar{3}$m
45°	422	4mm	$\bar{4}$2m	4/mmm
30°	622	6mm	$\bar{6}$2m	6/mmm

We can also draw the initial symmetry elements on a stereographic projection, apply them to deduce the images of a pole and then determine all the resulting symmetries.

2—Product of C_n and an inversion

$1 + \bar{1}$	$2 + \bar{1}$	$3 + \bar{1}$	$4 + \bar{1}$	$6 + \bar{1}$
$\bar{1}$	2/m	$\bar{3}$	4/m	6/m

3—Cubic groups

The matrix for rotation about Ox is:

$$\begin{pmatrix} 1 & 0 & 0 \\ 0 & \cos\varphi & -\sin\varphi \\ 0 & \sin\varphi & \cos\varphi \end{pmatrix} \quad \text{pour } \varphi = \frac{\pi}{2} \Rightarrow \begin{pmatrix} 1 & 0 & 0 \\ 0 & 0 & \bar{1} \\ 0 & 1 & 0 \end{pmatrix}$$

The matrices for rotation $+\pi/2$ about Oy and Oz are:

$$\text{O}y \Rightarrow \begin{pmatrix} 0 & 0 & 1 \\ 0 & 1 & 0 \\ \bar{1} & 0 & 0 \end{pmatrix} \quad \text{O}z \Rightarrow \begin{pmatrix} 0 & \bar{1} & 0 \\ 1 & 0 & 0 \\ 0 & 0 & 1 \end{pmatrix}$$

The matrices for rotation about threefold cube axes are:

$$\begin{array}{cccc} [111] & [\bar{1}11] & [1\bar{1}1] & [\bar{1}\bar{1}1] \\[4pt] \begin{pmatrix} 0 & 0 & 1 \\ 1 & 0 & 0 \\ 0 & 1 & 0 \end{pmatrix} & \begin{pmatrix} 0 & \bar{1} & 0 \\ 0 & 0 & 1 \\ \bar{1} & 0 & 0 \end{pmatrix} & \begin{pmatrix} 0 & \bar{1} & 0 \\ 0 & 0 & \bar{1} \\ 1 & 0 & 0 \end{pmatrix} & \begin{pmatrix} 0 & 0 & \bar{1} \\ 1 & 0 & 0 \\ 0 & \bar{1} & 0 \end{pmatrix} \end{array}$$

The product of a rotation through $+\pi/2$ about Oy and a rotation through $+\pi/2$ about Ox is therefore (non commutative product):

$$\begin{pmatrix} 0 & 0 & 1 \\ 1 & 0 & 0 \\ 0 & 1 & 0 \end{pmatrix} \quad \text{Rotation through } \frac{2\pi}{3} \text{ about } [111].$$

The product of a rotation through $+\pi/2$ about Ox and a rotation through $+\pi/2$ about Oy is:

$$\begin{pmatrix} 0 & 1 & 0 \\ 0 & 0 & \bar{1} \\ \bar{1} & 0 & 0 \end{pmatrix} \quad \text{Rotation through } \frac{2\pi}{3} \text{ about } [\bar{1}11].$$

Using the relation $\cos\gamma = \cos\alpha.\cos\beta - \sin\alpha.\sin\beta.\cos\Psi$, which gives half the rotation angle produced by multiplying two rotations of intersecting axes, we have:

$$\gamma = \pi/3 \qquad (\alpha = \beta = \pi/4, \ \Psi = \pi/2)$$

The presence of two fourfold axes $\perp$ implies the presence of four threefold axes oriented along the diagonals of the cube (and also of three fourfold axes normal to the faces of the cube).

Inverse case: the product of two threefold axes. We have to consider the case [$\bar{1}11$].[111] with $\Psi \approx 70°$ and the case [$\bar{1}\bar{1}1$].[111] with $\Psi = 109°28'$. The first case corresponds to a twofold axis oriented along [010] (or along [001] for the inverse product) and the second to a threefold axis oriented along [$1\bar{1}1$]. The presence of four threefold axes oriented along the diagonals of a cube only implies the presence of three twofold axes normal to the faces of the cube.

The product of a rotation about a threefold axis oriented along [111] followed by a rotation about a fourfold axis oriented along [001] is a twofold axis oriented along [011]. The inverse product corresponds to a twofold axis oriented along [101].

$$\gamma = \pi/2 \qquad (\alpha = \pi/4, \; \beta = \pi/3, \; \cos\varphi = 1/\sqrt{3})$$

$$[011] \Rightarrow \begin{pmatrix} \bar{1} & 0 & 0 \\ 0 & 0 & 1 \\ 0 & 1 & 0 \end{pmatrix} \qquad [101] \Rightarrow \begin{pmatrix} 0 & 0 & 1 \\ 0 & \bar{1} & 0 \\ 1 & 0 & 0 \end{pmatrix}$$

4—Product of C_n and a translation

The product of a pure rotation through an angle θ and a translation normal to the axis is a pure rotation about an axis situated on the mid-line of the vector **t** at a distance $h = t/2 . \tan\theta/2$.

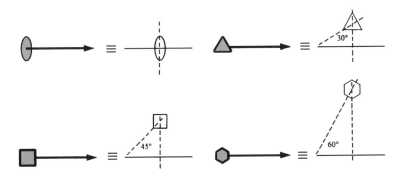

5—Products of symmetry operators

Point groups:

$(C2_y|\mathbf{0}).(C2_x|\mathbf{0}) = (C2_z|\mathbf{0})$ product of orthogonal twofold axes.
$(\sigma_x|\mathbf{0}).(\sigma_y|\mathbf{0}) = (C2_z|\mathbf{0})$ product of orthogonal mirror planes.
$(\sigma_z|\mathbf{0}).(I|\mathbf{0}) = (C2_z|\mathbf{0})$.
$(C2_z|\mathbf{0}).(I|\mathbf{0}) = (\sigma_z|\mathbf{0})$
$(C4_z|\mathbf{0}).(C2_x|\mathbf{0}) = (C2_{x\bar{y}}|\mathbf{0})$ twofold axis [$1\bar{1}0$].
$(C2_x|\mathbf{0}).(C4_z|\mathbf{0})$ ditto.
$(C6_z|\mathbf{0}).(C2_x|\mathbf{0}) = (C2_{30}|\mathbf{0})$ twofold along [21.0].
$(C2_x|\mathbf{0}).(C6_z|\mathbf{0}) = (C2_{-30}|\mathbf{0})$ twofold along [$1\bar{1}0$].
$(\sigma_{60}|\mathbf{0}).(\sigma_x|\mathbf{0}) = (C3_z|\mathbf{0})$.
$(\sigma_{30}|\mathbf{0}).(\sigma_x|\mathbf{0}) = (C6_z|\mathbf{0})$.

Space groups.

The translation is resolved into $\mathbf{t}_{\parallel}$ and $\mathbf{t}_{\perp}$. By a suitable choice of the position of the axes and mirror planes, the normal component of the translation can be eliminated. (For the axes see exercise 4, for the mirror planes, these must be placed on the mid-line of $\mathbf{t}_{\perp}$). In the present examples, $\mathbf{t}_{\parallel}$ is printed in *italics*.

Orthorhombic cell:

$(C2_x|\mathbf{0}).(E|\frac{1}{2}(\mathbf{a} + \mathbf{b})) = (C2_x|\frac{1}{2}\mathbf{a} + \frac{1}{2}\mathbf{b})$ 2_1[100] axis at $y = \frac{1}{4}$.
$(\sigma_x|\mathbf{0}).(E|\frac{1}{2}(\mathbf{a} + \mathbf{b})) = (\sigma_x|\frac{1}{2}\mathbf{b} + \frac{1}{2}\mathbf{a})$ mirror plane b(010) at $x = \frac{1}{4}$.
$(\sigma_y|\mathbf{0}).(E|\frac{1}{2}(\mathbf{a} + \mathbf{b})) = (\sigma_y|\frac{1}{2}\mathbf{a} + \frac{1}{2}\mathbf{b})$ mirror plane a(100) at $y = \frac{1}{4}$.
$(\sigma_x|\mathbf{0}).(\sigma_y|\frac{1}{2}(\mathbf{b} + \mathbf{c})) = (C2_z|\frac{1}{2}\mathbf{c} + \frac{1}{2}\mathbf{b})$ 2_1[001] axis at $y = \frac{1}{4}$.
$(\sigma_x|\frac{1}{2}(\mathbf{b} + \mathbf{a}).(\sigma_y|\frac{1}{2}(\mathbf{c} + \mathbf{b})) = (C2_z|\frac{1}{2}\mathbf{c} + \frac{1}{2}\mathbf{a})$ product (mirror plane c at $y = \frac{1}{4}$)
and (mirror plane b at $x = \frac{1}{4}$) = 2_1[001] axis at $x = \frac{1}{4}$.

Tetragonal cell:

$(C4_z|\mathbf{0}).(E|\mathbf{a}) = (C4_z|\mathbf{a})$ fourfold axis at $\frac{1}{2}, \frac{1}{2}, z$.
$(C2_z|\mathbf{0}).(E|\mathbf{a}) = (C2_z|\mathbf{a})$ twofold axis at $\frac{1}{2}, 0, z$.
$(C2_{xy}|\mathbf{0}).(E|\mathbf{a}) = (C2_{xy}|\frac{1}{2}(\mathbf{a} + \mathbf{b}) + \frac{1}{2}(\mathbf{a} - \mathbf{b}))$ 2_1 axis $//$ to [110] at $(\frac{1}{2}, 0, z)$
or $(\frac{1}{4}, -\frac{1}{4}, z)$.

$(C4_z^1|\mathbf{0}).(E|\frac{1}{2}(\mathbf{a} + \mathbf{b} + \mathbf{c})) = (C4_z^1|\frac{1}{2}\mathbf{c} + \frac{1}{2}(\mathbf{a} + \mathbf{b}))$ 4_2 axis at $0, \frac{1}{2}, z$.
$(C4_z^2|\mathbf{0}).(E|\frac{1}{2}(\mathbf{a} + \mathbf{b} + \mathbf{c})) = (C2_z|\frac{1}{2}\mathbf{c} + \frac{1}{2}(\mathbf{a} + \mathbf{b}))$ 2_1[001] axis at $\frac{1}{4}, \frac{1}{4}, z$.
$(C2_x|\mathbf{0}).(E|\frac{1}{2}(\mathbf{a} + \mathbf{b} + \mathbf{c})) = (C2_x|\frac{1}{2}\mathbf{a} + \frac{1}{2}(\mathbf{c} + \mathbf{b}))$ 2_1[100] axis at $x, \frac{1}{4}, \frac{1}{4}$.
$(C2_{xy}|\mathbf{0}).(E|\frac{1}{2}(\mathbf{a} + \mathbf{b} + \mathbf{c})) = (C2_{xy}|\frac{1}{2}(\mathbf{a} + \mathbf{b}) + \frac{1}{2}\mathbf{c})$ 2_1[110] axis height $\frac{1}{4}$.
$(\sigma_x|\mathbf{0}).(E|\frac{1}{2}(\mathbf{a} + \mathbf{b} + \mathbf{c})) = (\sigma_x|\frac{1}{2}(\mathbf{c} + \mathbf{b}) + \frac{1}{2}\mathbf{a})$ mirror plane n(010) at $x = \frac{1}{4}$.

$(\sigma_{xy}|0).(E|\frac{1}{2}(\mathbf{a}+\mathbf{b}+\mathbf{c})) = (\sigma_{xy}|\frac{1}{2}c+\frac{1}{2}(\mathbf{a}+\mathbf{b}))$ mirror plane c (110).

$(\sigma_z|0).(E|\frac{1}{2}(\mathbf{a}+\mathbf{b}+\mathbf{c})) = (\sigma_z|\frac{1}{2}(a+b)+\frac{1}{2}c)$ mirror plane n (001) height $\frac{1}{4}$.

This method can be used to generate space groups; first, a list is made of the symmetry operators for the point group (there are as many operators as there are equivalent directions), and they are then combined with non-integral and integral translations.

6—Groups derived from class m

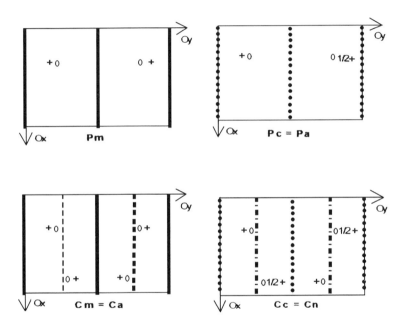

Pc is transformed into Pa by permuting the axes a and c, and by changing the origin (O′ at 0, $\frac{1}{4}$, 0) Cm is transformed into Ca and Cc into Cn.

Strictly speaking, the group Pn should also be considered, but changing the axes $\mathbf{a}' = \mathbf{a}/2$, $\mathbf{b}' = \mathbf{b}/2$, $\mathbf{c}' = \mathbf{c}/2$ brings us back to the group Cc.

7—The group Pmbm

Such notations are impossible; the symbol b in the second position corresponds to a mirror glide plane perpendicular to Oy with a translation b/2 parallel to Oy!

8—Identical results from different operations

Two different operations (a 2_1 screw axis and a mirror plane b) appearing to give the same result arises from the fact that the object transformed is plane and that the heights of the objects are not indicated. To obtain coherent representations, objects with relief must be used with the heights quoted in order to have a faithful representation of the symmetry operation.

9—The group Cmc2$_1$

This group is derived from the class mm2 with eight general equivalent positions.

$$x, y, z \qquad \tfrac{1}{2}+x, \tfrac{1}{2}+y, z$$
$$x, -y, \tfrac{1}{2}+z \qquad \tfrac{1}{2}+x, \tfrac{1}{2}-y, \tfrac{1}{2}+z$$
$$-x, y, z \qquad \tfrac{1}{2}-x, \tfrac{1}{2}+y, z$$
$$-x, -y, \tfrac{1}{2}+z \qquad \tfrac{1}{2}-x, \tfrac{1}{2}-y, \tfrac{1}{2}+z.$$

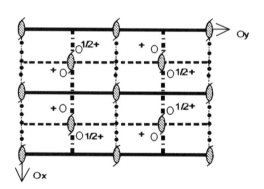

10—The group Amm2

This group is derived from the class mm2 with eight general equivalent positions:

$$x, y, z \qquad x, \tfrac{1}{2}+y, \tfrac{1}{2}+z$$
$$-x, -y, z \qquad -x, \tfrac{1}{2}-y, \tfrac{1}{2}+z$$
$$-x, y, z \qquad -x, \tfrac{1}{2}+y, \tfrac{1}{2}+z$$
$$x, -y, z \qquad x, \tfrac{1}{2}-y, \tfrac{1}{2}+z.$$

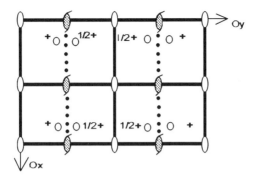

11—Product of a translation and a rotation

If we choose $Ox \parallel \mathbf{t}$, the coordinates of A'' will be:

$$x'' = x.\cos\theta - y.\sin\theta + t$$
$$y'' = x.\sin\theta + y.\cos\theta$$

In a frame centred on I (the centre of the equivalent rotation), we will have:

$$x''_I = x_I.\cos\theta - y_I.\sin\theta$$
$$y''_I = x_I.\sin\theta + y_I.\cos\theta \tag{1}$$

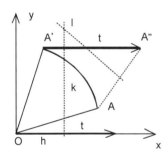

$$IA'' = IO + OA'' \qquad IA = IO + OA$$
$$x''_I = -h + x'' \qquad y''_I = -k + y''$$
$$x_I = -h + x \qquad y_I = -k + y$$

Solving system (1) above gives:

$$h = \frac{t}{2} \qquad k = \frac{t}{2.\tan\theta/2}$$

These two equations determine the position of the point I which is the centre of rotation equivalent to the product of a rotation and a subsequent translation.

Application to group I4

The relevant translations are $\mathbf{t1} = [100]$, $\mathbf{t2} = [010]$ and $\mathbf{t3} = \frac{1}{2}[111]$ (I lattice).

$$\mathcal{R}(O,\pi/2) + \mathbf{t1} \Rightarrow \mathcal{R}(A,\pi/2) \qquad \mathcal{R}(O,\pi/2) + \mathbf{t2} \Rightarrow \mathcal{R}(A,\pi/2)$$
$$\mathcal{R}(O,\pi/2) + \mathbf{t3} \Rightarrow \mathcal{R}(F,\pi/2) + \frac{1}{2}\mathbf{c}(4_2)$$

$\mathcal{R}(O,\pi) + \mathbf{t1} \Rightarrow \mathcal{R}(E,\pi)$ $\qquad\qquad$ $\mathcal{R}(O,\pi) + \mathbf{t2} \Rightarrow \mathcal{R}(F,\pi)$

$\mathcal{R}(O,\pi) + \mathbf{t3} \Rightarrow \mathcal{R}(G,\pi) + \frac{1}{2}\mathbf{c}(2_1)$

$\mathcal{R}(O,3\pi/2) + \mathbf{t1} \Rightarrow \mathcal{R}(A,3\pi/2)$ $\qquad$ $\mathcal{R}(O,3\pi/2) + \mathbf{t2} \Rightarrow \mathcal{R}(A,3\pi/2)$

$\mathcal{R}(O,3\pi/2) + \mathbf{t3} \Rightarrow \mathcal{R}(E,\pi/2) + \frac{1}{2}\mathbf{c}(4_2)$

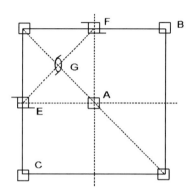

12—The group I4₁/a

The projection is deduced using the method of the previous exercise.

For this group, bear in mind the transformation of 4_1 into 4_3 and the presence of $\bar{4}$ axes (with their centre of inversion at a height 0).

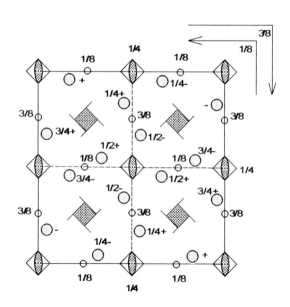

13.a—Tetragonal C = tetragonal P

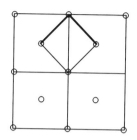

The transformation:

$$\mathbf{a}' = \tfrac{1}{2}(\mathbf{a} - \mathbf{b}),$$
$$\mathbf{b}' = \tfrac{1}{2}(\mathbf{a} + \mathbf{b}),$$
$$\mathbf{c}' = \mathbf{c}.$$

gives a primitive tetragonal cell.

13.b—Monoclinic F = Monoclinic C

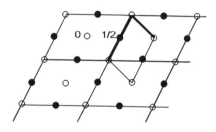

The transformation:

$$\mathbf{a}' = \mathbf{a},$$
$$\mathbf{b}' = \mathbf{b},$$
$$\mathbf{c}' = \tfrac{1}{2}(\mathbf{a} + \mathbf{c}).$$

gives a face-centred cell.

13.c—Hexagonal F = Orthorhombic I

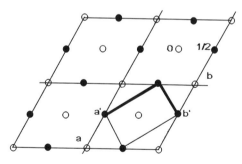

The transformation:

$$\mathbf{a}' = \tfrac{1}{2}(\mathbf{a} - \mathbf{b}),$$
$$\mathbf{b}' = \tfrac{1}{2}(\mathbf{a} + \mathbf{b}),$$
$$\mathbf{c}' = \mathbf{c}$$
$$O':0, \tfrac{1}{2}, \tfrac{1}{2}.$$

gives a centred orthorhombic cell.

14—Rotation matrix in a hexagonal lattice

We let $\mathbf{OA} = \mathbf{a}$, $\mathbf{OB} = \mathbf{b}$, $\{\mathbf{OA},\mathbf{OB}\} = 120°$

$$OD = A,\ OB = B\ \text{(orthonormal frame)}.$$

$$OD = OH + HD = \frac{2}{3}\sqrt{3}a + \frac{1}{3}\sqrt{3}b.$$

Consider a vector:

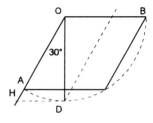

$$r = x.a + y.b = X.A + Y.B$$

in the plane.

After rotation through φ about an axis normal to the plane at O, this vector becomes:

$$r' = x'.a + y'.b = X'.A + Y'.B$$

$$\begin{pmatrix} A \\ B \end{pmatrix} = (H).\begin{pmatrix} a \\ b \end{pmatrix} = \begin{pmatrix} 2\sqrt{3}/3 & \sqrt{3}/3 \\ 0 & 1 \end{pmatrix}.\begin{pmatrix} a \\ b \end{pmatrix};$$

$$\begin{pmatrix} a \\ b \end{pmatrix} = (H)^{-1}.\begin{pmatrix} A \\ B \end{pmatrix} = \begin{pmatrix} \sqrt{3}/2 & -1/2 \\ 0 & 1 \end{pmatrix}.\begin{pmatrix} A \\ B \end{pmatrix}$$

$$\begin{pmatrix} X \\ Y \end{pmatrix} = (H^{-1})^{T}.\begin{pmatrix} x \\ y \end{pmatrix};\quad \begin{pmatrix} x' \\ y' \end{pmatrix} = (H)^{T}.\begin{pmatrix} X' \\ Y' \end{pmatrix};$$

$$\begin{pmatrix} X' \\ Y' \end{pmatrix} = (R_{\varphi}).\begin{pmatrix} X \\ Y \end{pmatrix} = \begin{pmatrix} \cos\varphi & -\sin\varphi \\ \sin\varphi & \cos\varphi \end{pmatrix}.\begin{pmatrix} X \\ Y \end{pmatrix}$$

$$\begin{pmatrix} x' \\ y' \end{pmatrix} = (H)^{T}.(R_{\varphi}).(H^{-1})^{T}.\begin{pmatrix} x \\ y \end{pmatrix} = R_{\varphi}^{H}.\begin{pmatrix} x \\ y \end{pmatrix}$$

$$(R_{\varphi}^{H}) = \begin{pmatrix} \cos\varphi + \frac{1}{3}\sqrt{3}\sin\varphi & -\frac{2}{3}\sqrt{3}\sin\varphi \\ \frac{2}{3}\sqrt{3}\sin\varphi & \cos\varphi - \frac{1}{3}\sqrt{3}\sin\varphi \end{pmatrix}$$

In particular:

$$\text{If } \varphi = \frac{\pi}{3}\ (R_{\varphi}^{H}) = \begin{pmatrix} 1 & -1 \\ 1 & 0 \end{pmatrix}\ \text{and if } \varphi = \frac{2\pi}{3}\ (R_{\varphi}^{H}) = \begin{pmatrix} 0 & -1 \\ 1 & -1 \end{pmatrix}$$

The coefficients of matrices representing rotations compatible with the notion of the lattice, are integers when these matrices are expressed in the frame of the base vectors.

STRUCTURE FACTOR

1—Diamond

0, 0, 0	$\frac{1}{2}, \frac{1}{2}, 0$	$0, \frac{1}{2}, \frac{1}{2}$	$\frac{1}{2}, 0, \frac{1}{2}$
$\frac{1}{4}, \frac{1}{4}, \frac{1}{4}$	$\frac{3}{4}, \frac{3}{4}, \frac{1}{4}$	$\frac{1}{4}, \frac{3}{4}, \frac{3}{4}$	$\frac{3}{4}, \frac{1}{4}, \frac{3}{4}$

$F_{hhh} = 1 + 3.\cos(2\pi h) + \cos(3\pi h/2) + 3.\cos(7\pi h/2)$. Hence:

$$F_{hhh} = 0 \text{ if } h = 4n + 2.$$

2—Fe$_3$Al

Phase A: Body-centred cubic, since all sites are equivalent.
Phase B: Simple cubic since the sites 0, 0, 0 and $\frac{1}{2}, \frac{1}{2}, \frac{1}{2}$ are not equivalent.
Phase C: Face-centred cubic.

Phase A: $f_A = \dfrac{3.f_{Fe} + f_{Al}}{4}$ for all sites.

Phase B: $f_B = \dfrac{f_{Fe} + f_{Al}}{2}$ for all sites $\frac{1}{2}, \frac{1}{2}, \frac{1}{2}$. (Fe and Al have equal probabilities).

$2\sin\theta/\lambda$	hkl_A	hkl_B	hkl_C	F_A	F_B	F_C
$2a/\sqrt{3}$	–	–	111			$j/2(f_{Fe} - f_{Al})$
a	–	100	200		$\frac{1}{2}(f_{Fe} - f_{Al})$	$\frac{1}{2}(f_{Fe} - f_{Al})$
$a/\sqrt{2}$	110	110	220	$\frac{1}{2}(3.f_{Fe} + f_{Al})$	$\frac{1}{2}(3.f_{Fe} + f_{Al})$	$\frac{1}{2}(3.f_{Fe} + f_{Al})$
$2a/\sqrt{11}$	–	–	311			$-j/2(f_{Fe} - f_{Al})$
$a/\sqrt{3}$	–	111	222		$\frac{1}{2}(f_{Fe} - f_{Al})$	$\frac{1}{2}(f_{Fe} - f_{Al})$

Extra lines of weak intensity are found in the ordered phases; these are the *superstructure* lines.

POWDER DIFFRACTION PATTERNS

1—Mixture of cubic species

For face-centred cubes the normal sequence of interplanar spacings is:

$$a/\sqrt{3} \quad a/\sqrt{4} \quad a/\sqrt{8} \quad a/\sqrt{11} \quad a/\sqrt{16} \quad a/\sqrt{19}\cdots$$

If the line at 3.24 Å is the 111 line of one species we have $a_1 = 5.62$ Å.

$d111 = 3.24$ Å $\quad d200 = 2.81$ Å $\quad d220 = 1.985$ Å $\quad d311 = 1.69$ Å $\quad d222 = 1.62$ Å
$d400 = 1.40$ Å

If the line with $d = 3.13$ Å is the line 111 of the other species, we have $a_2 = 5.42$ Å, but the series obtained does not fit.

If this line is the line 200 of the other species we have $a_2 = 6.26$ Å.

$d200 = 3.13$ Å $\quad d220 = 2.21$ Å $\quad d222 = 1.81$ Å $\quad d400 = 1.56$ Å $\quad d420 = 1.40$ Å

(It is, however, also possible to perform the indexing with a p cell of parameter 3.13 Å).

If we assume that $f_A = f_C$, and that the structure is of the NaCl type, the lattice appears to be type P. (See KCl in exercise 4).

2—Barium titanate

$$d^\alpha_{hkl} = \frac{a_\alpha}{(h^2 + k^2 + l^2)^{1/2}} \qquad d^\beta_{hkl} = \frac{a_\beta}{(h^2 + k^2 + l^2(a/c)^2)^{1/2}}$$

$$a_\alpha \approx a_\beta \qquad a/c = 1 - \varepsilon \Rightarrow d^\alpha_{hkl} \approx d^\beta_{hkl} \qquad (a/c)^2 \approx 1 - 2.\varepsilon$$

—For an hkl line of the cubic phase we have:

$$d^\alpha_{hkl} = K(h^2 + k^2 + l^2)^{-1/2}$$

The interplanar spacings are identical for the lines klh, lhk, hlk, lkh, khl.

—For the tetragonal phase, on the other hand:

$$d^\beta_{hkl} = K(h^2 + k^2 + l^2 - 2.\varepsilon.l^2)^{-1/2} \text{ (ditto for } khl),$$

$$d^\beta_{klh} = K(h^2 + k^2 + l^2 - 2.\varepsilon.h^2)^{-1/2} \text{ (ditto for } lkh),$$

$$d^\beta_{hkl} = K(h^2 + k^2 + l^2 - 2.\varepsilon.k^2)^{-1/2} \text{ (ditto for } hlk).$$

If $h \neq k \neq l$, a line of the α phase gives 3 lines in the β phase.

If $h = k \neq l$, a line of the α phase gives 2 lines in the β phase.

If $h = k = l$, a line of the α phase gives 1 line in the β phase.
Tetragonalisation results in a lifting of the degeneracy in the values of $h\,k\,l$ for certain cubic lines.

3—Cubic crystals

Barium: $a = 5.025\,\overset{\circ}{\text{A}}$ I lattice.
CsCl: $a = 4.123\,\overset{\circ}{\text{A}}$ P lattice.
Diamond: $a = 3.5667\,\overset{\circ}{\text{A}}$ F lattice (Note the absence of $h + k + l = 4n + 2$).
Copper: $a = 3.615\,\overset{\circ}{\text{A}}$ F lattice.

Barium (*Ba*)			CsCl			Diamond			Copper		
$D/\overset{\circ}{\text{A}}$	I	hkl	$D/\overset{\circ}{\text{A}}$	I	hkl	$D/\overset{\circ}{\text{A}}$	I	hkl	$D/\overset{\circ}{\text{A}}$	I	hkl
3.55	100	110	4.12	45	100	2.06	100	111	2.088	100	111
2.513	20	200	2.917	100	110	1.261	25	220	1.808	46	200
2.051	40	211	2.380	13	111	1.0754	16	311	1.278	20	220
1.776	18	220	2.062	17	200	0.8916	8	400	1.090	17	311
1.590	12	310	1.844	14	210	0.8182	16	331	1.0436	5	222
1.451	6	222	1.683	25	211				0.9038	3	400
1.343	14	321	1.457	6	220				0.8293	9	331
1.1852	6	330	1.374	5	330				0.8083	8	420
1.1236	4	420	1.304	8	310						

4—Cubic crystals

α *iron*: As there are only six lines in the photograph we might assume a P lattice with a parameter of 2.0268 Å. In fact the lattice is I and the parameter is 2.8664 Å.
NaCl: $a = 5.6402\,\overset{\circ}{\text{A}}$ F lattice.
KCl: $a = 6.2917\,\overset{\circ}{\text{A}}$ F lattice.
The fractional coordinates are: Cl: 0. 0. 0 + FCC; Na: $\frac{1}{2}$. 0. 0 + FCC.
For lines with odd indices the structure factor is:

$$F_{hkl} = 4.f_{\text{Cl}} - 4.f_{\text{Na}}. \quad (F_{hkl} \text{ is therefore weak.})$$

For KCl the scattering factors of K^+ and Cl^- are practically the same and the intensities of lines with odd indices are close to zero.
If: $a = 5.4309\,\overset{\circ}{\text{A}}$ F lattice.
The fractional coordinates of the atoms are:

$$\text{Si: } 0, 0, 0 + \text{FCC}; \quad\quad \tfrac{1}{4}, \tfrac{1}{4}, \tfrac{1}{4} + \text{FCC}.$$

Calculation of the structure factor shows the absence of lines such as $h + k + l = 4n + 2$.

α iron			NaCl			KCl			Si		
$D/\text{Å}$	I	hkl	$D/\text{Å}$	I	hkl	$D/\text{Å}$	I	hkl	$D/\text{Å}$	I	hkl
2.0268	100	110	3.26	13	111	3.633	1	111	3.1355	100	111
1.4332	20	200	2.821	100	200	3.146	100	200	1.9201	55	220
1.1702	30	211	1.994	55	220	2.225	37	220	1.6375	30	311
1.0134	10	220	1.701	2	311	1.897	<1	311	1.3577	6	400
0.9064	12	310	1.628	15	222	1.817	10	222	1.2459	11	331
0.8275	6	222	1.410	6	400	1.573	5	400	1.1086	12	422
			1.294	1	331	1.407	9	420	1.0452	6	511
			1.261	11	420	1.284	5	422	0.9600	3	440
			1.1515	7	422	1.112	1	440	0.9180	7	531

5—Time-of-flight method

Let L be the distance travelled up to the detector. We have: $mv = mL/t = h/\lambda$. The detector receives neutrons scattered at the fixed diffraction angle θ_0. A family of planes (hkl) diffracts the wavelength $\lambda_{hkl} = 2d_{hkl}.\sin\theta_0$. For this family the time of flight will be:

$$t_{hkl} = \frac{m}{h} L.\lambda_{hkl} = 2\frac{m}{h} L.d_{hkl}.\sin\theta_0$$

peak	t/ms	$d_{hkl}\text{Å}$	hkl
6	13.503	1.916	220
5	11.515	1.634	311
4	9.549	1.355	400
3	8.760	1.243	331
2	7.796	1.106	422
1	7.351	1.043	511/333

The cell parameter is 5.419 Å.

The Structure of Rutile

(For the diagram see page 237 of chapter 16).
The fractional coordinates are:

O1: $x, x, 0$ O4: $\frac{1}{2} - x, \frac{1}{2} + x, \frac{1}{2}$

O2: $1 - x, 1 - x, 0$ O5: $x, x, 1$

O3: $\frac{1}{2} + x, \frac{1}{2} - x, \frac{1}{2}$ O6: $1 - x, 1 - x, 1.$

Ti1: $0, 0, 0$ Ti4: $1, 1, 0$

Ti2: $0, 1, 0$ Ti5: $\frac{1}{2}, \frac{1}{2}, \frac{1}{2}$

Ti3: $1, 0, 0$ Ti6: $0, 1, 1.$

The oxygens are at the centre of a triangle of titaniums and the titaniums are at the centre of an octahedron of oxygens. The interatomic distances are:

$$d^2 = (x - x')^2.a^2 + (y - y')^2.a^2 + (z - z')^2.c^2$$

Ti–O bonds:

Ti5–O4 (Ti1–O1) $d = \sqrt{2}.x.a \approx 1.98$ Å.

Ti5–O1(Ti2–O4) $d = (2.(\frac{1}{2} - x)^2.a^2 + \frac{1}{4}.c^2)^{\frac{1}{2}} \approx 1.947$ Å.

O–O bonds:

O1–O5 $d = c \approx 2.96$ Å.

O1–O2 $d = \sqrt{2}.(1 - 2x).a \approx 2.53$ Å.

O1–O4 $d = ((\frac{1}{2} - 2x)^2.a^2 + \frac{1}{4}.a^2 + \frac{1}{4}.c^2)^{1/2} \approx 2.78$ Å.

The TiO_6 octahedra have orthorhombic symmetry.

Calcium Titanate

On the projection of the structure on (010), the values of x, y and z close to zero have been doubled to increase the distortion.
Parameters of the pseudo-cubic cell:

$\mathbf{a}' = \frac{1}{2}(\mathbf{a} + \mathbf{c})$ $\mathbf{b}' = \frac{1}{2}\mathbf{b}$ $\mathbf{c}' = \frac{1}{2}(\mathbf{c} - \mathbf{a})$

$a' = c' = 3.822$ Å. $b' = 3.82$ Å. $\tan(\beta/2) = c/a \Rightarrow \beta = 90°44'.$

Axis transformation matrix:

$$\begin{pmatrix} \frac{1}{2} & 0 & \frac{1}{2} \\ 0 & \frac{1}{2} & 0 \\ \frac{\bar{1}}{2} & 0 & \frac{1}{2} \end{pmatrix}$$

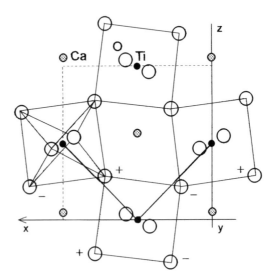

This structure differs from that of ideal perovskite in the displacement of the calcium and the rotations of the TiO$_6$ octahedra about x and about z. Note the alternating heights of the oxygens above and below the (010) plane.

Solutions to the Problems

Cuprite

Indexing:

d_{hkl}	3.020	2.465	2.135	1.743	1.510	1.287	1.233
Indices	110	111	200	211	220	311	222

$$a = 4.27 \text{ Å} \quad \text{P lattice.}$$

Group Pn3m (only symmetry elements normal to (001) are shown; for a complete projection, see the International Tables).

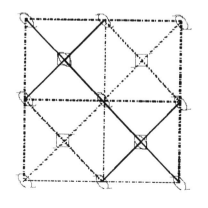

Structure factor:

$$F_{hkl} = f_0 . e^{j(\pi/2)(h+k+l)}(1 + e^{j\pi(h+k+l)})$$

$$+ f_{Cu}(1 + e^{j\pi(h+k)} + e^{j\pi(h+l)} + e^{j\pi(l+k)})$$

$$F_{hkl} = 0 \text{ if } h + k + l = 2n + 1 \text{ with}$$

h, k, l of different parities (e.g.: 300).

$$n[110] = 6.038 \text{ Å}. \quad l_{2,-2} \approx 6.8 \text{ cm.}$$

$$h.u + k.v + l.w = K. \quad \text{Here: } h + k = K.$$

For the zero layer $K = 0$: the nodes in the reciprocal plane are of the form $h\bar{h}l$. The plane can be generated from rows $[001]$ and $[1\bar{1}0]$.

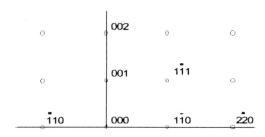

Trigonal $\Leftrightarrow$ Cubic Phase Transition

1—See chapter 2 for definitions of the reciprocal lattice.

$2 - V = (\mathbf{a, b, c}) = a^3 . \sin \alpha . \sin \varphi$

$\mathbf{a}.(\mathbf{b} + \mathbf{c}) = \mathbf{a}.\mathbf{b} + \mathbf{a}.\mathbf{c} = |\mathbf{a}|.|\mathbf{b} + \mathbf{c}|. \cos \varphi \Rightarrow \cos \varphi = \cos \alpha / \cos \alpha/2$

$3 - \dfrac{1}{d_{hkl}^2} = h^2 A^2 + k^2 B^2 + l^2 C^2 + 2h.k.\mathbf{A.B} + 2h.l.\mathbf{A.C} + 2k.l.\mathbf{B.C}$

Now : $\mathbf{A.A} = \dfrac{a^4}{V^2} \sin^2 \alpha$ and $\mathbf{A.B} = \dfrac{a^4(\cos^2 \alpha - \cos \alpha)}{V^2}$

from which the equation in the problem follows.

4 – See the chapter on lattices.

5 – The hkl lines conserved by the threefold axis remain degenerate; the others are resolved.

$$\alpha = \pi/2 - \varepsilon \Rightarrow \sin \alpha = \cos \varepsilon \approx 1 \quad \text{and} \quad \cos \alpha = \sin \varepsilon \approx \varepsilon$$

$$\dfrac{1}{d_{hkl}^2} \text{Trig} \approx \dfrac{h^2 + k^2 + l^2 - 2.\varepsilon.(hk + kl + hl)}{a^2} \text{and} \dfrac{1}{d_{hkl}^2} \text{Cub} \approx \dfrac{h^2 + k^2 + l^2}{a^2}$$

Cubic line (111):

◆ 111 and $\bar{1}\bar{1}\bar{1}$ $\dfrac{1}{d^2} = \dfrac{3 - 6\varepsilon}{a^2}$ Intensity = 2

◆ $\bar{1}11, 1\bar{1}1, 11\bar{1}, \bar{1}\bar{1}1, 1\bar{1}\bar{1}, \bar{1}1\bar{1}$ $\dfrac{1}{d^2} = \dfrac{3 + 2\varepsilon}{a^2}$ Intensity = 6

Cubic line (200) unchanged.

Cubic line (110):

◆ 110, 011, 101, $\bar{1}\bar{1}0$, $0\bar{1}\bar{1}$, $\bar{1}0\bar{1}$ $\dfrac{1}{d^2} = \dfrac{2-2\varepsilon}{a^2}$ Intensity $= 6$

◆ $\bar{1}10$, $0\bar{1}1$, $\bar{1}01$, $1\bar{1}0$, $01\bar{1}$, $10\bar{1}$ $\dfrac{1}{d^2} = \dfrac{2+2\varepsilon}{a^2}$ Intensity $= 6$

$6 - \quad d_{100} = a = 3.542\,\text{Å}.$ $1 - \varepsilon = 0.9917$ $\varepsilon = 89°30'.$

$d_{111} = 2.062\,\text{Å}.$ $d_{2\bar{1}0} = 1.580\,\text{Å}.$

1.1 Structure of NiO

After indexing, one parameter $a = 4.183\,\text{Å}$ is found.

$2.\theta$	Intensities	$d_{hkl}/\text{Å}$	$h\,k\,l$
37.283	57	2.4090	111
43.279	100	2.0880	200
62.729	45	1.4796	220
75.265	14	1.2612	311
79.310	10	1.2067	222
94.887	4.5	1.0454	400
106.592	6.3	0.9605	331
110.636	14.9	0.9365	420
128.423	13.2	0.8552	422

$h\,k\,l$	Int	θ	LP	m	$f\,O^{2-}$	$f\,Ni^{2+}$
1 1 1	57	18.641	16.868	8	5.98	20.709
2 0 0	100	21.640	12.104	6	5.28	19.383
2 2 0	45	31.364	5.231	12	3.657	15.984
3 1 1	14	37.632	3.606	24	3.016	14.460
2 2 2	10	39.655	3.299	8	2.855	14.049

$h\,k\,l$	Type NaCl			Type ZnS		
	F_{hkl}	I_{hkl}	kI_{hkl}	F_{hkl}	I_{hkl}	kI_{hkl}
1 1 1	−58.889	467974	66	23.9–82.8.j	1003130	100
2 0 0	98.655	706835	100	−56.41	231095	23
2 2 0	78.566	387467	55	78.56	387467	38
3 1 1	−45.778	181363	25	12 + 57.8.j	302125	30
2 2 2	67.616	120662	17	−44.77	52899	5

The structure is of the NaCl type. The differences between the calculated and measured intensities arise from terms neglected in the calculations (thermal agitation and absorption).

Oxides of Iron

$\mu = n.M/V.N$

$V_A = a^3 = 80.06 \text{ Å}^3$

$V_B = a^3 = 586.4 \text{ Å}^3$

$V_C = a^2.C.\sin(60°) = 301.06 \text{ Å}^3$.   Primitive cell $V_C = 100.35 \text{ Å}^3$.

Possible formulæ:

	FeO	Fe_2O_3	Fe_3O_4
M	71.85	159.7	231.55
$M/N/(10^{-22})$	1.193	2.652	3.846
$\mu V/(g/\text{Å}^3 \times 10^{-22})$	4.78	5.28	30.49
n	4 (A)	2 (C)	8 (B)

Compound A FeO	4 formula units per cell	F lattice.
Compound B Fe_3O_4	8 formula units per cell	F lattice.
Compound C Fe_2O_3	2 formula units per cell	R lattice.

For a cubic F lattice, we have $I_{hkl} \neq 0$ if h, k, l are of the same parity.

Possible lines	$d_{hkl}/\text{Å}$	$\theta(\sin \theta = \gamma/2d_{hkl})$
111	2.4883	18
200	2.155	20.9°
220	1.5238	30.35°

If the rotation row is [001], the spot 111 is on layer 1.

$y = R.\tan \alpha$ where $\alpha = \lambda/c$ $y = 29.6$ mm

There are two possible methods of determining x:

—use the method explained in the text: $\cos 2\theta = \cos \beta . \cos \alpha$ where $x = R.\beta$

—use a graphic method based on the reciprocal lattice.

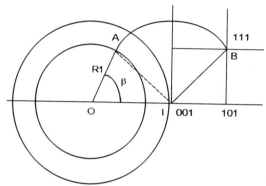

Layer 1 cuts the Ewald sphere in a circle of radius R_1 such that $R_1^2 = R^2 - D_{001}^2$ i.e.:

$$R_1 = R\sqrt{\frac{n_{001}^2 - \lambda^2}{n_{001}^2}} = 51.16 \text{ mm}$$

The node 111 penetrates the Ewald sphere at A, hence:

$$IA = IB = A^*.\sqrt{2}$$

Trigonometric relations in triangle OAI give $\cos \beta = 0.7796$

hence $\beta = 0.677$ Rd and $x = 38.78$ mm.

The structure is thus of the NaCl type.

KIO$_2$F$_2$

$$n = K.\lambda.\sqrt{\frac{R^2 + y^2}{y}} \qquad \frac{\Delta n}{n} = \left| \frac{y}{R^2 + y^2} - \frac{1}{y} \right| \Delta y$$

$a = 8.38 \pm 0.025 \text{ Å}; \qquad c = 8.41 \pm 0.025 \text{ Å}.$

The crystal is thus either orthorhombic or tetragonal; in fact it is orthorhombic since it is biaxial. $\mu = n.M/V.N$, $n = 4$.

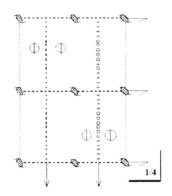

Group Pbcm.
Eight general equivalent positions:
K and I occupy special positions.

x, y, z	$-x, 1/2 + y, z$
$-x, -y, -z$	$x, 1/2 - y, -z$
$x, y, 1/2 - z$	$-x, 1/2 + y, 1/2 - z$
$-x, -y, 1/2 + z$	$x, 1/2 - y, 1/2 + z.$

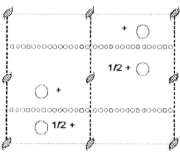

Group Pca2$_1$.
Four general equivalent positions.
x, y, z	$x + 1/2, -y, z$
$1/2 - x, y, 1/2 + z$	$-x, -y, 1/2 + z$

hkl: no condition.
$0kl$: $l = 2n$. $h0l$: $h = 2n$.
$hk0$: no condition.
$h00$: $h = 2n$.
$0k0$: no condition.
$00l$: $l = 2n$.

Pseudosymmetry

$\tan \alpha_{Kmax} = 40/R = \pi/3 \quad \alpha_{Kmax} = 46.4°.$

$n_{uvw} \cdot \sin \alpha_K = K\lambda \qquad K_{max} = n_{uvw} \cdot \sin \alpha_K / \lambda.$

Rotation about [001] $K_{max} = 4.05 \Rightarrow 9$ layers (including the zero layer).

Rotation about [010] $K_{max} = 6.7 \Rightarrow 13$ layers.

General method:

$$\cos \theta = \frac{\mathbf{N}^*_{\bar{2}01} \cdot \mathbf{N}^*_{100}}{N^*_{\bar{2}01} \cdot N^*_{100}} = \frac{-2\mathbf{A} \cdot \mathbf{C} + C^2}{\sqrt{(4A^2 - 4\mathbf{A} \cdot \mathbf{C} + C^2) \cdot C^2}} \approx 0$$

Direct method

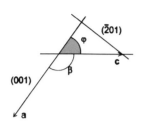

Vectors **a** and **c** are drawn in the plane (010). The plane $\bar{2}01$ cuts a length $-a/2$ on the Ox axis and a length c on the Oz axis.

$$a/2c = 0.582$$

$$\cos\varphi = \cos 54°20' = 0.582.$$

The two planes are orthogonal.

The suggested cell is pseudo-orthorhombic.
The actual symmetry is deduced from diffraction photographs and physical properties.
The systematic absences for $0k0$ where $k = 2n + 1$ are compatible with 2_1 axis parallel to [010]. The group is $P2_1/m$.

General Positions:

$$x, y, z \qquad -x, 1/2+y, -z \qquad x, -y, z \qquad -x, 1/2-y, -z$$

Special positions:

$$0, 0, 0 \qquad 0, 1/2, 0 \qquad 1/2, 0, 0 \qquad 1/2, 1/2, 0$$

AuCu

On indexing we find: $a = 3.97$ Å.

2.293 Å	1.982 Å	1.405 Å	1.195 Å	1.146 Å	0.992 Å	0.912 Å
111	200	220	311	222	400	331

F lattice $\mu = 14 \text{g/cm}^3$

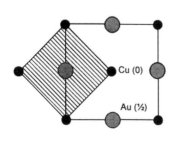

The alternating layers of copper and gold imply a tetragonal lattice with a primitive unit-cell such that the coordinates of the atoms are: Cu 0, 0, 0; Au 1/2, 1/2, 1/2.

$$\mathbf{a}_1 = 1/2(\mathbf{a}_F + \mathbf{b}_F)$$

$$\mathbf{b}_1 = 1/2(\mathbf{a}_F - \mathbf{b}_F)$$

$$\mathbf{c}_1 = \mathbf{c}_F$$

$$(c/a)_1 = a_F.\sqrt{2}/a_F = \sqrt{2}.$$

Cu (0)

Au (½)

To establish the new indexing, the covariance of the Miller indices is used:

Cubic F	111	200	220	311	222	400	331
Tetragonal I	101	110	200	211	202	220	301

Cubic phase: $(1/d)^2_{hkl} \, \mathrm{Cub} = (h^2 + k^2 + l^2/a^2)$

For phase B, $c/a = \sqrt{2} - \varepsilon \Rightarrow (a/c^2) = 1/2 + \varepsilon'$

Tetragonal phase: $\dfrac{1}{d^2_{hkl}}\,\mathrm{Tetra} = \dfrac{h^2 + k^2 + l^2(a/c)^2}{a^2} = \dfrac{h^2 + k^2 + l^2/2 - 21^2\varepsilon'}{a^2}$

hhh lines unchanged, *hhl* doubled, *hkl* tripled.

Cubic	100	110	111	200	210	211	220	221–300
Tetragonal	100	110	111	200	210	211	220	221–212
	001	101		002	201	112	202	300–003
					012			

Structure factors calculated for <u>one</u> unit-cell:

Phase A: $F_{hkl} = 1/2(f_{Cu} + f_{Au}).(1 + e^{j\pi(h+k+l)})$

$h + k + l = 2n$ $\qquad$ $F_{hkl} = f_{Cu} + f_{Au}$

$h + k + l = 2n + 1$ $\quad$ $F_{hkl} = 0.$

Phase B: $F_{hkl} = f_{Cu} + f_{Au}.e^{j\pi(h+k+l)}$

$h + k + l = 2n$ $\qquad$ $F_{hkl} = f_{Cu} + f_{Au}$

$h + k + l = 2n + 1$ $\quad$ $F_{hkl} = f_{Cu} - f_{Au}.$

Setting $c/a = \sqrt{2}$ we obtain the values of d_{hkl} for the various lines:

100 a	100 $a/\sqrt{2}$	111$a\sqrt{\frac{2}{5}}$	200 $a/2$	210 $a/\sqrt{5}$	211 $a\sqrt{\frac{2}{11}}$	220 $\dfrac{a}{2\sqrt{2}}$
001 $a\sqrt{2}$	101 $a\sqrt{\frac{2}{5}}$		002 $a/\sqrt{2}$	201 $a/\frac{\sqrt{2}}{3}$	112 $a/2$	202 $\dfrac{a}{\sqrt{6}}$
				102 $a/\sqrt{3}$		

Cesium Bromate

In the hexagonal cell the Miller indices are:

$H = h - k$; $K = k - l$; $L = h + k + l$. $(111)_R \Rightarrow (00.1)_H$ and $(100)_R \Rightarrow (10.1)_H$

$\mathbf{N}^{*}_{00.1}.\mathbf{N}^{*}_{10.1} = |\mathbf{N}^{*}_{00.1}|.|\mathbf{N}^{*}_{10.1}|.\cos \rho_b$

$C^* = 1/c$ $A^* = 2/a.\sqrt{3}$. $(c/a)_H = 1.215$.

The row normal to face a is $[001]_H$ $n_{001} = 8.228$ Å.

The row normal to face f is $[110]_H$ $n_{110} = n_{100} = 6.773$ Å.

$(c/a)_H = 1.215$.

Tetrahedron and Octahedron

Class $\bar{4}2m$.

$p = (111) \Rightarrow c/a \approx 0.98$.

$t = (0kl)$ $ck/al = \tan(63°05')$ $k/l = 2$ $t = (021)$.

The condition for systematic absences indicates an I lattice.

AB is the edge between (111) and $(\bar{1}\bar{1}1)$ i.e. $[1\bar{1}0]$.

BC is the edge between (111) and (021) i.e. $[\bar{1}\bar{1}2]$.

We deduce $a = b = 5.27$ Å.

$[\bar{1}\bar{1}2] = 6.33$ Å is incompatible with the value of c/a.

We try: $c/a \approx 2$.

with this hypothesis $(111) \Rightarrow (112)$ and $(021) \Rightarrow (011)$.

AB remains $[1\bar{1}0]$ and $a = b = 5.27\,\text{Å}$. BC becomes $[\bar{1}\bar{1}1]$.

The lattice is body-centred, hence the parameter of this row is $2 \times 6.33\,\text{Å} = 12.66\,\text{Å}$.

$|n_{\bar{1}\bar{1}1}| = 2a^2 + c^2 \Rightarrow c = 10.2\,\text{Å}$.

Sodium Chlorate

Class 23.

The rotation row is [001]: the cell parameter is $6.56\,\text{Å}$.

$d_{210} = a/\sqrt{5}$.

$\lambda = 2.d_{hkl}.\sin\theta.$ $\theta = 15°13'$.

The row is $[1\bar{2}\bar{1}]$ and its parameter is $a.\sqrt{6} = 16.07\,\text{Å}$.

Hexagonal Close Packed

Hexagonal lattice: $a = b \neq c,,\ \alpha = \beta = \pi/2,\ \gamma = 2\pi/3$.

$$d_{hkl} = \frac{a}{\sqrt{\frac{4}{3}(h^2 + k^2 + hk) + l^2.(a/c)^2}}$$

The line 002 enables us to determine the parameter $c = 2.2 \times 605 = 5.21\,\text{Å}$.

The line 110 enables us to determine the parameter $a = 2.1 \times 6047 = 3.2094\,\text{Å}$. $c/a = 1.62335$.

d_{hkl} Å	2.778	2.605	2.452	**1.9006**	1.6047	**1.389**
$hk.1$	**10.0**	00.2	10.1	10.2	11.0	20.0

—With the rigid sphere model of the hexagonal close packed structure, the ratio c/a is i.e.: $2.\sqrt{2/3}$ i.e. 1.63299. This model fits the data for magnesium.

The fractional coordinates of the atoms in the HCP cell (assemblage ABAB..) are: 0, 0, 0; (type A) and 1/3, 2/3, 1/2 (type B). The choice suggested (2/3, 1/3, 1/4) (1/3, 2/3, 3/4) i.e. ±(2/3, 1/3, 1/4) involves a change of origin to conform to the coordinates chosen in the Tables.

There are as many octahedral vacancies as there are atoms (one atom is surrounded by six vacancies and one vacancy is surrounded by six atoms); however, there are twice as many tetrahedral vacancies as atoms.

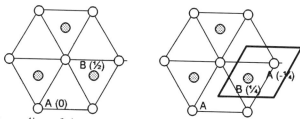

If R is the radius of the atoms in the octahedral vacancies, we can introduce atoms i such that $Ri/R < 0.414$. For the tetrahedral vacancies we have $Ri/R < 0.2247$ (see the section on coordination in ionic structures).

The screw axis 6_n at the origin is a 6_3 axis (a translation of $c/2$ between two adjacent atoms). The group is $P\frac{6_3}{m}mc$.

The mirror planes normal to Ox are types m and a; diagonal mirror planes ($\parallel$ to Ox) are types c and n. The pattern of 2 and 2_1 axes is repeated around the lines of each of the vertical axes 6_3 and 2_1. The centre of inversion of the $\bar{6}$ axes is at a height of 1/4 (in the horizontal mirror plane).

General conditions for systematic absences in the group:

$hh.l$ where $l = 2n$. (mirror plane c: $x, y, z \Rightarrow y, x, z + 1/2$).

$00.l$ where $l = 2n$. (6_3 axis)

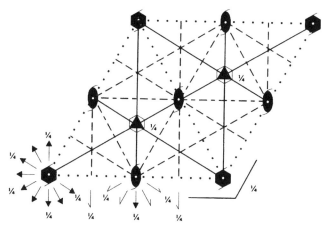

BPO$_4$

The parameter c is 6.642 Å.

Since the lattice is type I we have $h + k + l = 2n$.

The line where $d_{hkl} = 3.3207$ Å is therefore (002).

If we assume that the line of $d_{hkl} = 3.0699$ Å is (110), we have:

$a = 4.342$ Å. The line where $d_{hkl} = 3.6351$ Å is (101) and $c/a = 1.5297$.

Coordinates of the oxygens with the inverse fourfold axis:

$$x, y, z \qquad \overline{x}, \overline{y}, z; \qquad \overline{y}, \overline{x}, \overline{z}; \qquad \overline{y}, x, \overline{z};$$

the translations of the I lattice must then be added.

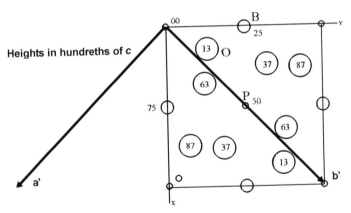

The coordinate changing matrix (X) is the transpose of the inverse of the coordinate changing matrix (A).

$$(A) = \begin{pmatrix} 1 & \overline{1} & 0 \\ 1 & 1 & 0 \\ 0 & 0 & 1 \end{pmatrix} \quad (X) = \begin{pmatrix} 1/2 & \overline{1/2} & 0 \\ 1/2 & 1/2 & 0 \\ 0 & 0 & 1 \end{pmatrix}$$

In the new cell, the sub-lattices of boron and potassium are type F. In this cell $c/a = 1.0816 \approx 1$ (Close to a cubic lattice).

The assemblage of boron and potassium atoms is identical to that of sulphur and zinc atoms in zinc blende.

Group cmca

The parameter is 13.20 Å.

The symmetry in the diffraction photographs corresponds to the Laue class mmm. Since the lattice is triorthogonal, the reciprocal vectors are in the same directions as the direct vectors. In the photographs with the beam ∥ to **b**, reciprocal planes of the type $(h01)^*$ are obtained; with the beam ∥ to **c**, $(hk0)^*$ planes are obtained.

In both types of photograph the vector **A*** is horizontal and normal either to **B*** or to **C***. To index the photographs correctly the systematic absences must be taken into account.

On the photographs, from the measurements:

$$18A^* = 58.2 \text{ mm}. \quad 8B^* = 56 \text{mm}. \quad 6C^* = 50 \text{ mm}.$$

we deduce:

$$a \approx 13.2 \text{ Å}; \qquad b \approx 6.1 \text{ Å}; \qquad c \approx 5.1 \text{ Å}.$$

From the density, the number of formula units is close to 4: $Z = 4$.

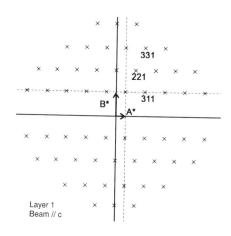

Systematic absences:

Lattice C: $x, y, z \to 1/2 + x, 1/2 + y, z$ hkl $h + k = 2n$.

mirror c∥(010): $x, y, z \to x, -y, 1/2 + z$ $h0l$ $l = 2n$ and $h = 2n$.

mirror a∥(100): $x, y, z \to -x, 1/2 + y, z$ $0kl$ $k = 2n$.

mirror a∥(001): $x, y, z \to 1/2 + x, y, -z$ $hk0$ $h = 2n$ and $k = 2n$.

2_1 axis $\parallel$ Ox: $x, y, z \rightarrow 1/2 + x, -y, -z$ $h00 \quad h = 2n.$

2_1 axis $\parallel$ Oy: $x, y, z \rightarrow -x, 1/2 + y, -z$ $0k0 \quad k = 2n.$

2_1 axis $\parallel$ Oz: $x, y, z \rightarrow -x, -y, 1/2 + z$ $00l \quad l = 2n.$

Group **Cmca**

There are sixteen general equivalent positions:

$x, y, z;$ $\qquad\qquad -x, -y, -z;$

$-x, 1/2-y, 1/2+z;$ $\quad x, 1/2+y, 1/2-z;$

$-x, 1/2+y, 1/2-z;$ $\quad x, 1/2-y, 1/2+z;$

$x, -y, -z;$ $\qquad\qquad -x, y, z.$

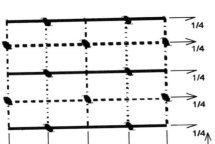

Since the lattice is type C, the translation:

$$t = (1/2, 1/2, 0)$$

must be added.

Crystallographic point groups and their subgroups after Hermann

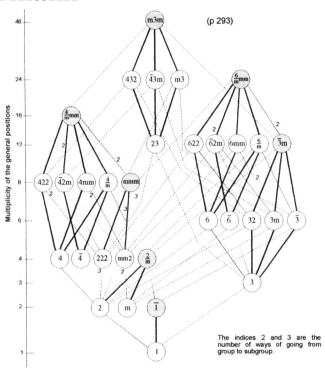

Figure 1. Point groups and subgroups

The holohedral classes are in grey. A connecting line between two groups means that the lower group is a subgroup of the upper group. The links between classes in the same system are shown by bold lines. The vertical scale shows the number of general equivalent directions (the order of the group).

Appendix A

Atlas of Crystallographic Forms

This atlas contains the crystalline forms (the set of equivalent faces in {*hkl*} notation) of the 32 **classes** of symmetry point groups; the classes are arranged into the seven crystal **systems**, each system beginning with the holohedral class (i.e. the class which has the symmetry of the lattice). First the *general* form, then the *special* forms are given (the pole of a face coinciding with a symmetry element). For the holohedral classes, all the stereographic representations and all the corresponding solids are given, but for the merihedral classes, only those forms are given which differ from the holohedral forms.

The dotted lines on the stereographic projections are simply visual aids and should not be confused with projections of the symmetry elements. Hidden faces of solids are only represented if there is no resulting confusion. Certain forms do not enclose space; in a real crystal these forms cannot exist alone. In order to show symmetries more clearly and to simplify the art work, solids have been represented with the same development for all faces; the habits of actual crystals are often very different from those of the ideal solids shown in this atlas. Crystals usually possess several associated habits and different faces often have different developments according to how the crystal was grown.

In the nomenclature, common terms are used for well-known forms such as the cube, tetrahedron, octahedron and rhombohedron. The pinacoid involves two parallel planes. For the cubic forms the following systematic system is used: the suffix 'hedron' (face) is preceded by a numerical prefix (Greek not Latin) indicating the number of faces. Thus we have the tetrahedron, hexahedron (cube), octahedron and dodecahedron. To this base are added the prefixes bi, tri, tetra, hexa etc. to indicate that the number of faces is doubled, tripled etc. Thus a trioctahedron is a solid in which each of the faces of an octahedron has been replaced by a triangular pyramid. A second prefix specifies the shape of the faces; e.g. for the pentagontrioctahedron the faces in question are pentagons.

The frame used is indicated on the first projection in each system. Axes out of the plane of projection are dotted, and the axis normal to the plane is shown by a dot in a circle.

Triclinic System

Class $\bar{1}$ (C$_i$)   Element: C

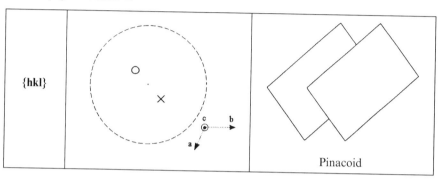

{hkl}

Pinacoid

Class 1 (C1)   Element: none

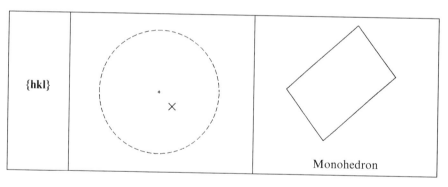

{hkl}

Monohedron

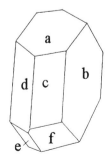

Crystal of albite NaAlSi$_3$O$_8$

Class $\bar{1}$

a: {001}   Pinacoids
b: {010}
c: {110}
d: {1$\bar{1}$0}
e: {1$\bar{1}$$\bar{1}$}
f: {11$\bar{1}$}

Monoclinic System

Class 2/m (C$_{2h}$)   Elements: $\dfrac{A_2}{M}$ C

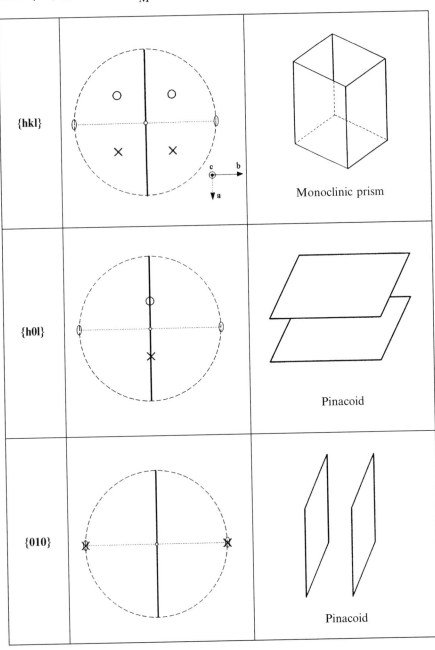

Monoclinic prism

Pinacoid

Pinacoid

Class 2 (C₂) Element: A₂

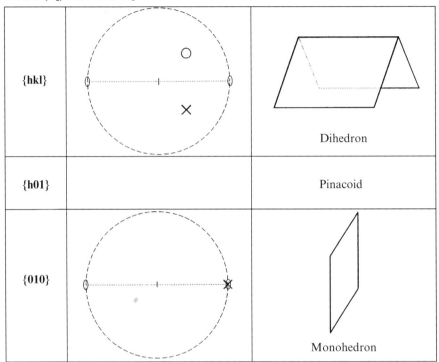

{hkl}		Dihedron
{h01}		Pinacoid
{010}		Monohedron

Class m (Cₛ) Element: M

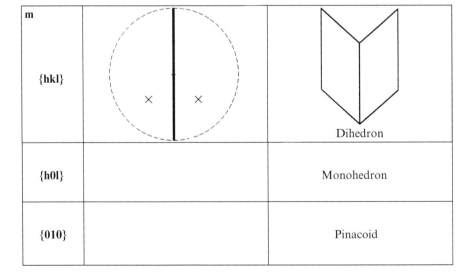

m {hkl}		Dihedron
{h0l}		Monohedron
{010}		Pinacoid

Orthorhombic System

Class mmm (D$_{2h}$)   Elements: $\frac{3A_2}{3M}$ C

{hkl}

Orthorhomic octahedron

{0kl}

{h0l}
{hk0}

Orthorhomic prisms

{100}

{010}
{001}

Pinacoids

Class mm2 (C$_{2v}$) Elements: A$_2$ M' M"

{hkl}		 Orthorhombic pyramid
{h0l} {0kl}		 Dihedron
{hk0}		 Orthorhombic prism
{100} {010}		Pinacoids
{001}		 Monohedron

Class 222 (D₂) Elements: 3A₂

{hkl}	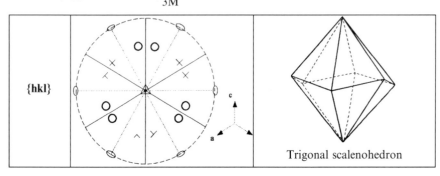	
		Orthorhombic tetrahedron
{0kl} {h0l} {hk0}		Orthorhombic prisms
{100} {010} {001}		Pinacoids

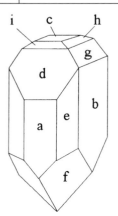

Crystal of calamine: Zn₄(OH)₂Si₂O₇,H₂O

Class mm2

a: {100} Pinacoid
b: {010} Pinacoid
c: {001} Monohedron
d: {301} Dihedron
e: {110} Prism
f: {12$\bar{1}$} Pyramid
g: {031} Dihedron
h: {011} Dihedron
i: {101} Dihedron

Trigonal system

Classes in the trigonal system are compatible with a hexagonal lattice.

Class $\bar{3}$ m (D₃d) Elements: $A_3 \dfrac{3A_2}{3M} C$

{hkl}		Trigonal scalenohedron

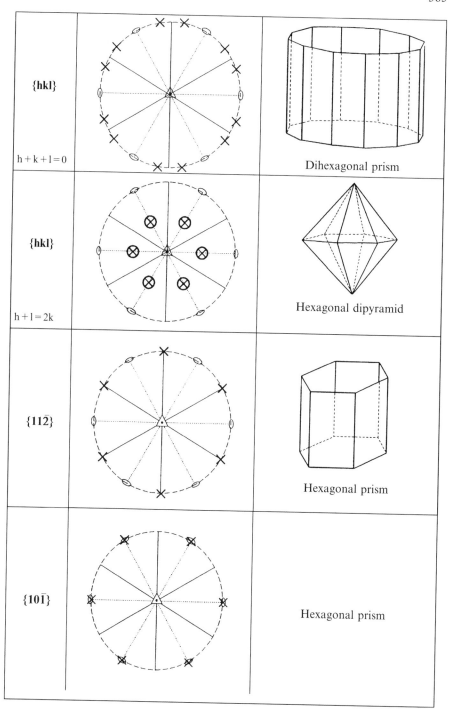

{hkl}

h + k + l = 0

Dihexagonal prism

{hkl}

h + l = 2k

Hexagonal dipyramid

{11$\bar{2}$}

Hexagonal prism

{10$\bar{1}$}

Hexagonal prism

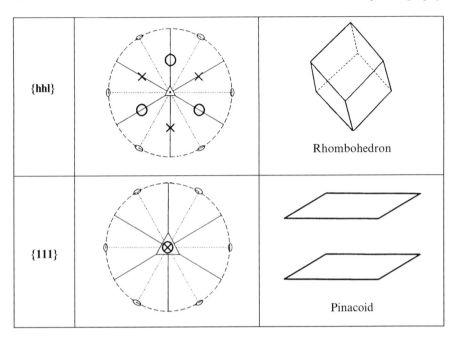

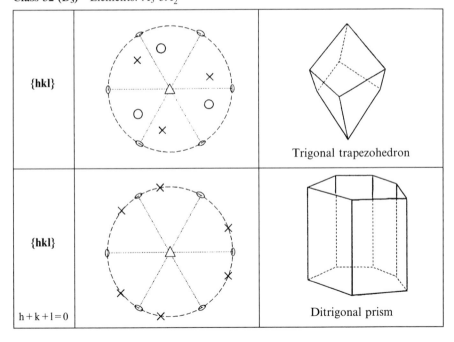

{hkl} h + l = 2k	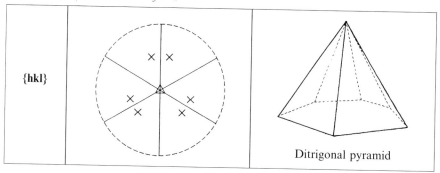	Trigonal dipyramid
{11$\bar{2}$}		Hexagonal prism
{10$\bar{1}$}		Trigonal prism
{hhl}		Rhombohedron
{111}		Pinacoid

Class 3 m (C$_{3v}$) Elements: A$_3$ 3M$'$

{hkl}		Ditrigonal pyramid

{hkl} h + k + l = 0		Ditrigonal prism
{hkl} h + l = 2k		Hexagonal pyramid
{112̄}		Trigonal prism
{101̄}		Hexagonal prism
{hhl}		Trigonal pyramid
{111}		Monohedron

Class $\bar{3}$ (S$_6$)   Element: A$_3$C

{hkl}	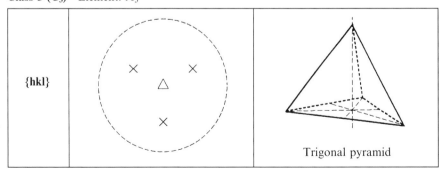	Rhombohedron
{hkl} $h+k+l=0$		Hexagonal prism
{hkl} $h+l=2k$		Rhombohedron
{11$\bar{2}$}		Hexagonal prism
{10$\bar{1}$}		Hexagonal prism
{hhl}		Rhombohedron
{111}		Pinacoid

Class 3 (C$_3$)   Element: A$_3$

{hkl}		Trigonal pyramid

{hkl} h + k + l = 0		Trigonal prism
{hkl} h + l = 2k		Trigonal pyramid
{11$\bar{2}$}		Trigonal prism
{10$\bar{1}$}		Trigonal prism
{hhl}		Trigonal pyramid
{111}		Monohedron

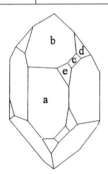

Crystal of quartz, right-hand.

Class 32 (hexagonal lattice)

	Hexag.	Trig.	Forms
a:	{10.0}	{11$\bar{2}$}	Hexagonal prism
b:	{10.1}	{100}	Rhombohedron
c:	{11.1}	{41$\bar{2}$}	Ditrigonal pyramid
d:	{01.1}	{22$\bar{1}$}	Rhombohedron
e:	{51.1}	{4$\bar{1}\bar{2}$}	Trapezohedron

Tetragonal system

Class 4/mmm (D$_{4h}$)   Elements: $\dfrac{A_4}{M}\dfrac{2A_2'}{2M'}\dfrac{2A_2''}{2M''}C$

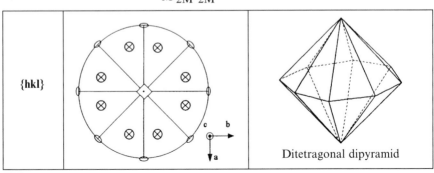

{hkl}		Ditetragonal dipyramid

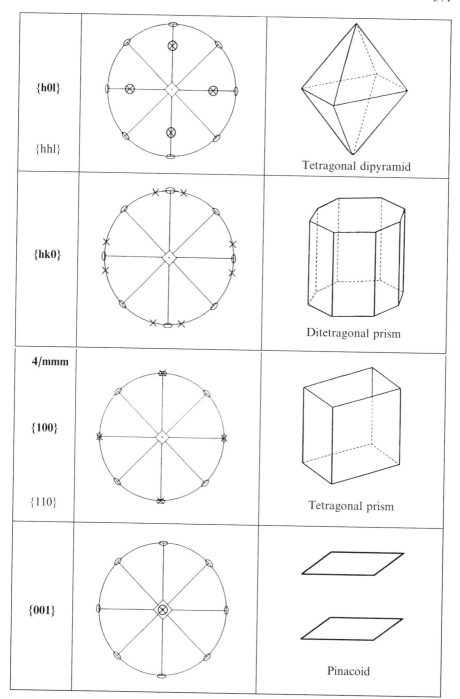

{h0l} {hhl}		Tetragonal dipyramid
{hk0}		Ditetragonal prism
4/mmm **{100}** {110}		Tetragonal prism
{001}		Pinacoid

Class 422 (D₄)   Elements: A_4 $2A_2'$ $2A_2''$

{hkl}		 Tetragonal trapezohedron
{h0l} {hhl}		Tetragonal dipyramid
{hk0}		Ditetragonal prism
{100} {110}		Tetragonal prism
{001}		Pinacoid

Class 4 mm (C₄ᵥ)   Elements: A_4 $2M'$ $2M''$

4 mm		
{hkl}		 Ditetragonal pyramid

{h0l}		Tetragonal pyramid
{hhl}		
{hk0}		Ditetragonal prism
{100}		Tetragonal prism
{110}		
{001}		Monohedron

Class 4/m (C$_{4h}$)   Elements: $\frac{A_4}{M}$ C

{hkl}		Tetragonal dipyramid
{h0l}		Tetragonal dipyramid
{hhl}		

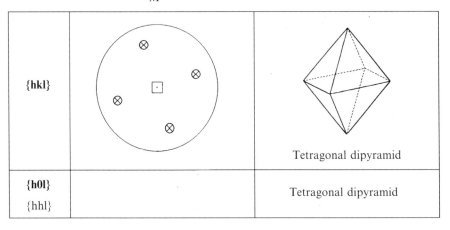

4/m {hk0}	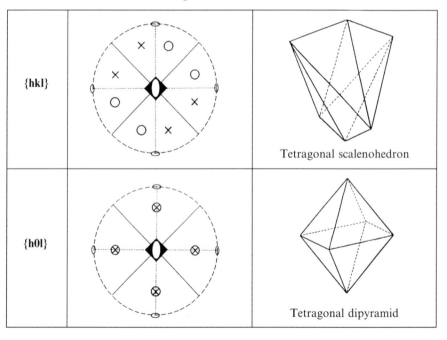	Tetragonal prism
{100} {110}		Tetragonal prism
{001}		Pinacoid

Class $\bar{4}2\,m$ (D$_{2d}$)   Elements: $\bar{A}_4$ $2A'_2$ $2M''$

{hkl}		Tetragonal scalenohedron
{h0l}		Tetragonal dipyramid

{hhl}		Tetragonal tetrahedron
{hk0}		Ditetragonal prism
{100} {110}		Tetragonal prism
{001}		Pinacoid

Class 4 (C₄) Element: A₄

{hkl}		Tetragonal pyramid
{h0l} {hhl}		Tetragonal pyramid
{hk0}		Tetragonal prism
{100} {110}		Tetragonal prism
{001}		Monohedron

Class $\bar{4}$ (S₄)   Element: $\bar{A}_4$

{hkl}		Tetragonal tetrahedron
{h0l} **{hhl}**		Tetragonal tetrahedron
{hk0}		Tetragonal prism
{100} **{110}**		Tetragonal prism
{001}		Pinacoid

Hexagonal System

For crystals with a Hexagonal lattice, the five classes of the trigonal system must be added to the Hexagonal system.

Class 6/mmm (D₆ₕ)   Elements: $\dfrac{A_6}{M} \dfrac{3A_2'}{3M'} \dfrac{3A_2''}{3M''} C$

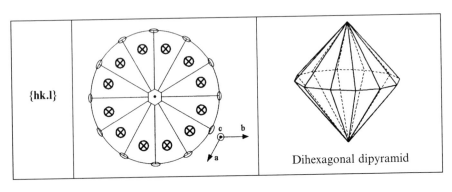

{hk.l}		Dihexagonal dipyramid

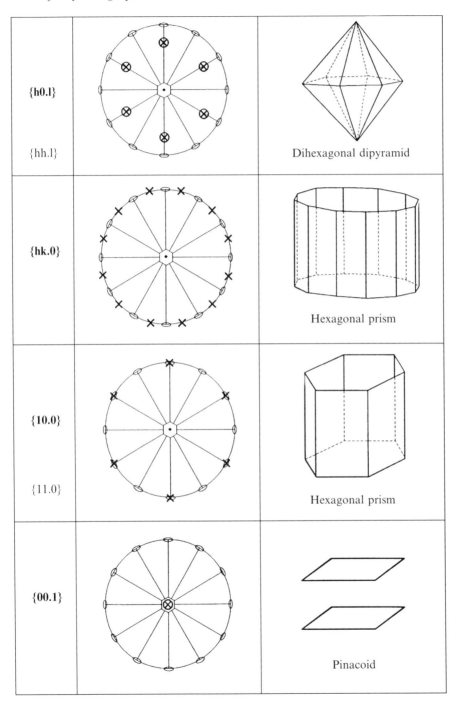

{h0.l} {hh.l}		Dihexagonal dipyramid
{hk.0}		Hexagonal prism
{10.0} {11.0}		Hexagonal prism
{00.1}		Pinacoid

Class 622 (D_6)   Elements: $A_6\ 3A_2'\ 3A_2''$

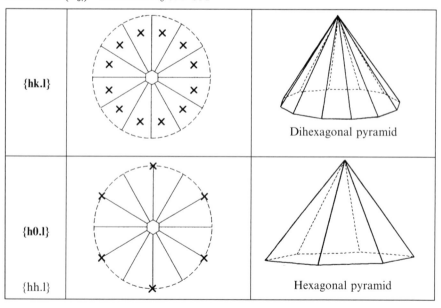

{hk.l}		Hexagonal trapezohedron
{h0.l} {hh.l}		Hexagonal dipyramid
{hk.0}		Dihexagonal prism
{10.0} {11.0}		Hexagonal prism
{00.1}		Pinacoid

Class 6 mm (C_{6v})   Elements: $A_6\ 3M'\ 3M''$

{hk.l}		Dihexagonal pyramid
{h0.l} {hh.l}		Hexagonal pyramid

{hk.0}		Dihexagonal prism
{10.0} {11.0}		Hexagonal prism
{00.1}		Monohedron

Class 6/m (C$_{6h}$)   Elements: $\dfrac{A_6}{M}$ C

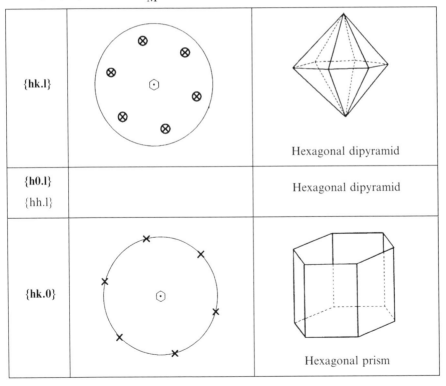

{hk.l}		Hexagonal dipyramid
{h0.l} {hh.l}		Hexagonal dipyramid
{hk.0}		Hexagonal prism

{10.0} {11.0}		Hexagonal prism
{00.1}		Pinacoid

Class $\bar{6}2\,m$ (D$_{3h}$)   Elements: $\frac{A_3}{M}\,3A_2'\,3M''$

{hk.l}		 Ditrigonal dipyramid
{h0.l}		 Hexagonal dipyramid
{hh.l}		 Trigonal dipyramid

{hk.0}		Ditrigonal prism
{10.0}		Hexagonal prism
{11.0}		Trigonal prism
{00.1}		Pinacoid

Class 6 (C₆) Element: A₆

{hk.l}		Hexagonal pyramid
{h0.l} {hh.l}		Hexagonal pyramid
{hk.0}		Hexagonal prism
{10.0} {11.0}		Hexagonal prism
{00.1}		Monohedron

Class $\bar{6}$ (C$_{3h}$)   Elements: $\dfrac{A_3}{M}$

{hk.l}		Trigonal dipyramid
{h0.l} {hh.l}		Trigonal dipyramid
{hk.0}		Trigonal prism
{10.0} {11.0}		Trigonal prism
{00.1}		Pinacoid

Cubic System

Class m3m (O$_h$)   Elements: $\dfrac{3A_4}{3M}$ $4A_3$ $\dfrac{6A_2'}{6M}$ C

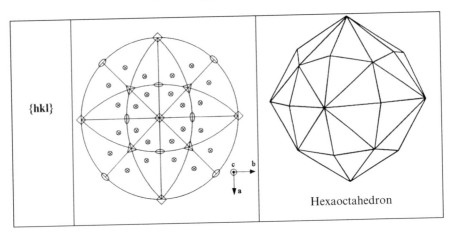

{hkl}		Hexaoctahedron

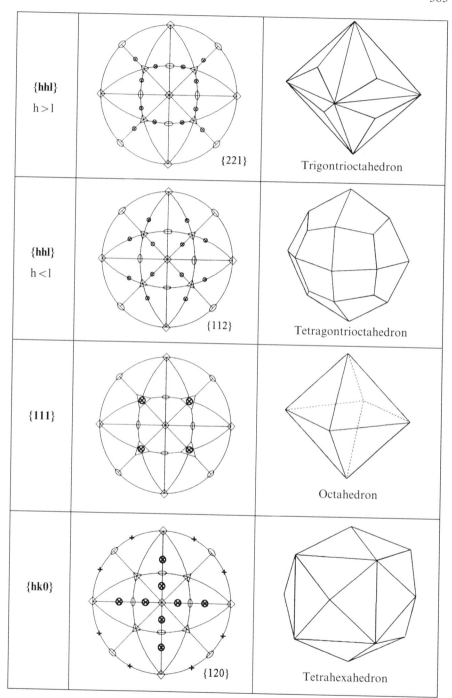

{hhl}
h > l

{221}

Trigontrioctahedron

{hhl}
h < l

{112}

Tetragontrioctahedron

{111}

Octahedron

{hk0}

{120}

Tetrahexahedron

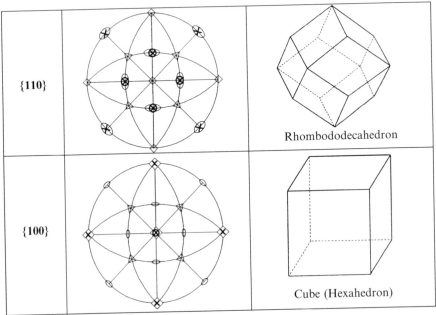

{110}		Rhombododecahedron
{100}		Cube (Hexahedron)

Class 432 (O)   Elements: $3A_4$ $4A_3$ $6A'_2$

{hkl}	{123}	Pentagontrioctahedron (righthand)
{hhl} h < l		Trigonotrioctahedron
{hhl} h > l		**Tetragontrioctahedron**
{111}		Octahedron
{hk0}		Tetrahexahedron
{110}		Rhombododecahedron
{100}		Cube

Class $\overline{4}$3m (T$_d$)   Elements: $3\overline{A}_4$ $4A_3$ $6M'$

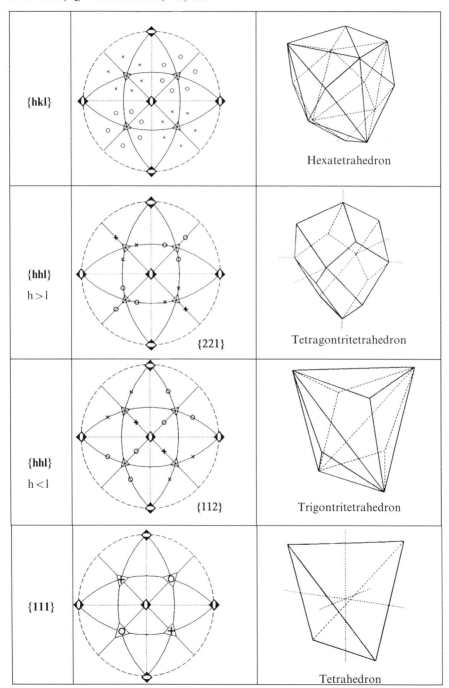

{hkl}		Hexatetrahedron
{hhl} h > 1	{221}	Tetragontritetrahedron
{hhl} h < 1	{112}	Trigontritetrahedron
{111}		Tetrahedron

{hk0}		Tetrahexahedron
{110}		Rhombododecahedron
{100}		Cube

Class m3 (T$_h$)   Elements: $\frac{3A_2}{3M}$ 4A$_3$ C

{hkl}		Didodecahedron
{hhl} h > l		Trigontrioctahedron
{hhl} h < l		Tetragontrioctahedron
{111}		Octahedron
{hk0}	{120}	Pentagondodecahedron

{110}		Rhombododecahedron
{100}		Cube

Class 23 (T)   Elements: $3A_2$ $4A_3$

{hkl}		
		Pentagontritetrahedron (lefthand)
{hhl} h>l		Tetragontritetrahedron
{hhl} h<l		Trigontritetrahedron
{111}		Tetrahedron
{hk0}		Pentagondodecahedron
{110}		Rhombododecahedron
{100}		Cube

Non-crystallographic Point Groups

These are listed by the same method as for crystallographic point groups, but free of lattice-related constraints.

Cyclic groups	n	$5,7,8,9,10\ldots,\infty$
Dihedral groups	n2	$52, 72, 822, 92, 10\,2\,2\ldots,\infty2$
Improper groups	$\bar{n}$	$\bar{5}, \bar{7}, \bar{8}, \bar{9}, \overline{10}\ldots$
	n/m	$8/m, 10/m\ldots,\infty/m$ (n even)
	nm	$5m, 7m, 8mm, 9m, 10mm\ldots,\infty m$
	$\bar{n}2$	$\bar{5}2, \bar{7}2, \bar{8}2m, \bar{9}2, \overline{10}m2\ldots$
	$\frac{n}{m}m$	$\frac{8}{m}mm, \frac{10}{m}mm\ldots, \frac{\infty}{m}m = \frac{\infty}{m}\frac{2}{m}$ (n even)
Icosahedral groups		$532, \bar{5}3\frac{2}{m}$ (several principal axes)

Continuous groups have one isotropy axis (∞). There also exist spherical groups (with several isotropy axes) $\infty\,\infty$ and $\infty/m\,\infty/m$. The symmetry of objects in the group $\infty\,\infty$ is that of a sphere filled with an optically active liquid; objects in the group $\infty/m\,\infty/m$ have the symmetry of a sphere.

The five continuous groups with an isotropy axis may be represented by the following objects (which can be used in applying the Curie laws):

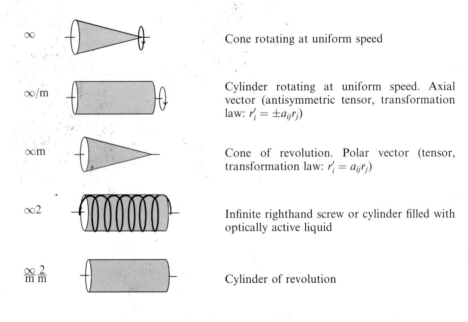

∞ Cone rotating at uniform speed

∞/m Cylinder rotating at uniform speed. Axial vector (antisymmetric tensor, transformation law: $r'_i = \pm a_{ij}r_j$)

∞m Cone of revolution. Polar vector (tensor, transformation law: $r'_i = a_{ij}r_j$)

$\infty2$ Infinite righthand screw or cylinder filled with optically active liquid

$\frac{\infty}{m}\frac{2}{m}$ Cylinder of revolution

Appendix B

Gnomonic and Orthographic Projections

1 GNOMONIC PROJECTIONS

Because it conserves angles, the stereographic projection is the most widely used method of representing crystals; however, a highly developed visual imagination is certainly required. To overcome this drawback, other methods of representation have been developed. One of these is the gnomonic projection (Laue's method); although angles are not conserved, it is nevertheless possible to deduce from this orthographic projections which are fairly faithful representations of the crystal.

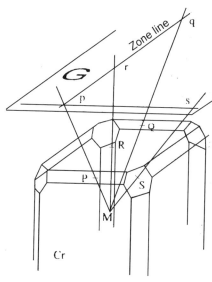

Figure 24.1

We consider a crystal Cr, a point M inside the crystal, the bundle of normals from M: MP, MQ, MR, MS..., and the plane of projection G.

The traces of these normals in G are the poles of the of the faces and the set of these poles constitutes the gnomonic projection of the crystal.

Let P, Q, R be planes in zone. Their normals will be contained in a single plane which is the zone plane. (This plane is normal to the edge common to the zone planes). The poles p, q, r of the faces of the zone lie on the straight line which is the intersection of the zone plane with the plane of the projection. This line is the zone line. The direction of the edges of the faces in zone is the zone axis.

❒ PRINCIPLE

In a gnomonic projection, the faces in zone have their poles aligned on a zone line.

⧄ CHARACTERISING A FACE

The angles φ and ρ are measured with a two-circle goniometer.

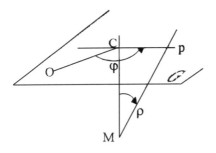

Figure 24.2

In a gnomonic projection, the elements which characterise a face are:

♦ The plane of projection G,
♦ the origin M of the normals to the faces,
♦ the normal MC to the plane of projection, from M.

We let MC = D. The normal to the face P, which penetrates the plane of projection at p, will be defined by the angle $\{CMP\} = \rho$ (inclination) and $\{Ocp =\}\varphi$ (azimuth).

CO, an arbitrary direction in the plane G, is chosen as the origin of the azimuths. The poles of faces which have an inclination of 90° lie at infinity.

⧄ CONSTRUCTING THE POLE OF A FACE

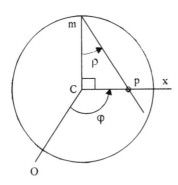

Figure 24.3

In the plane G, draw a circle of radius CD = MC = d; this is called the *gnomonic circle*.
Take the line of origin CO; the pole of face P lies on the line Cx which makes an angle φ with Ox.
To obtain the position of p on Cx, the plane CPM is folded onto plane G , taking Cx as hinge line. The angle of inclination $\{CMP\} = \rho$ applied to $\{Cmp\}$ then gives the position of p.

⧄ GEOMETRICAL CONSTRUCTIONS

A few geometrical constructions can be made from a gnomonic projection. Here we shall describe the construction of the *alignment point*, and of the edge between two faces.

Angles between two faces

Let p_1 and p_2 be the poles of the faces and LZ their zone line.

What we actually determine is the angle $\{P_1MP_2\} = \{p_1Mp_2\}$ between the normals to the faces. The plane normal to the zone line passing through M intersects this line at H and also contains the centre C of the gnomonic circle. By folding the zone plane about LZ onto the plane G, $\{p1Mp2\}$ can be measured. We proceed in two steps:

— In the plane G draw the normal HC to the zone line LZ, passing through the centre of the circle, and then fold the plane HMC about HC up onto the plane G (M arrives at m and HM = Hm).

—In G, the normal to HC at C intersects the gnomonic circle at the point m; from this deduce the point A such that Hm = HA.

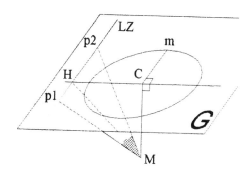

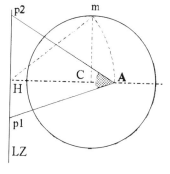

Figure 24.4a

Figure 24.4b

The required angle is $\{p_1Ap_2\}$. The point A is the pole under which all the poles of LZ are seen at their real angles, and is called the *alignment point* of the zone line.

The edge between two faces

The edge between two faces P_1 and P_2 is parallel to their zone axis: it is therefore normal to the zone plane P_1MP_2 and to HM.

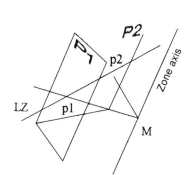

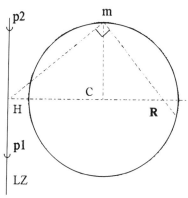

Figure 24.5a

Figure 24.5b

We represent the edge by its pole R. To construct this pole, rotate about HC; since MR is normal to MH, R lies on the intersection of HC and the normal to Hm.

2 FRONTAL PROJECTION

This is a projection onto a plane parallel to the gnomonic projection plane. The edges which define the crystal faces are perpendicular to the zone lines of intersecting planes and the gnomonic projection enables the direction of these edges to be determined—but the lengths of the edges depend on the conditions under which crystal growth took place. In the frontal projection, the faces can be more or less developed so that an accurate representation of the crystal is obtained.

Our example is zinc bromate, $Zn(BrO_3)2.6H_2O$ (cubic) which, when grown in aqueous conditions, exhibits the forms $\{100\}$ and $\{111\}$.

❑ CONSTRUCTING THE GNOMOGRAM

This construction is carried out from measurements of the interfacial angles.

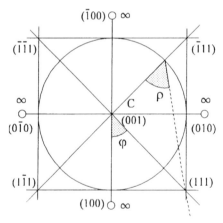

Figure 24.6

The faces (100) and (010) have their poles at infinity. ($\rho = 90°$). The face (111) has angles $\varphi = 45°$ and $\rho = 45°44'$.

The poles of the $\{111\}$ form faces lie at the corners of the square circumscribing the gnomonic circle, since:

$$\tan\rho = R\sqrt{2}/R = \sqrt{2} = \tan54°44'.$$

The various zone lines are drawn on the gnomogram, but poles with an angle p > 90° are omitted so as not to encumber the diagram.

❑ CONSTRUCTING THE FRONTAL PROJECTION

To do this it is only necessary to construct the normals to all those zone lines which correspond to an edge.

In figure 24.7, the construction lines are shown for the edge between the faces (001) and ($1\bar{1}1$); similarly, the edge between the faces (111) and ($1\bar{1}1$) and the corresponding zone line are shown in bold lines.

In this example, the relative developments of faces having the same form have been made identical.

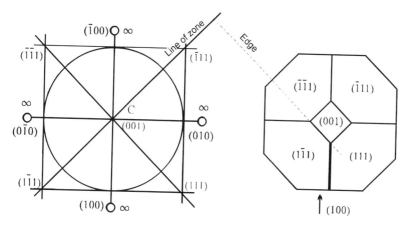

Figure 24.7

3 ORTHOGONAL PROJECTIONS

◻ CONSTRUCTING THE GUIDE LINE

To draw the crystal, a normal projection is made onto a plane which is oblique to the plane of the drawing. To visualise correctly all the faces, the plane on which the projection is to be made must be chosen to be neither parallel not perpendicular to any of the crystal faces.

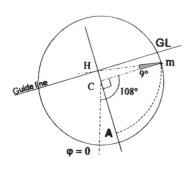

Figure 24.8

A projection plane which is often used is that passing through M whose trace GL on the gnomonic plane makes an angle $\varphi = 108°$ and an angle $\rho = 9°$. Figure 24.8 shows how to construct the *guide line* GL and its alignment point A. (CA since Hm = HA).

◻ CONSTRUCTING THE LINES OF PROJECTION

The normals to the frontal plane are normal to the guide line GL of this plane. In the oblique projection, the corners of the crystal therefore lie on the perpendiculars to GL from the corresponding corners of the frontal projection. These perpendiculars are the *lines of projection*. Note that edges normal to the frontal plane coincide with the projection lines.

◻ CONSTRUCTING THE EDGES

In this construction we seek the edge between two faces P_1 and P_2 whose poles p_1 and p_2 are on the zone line LZ. This edge is normal to the zone plane p_1Mp_2 and hence to all lines in this plane. If B is the intersection of the zone line LZ and the guide line GL, the edge will be normal to MB.

When the oblique projection plane is swung about GL onto the plane of the drawing, the point M arrives at A (the alignment point of GL) and the point B remains stationary since it lies on the hinge GL.

In the plane of the figure, the edge being sought is thus normal to the line AB. To construct, for example, the edge between (111) and ($\bar{1}$11) we determine the point B_0 and draw the straight line AB_0. In the frontal projection the edge will be bounded by F and G; in the oblique projection the edge will be normal to AB and bounded by the lines of projection from F and from G. Similarly, the edge EF between the faces (001) and (111) is normal to AB_1, and the edge GK between the faces (010) and ($\bar{1}$11) is normal to AB_2.

To draw the orthographic projection, we begin by drawing consecutively all the edges of one face, followed by the adjacent faces. If the figure obtained is not satisfactory, the relative development of the faces can be modified in the frontal projection, or the angles of the guide line can be modified

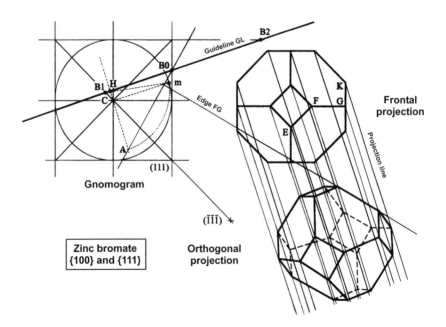

Figure 24.9

Appendix C

The Seventeen Plane Groups

Apart from their academic interest, plane groups find application in surface physics. The groups relate to the regular tiling of a plane surface, and the Dutch engraver Maurits Cornelius Escher[1] made numerous illustrations involving such tiling. An excellent exercise is to take his engravings and look for the groups used in them and the symmetry elements.

1 AXES OF ROTATION AND PLANE LATTICES

The lattice must remain unchanged during tiling operations, and the only direct axes possible are therefore 1, 2, 3, 4 and 6. Only axes normal to the plane are taken into account in two-dimensional lattices. For plane symmetries with one twofold axis, (centre of symmetry), the inversion operation is identical. In plane geometry, we must also consider symmetry about a straight line. This line may be considered as the trace of a mirror plane normal to the plane; this 'mirror-line' is denoted m.

Only **four systems** are possible, given the restrictions on symmetry operations compatible with the periodicity of a plane lattice.

◻ THE HEXAGONAL SYSTEM

The cell is such that: $a = b$ and $\gamma = 2\pi/3$. It is compatible with sixfold or threefold axes.

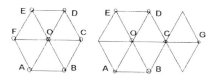

Figure 25.1

[1]M.C.ESCHER, The Graphic Work, Oldbourne, London (1960).

A sixfold axis at O will generate nodes B, C, D, E and F starting from node A, while a threefold axis at O will only generate nodes C and E. But if the cell is part of a lattice, there also exists a threefold axis at C which will generate nodes B and D starting from node G; the final result is thus identical.

❐ THE SQUARE SYSTEM

The base vectors are: $a = b$ and $\gamma = \pi/2$. The cell is compatible with fourfold axes.

❐ THE OBLIQUE SYSTEM

The base vectors are general: $a \neq b$ and $\gamma \neq \pi/2$. The cell is compatible with twofold axes.

❐ THE RECTANGULAR SYSTEM

The cell has $a \neq b$ and $\gamma = \pi/2$.

For the lattice to remain unchanged, any mirror-line must be parallel to one of the base vectors and normal to the other; the lattice is thus rectangular.

2 THE BRAVAIS LATTICES

Non-integral lattice translations which conserve lattice symmetry are sought. For the hexagonal system, translations of the type: $\mathbf{t1} = 1/3\mathbf{a} + 2/3\mathbf{b}$ and $\mathbf{t2} = 2/3\mathbf{a} + 1/3\mathbf{b}$ are possible, but there exists a simple cell with the same symmetry. In all the lattices, the translations $\mathbf{a}/2$ and $\mathbf{b}/2$ must be excluded either because they do not conserve symmetry (hexagonal and square systems) or because they enable smaller cells to be defined (oblique and rectangular systems).

The translation $\frac{1}{2}(\mathbf{a} + \mathbf{b})$ does not conserve hexagonal symmetry and leads to a smaller cell in the square and oblique lattices, but it does conserve symmetry in a rectangular cell which will then be centred.

Thus there exist five plane Bravais lattices:

primitive oblique **p**,

primitive rectangular **p** and centred rectangular **c**,

primitive square **p**,

primitive hexagonal **p**.

To distinguish plane from three-dimensional lattices, Bravais lattices are denoted by lower-case letters.

3 PLANE POINT GROUPS

The combination of plane symmetry operators (axes and mirror-lines) leads to the ten plane groups (with notations identical to those of the point groups). In table 1, the four holohedral classes (which have the symmetry of the lattice) are in bold letters and the six Laue classes in italics.

Table 25.1. The ten plane classes

Oblique	1	*2*		
Rectangular	m	**2mm**		
Square	4	**4mm**		
Hexagonal	3	3m	6	**6mm**

4 PLANE GROUPS

The product of point symmetry operations and the group of translations is a single operation of new symmetry, the mirror glide line, denoted g, and shown by dashed lines.

Combining all the possible operations leads to seventeen plane groups sharing the same construction principals as for space groups. The notation used is also identical.

Table 25.2. The seventeen plane groups

Classes	Groups				
1	p1				
2	p2				
m	pm	pg	cm		
2mm	p2mm	p2mg	p2gg	c2mm	
4	p4				
4mm	p4mm	p4gm			
3	p3				
3m	p3m1	p31m			
6	p6				
6mm	p6mm				

The seventeen plane groups are illustrated below. The whole set of equivalent objects within the cell is shown wherever clarity is not thus compromised.

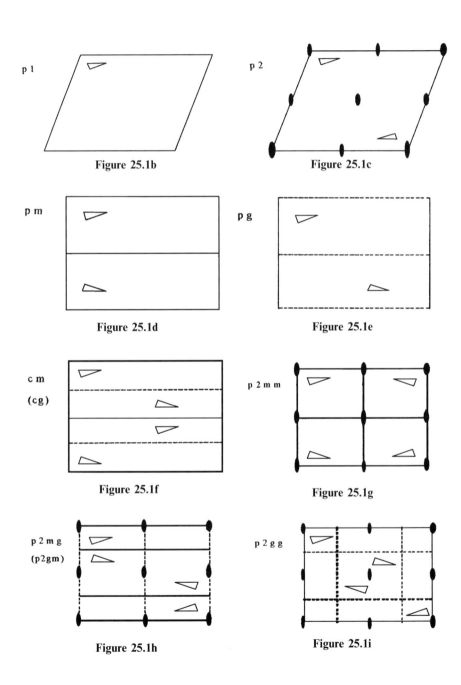

Figure 25.1b

Figure 25.1c

Figure 25.1d

Figure 25.1e

Figure 25.1f

Figure 25.1g

Figure 25.1h

Figure 25.1i

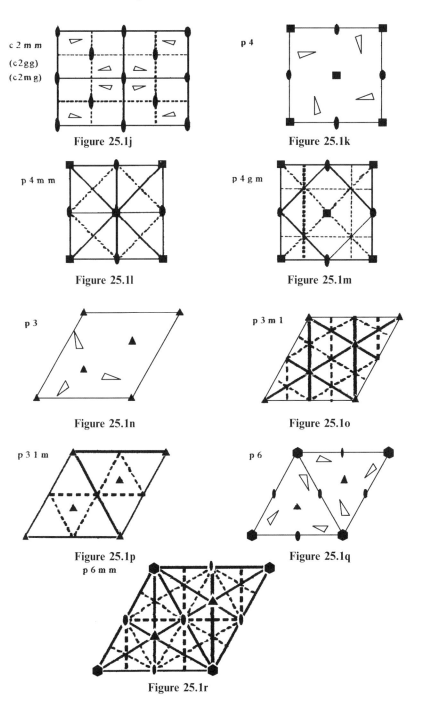

Figure 25.1j

Figure 25.1k

Figure 25.1l

Figure 25.1m

Figure 25.1n

Figure 25.1o

Figure 25.1p

Figure 25.1q

Figure 25.1r

Appendix D
The 230 Space Groups

Classes	Space groups						
Triclinic 1 $\bar{1}$	2 groups (1 to 2): $P1$ $P\bar{1}$						
Monoclinic 2 m 2/m	13 groups (3 to 15): $P2$ Pm $P2/m$	$P2_1$ Pc $P2_1/m$	$C2$ Cm $C2/m$	 Cc $P2/c$	 $P2_1/c$	 $C2/c$	
Orthorhombic 222 mm2 mmm	59 groups (16 to 74): $P222$ 1222 $Pmm2$ $Pba2$ $Abm2$ $Ima2$ $Pmmm$ $Pcca$ $Pbca$ $Ccca$	 $P222_1$ $12_12_12_1$ $Pmc2_1$ $Pna2_1$ $Ama2$ $Pnnn$ $Pbam$ $Pnma$ $Fmmm$	 $P2_12_12$ $Pcc2$ $Pnn2$ $Aba2$ $Pccm$ $Pccn$ $Cmcm$ $Fddd$	 $P2_12_12_1$ $Pma2$ $Cmm2$ $Fmm2$ $Pban$ $Pbcm$ $Cmca$ $Immm$	 $C222_1$ $Pca2_1$ $Cmc2_1$ $Fdd2$ $Pmma$ $Pnnm$ $Cmmm$ $Ibam$	 $C222$ $Pnc2$ $Ccc2$ $Imm2$ $Pnna$ $Pmmm$ $Cccm$ $Ibca$	 $F222$ $Pmn2_1$ $Amm2$ $Iba2$ $Pmna$ $Pbcn$ $Cmma$ $Imma$
Tetragonal 4 $\bar{4}$ 4/m 422 4mm $\bar{4}$2m	68 groups (75 to 142): $P4$ $P\bar{4}$ $P4/m$ $P422$ $P4_32_12$ $P4mm$ $P4_2bc$ $P\bar{4}2m$ $P\bar{4}n2$	$P4_1$ $I\bar{4}$ $P4_2/m$ $P42_12$ $I422$ $P4bm$ $I4mm$ $P\bar{4}2c$ $I\bar{4}m2$	$P4_2$ $P4/n$ $P4_122$ $I4_122$ $P4_2cm$ $I4cm$ $P\bar{4}2_1m$ $I\bar{4}2c$	$P4_3$ $P4_2/n$ $P4_12_12$ $P4_2nm$ $I4_1md$ $P\bar{4}2_1c$ $I\bar{4}2m$	$I4$ $I4/m$ $P4_222$ $P4cc$ $I4_1cd$ $P\bar{4}m2$ $I\bar{4}2d$	$I4_1$ $I4_1/a$ $P4_22_12$ $P4nc$ $P\bar{4}c2$	 $P4_322$ $P4_2mc$ $P\bar{4}b2$

(*continued*)

(*continued*)	
4/mmm	P4/mmm P4/mcc P4/nbm P4/nnc P4/mbm P4/mnc P4/nmm P4/ncc $P4_2$mmc $P4_2$/mcm $P4_2$/nbc $P4_2$/nnm $I4_1$/amd $I4_1$/acd $P4_2$/nmc $P4_2$/mbc $P4_2$/mnm I4/mmm I4/mcm $P4_2$/ncm

Trigonal   25 groups (143 to 167):

3	P3	$P3_1$	$P3_2$	R3		
$\bar{3}$	$P\bar{3}$	$R\bar{3}$				
32	P312	P321	$P3_112$	$P3_121$	$P3_212$	$P3_221$	R32
3m	P3m1	P31m	P3c1	P31c	R3m	R3c
$\bar{3}$m	$P\bar{3}1$m	$P\bar{3}1$c	$P\bar{3}$m1	$P\bar{3}$c1	$R\bar{3}$m	$R\bar{3}$c

Hexagonal   27 groups (168 to 194):

6	P6	$P6_1$	$P6_5$	$P6_2$	$P6_4$	$P6_3$
$\bar{6}$	$P\bar{6}$					
6/m	P6/m	$P6_3$/m				
622	P622	$P6_122$	$P6_522$	$P6_222$	$P6_422$	$P6_322$
6mm	P6mm	P6cc	$P6_3$cm	$P6_3$mc		
$\bar{6}$2m	$P\bar{6}$m2	$P\bar{6}$c2	$P\bar{6}$2m	$P\bar{6}$2c		
6/mmm	P6/mmm		P6/mcc	$P6_3$/mcm		$P6_3$/mmc

Cubic   36 groups (195 to 230):

23	P23	F23	I23	$P2_13$	$I2_13$	
$m\bar{3}$	$Pm\bar{3}$	$Pn\bar{3}$	$Fm\bar{3}$	$Fd\bar{3}$	$Im\bar{3}$	$Pa\bar{3}$	$Ia\bar{3}$
432	P432	$P4_232$	F432	$F4_132$	I432	$P4_332$	$P4_132$
	$I4_132$					
$\bar{4}$3m	$P\bar{4}3$m	$F\bar{4}3$m	$I\bar{4}3$m	$P\bar{4}3$n	$F\bar{4}3$c	$I\bar{4}3$d
$m\bar{3}$m	$Pm\bar{3}$m	$Pn\bar{3}$n	$Pm\bar{3}$n	$Pn\bar{3}$m	$Fm\bar{3}$m	$Fm\bar{3}$c	$Fd\bar{3}$m
	$Fd\bar{3}$c	$Im\bar{3}$m	$Ia\bar{3}$d				

Groups in this table are classified into 'standard groups' as in the International Tables (i.e. base vectors are chosen in accordance with conventions). If alternative base vectors are chosen, this normally involves modifying the name of the group. In order to follow the notation in the International Tables, the cubic classes m3 and m3m are denoted $m\bar{3}$ and $m\bar{3}$m. Note also the nomenclature of trigonal groups which have a hexagonal cell.

Making three-dimensional models
Tetrahedron, octahedron, pentagonal dodecahedron and rhombohedron

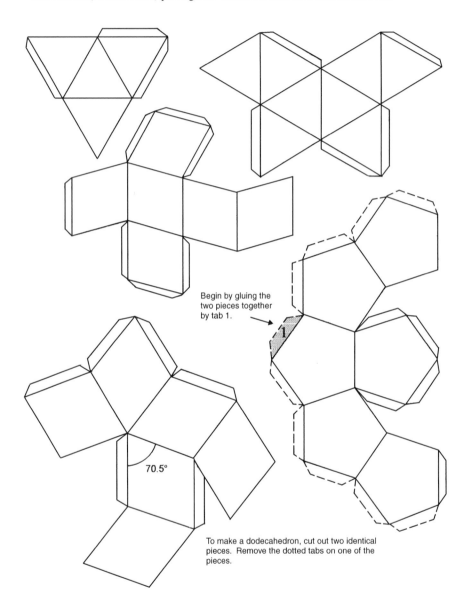

Begin by gluing the
two pieces together
by tab 1.

1

70.5°

To make a dodecahedron, cut out two identical
pieces. Remove the dotted tabs on one of the
pieces.

Figure 26.1. Copy figures (after enlarging if desired) onto thin card. Score the lines with
a ball pen to make folding easier. Cut out, bend and assemble using universal glue. The
cube is made in a similar way to the rhombohedron.

WULFF NET

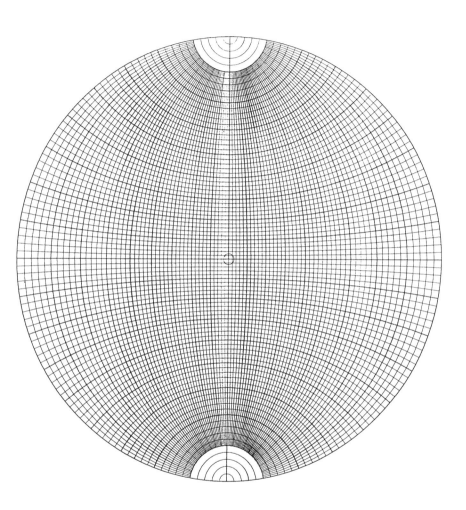

Strengthen the centres of the net and the transparent paper with adhesive tape. On the transparent paper, draw a circle with the same radius as the net. Use a pin as common axis of rotation for paper and net.

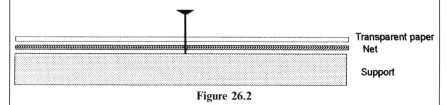

Transparent paper

Net

Support

Figure 26.2

Spherical trigonometry relations

Angles of the triangle: $\hat{A}, \hat{B}, \hat{C}.$ Sides of the triangle: a, b, c

$$\cos a = \cos b. \cos c + \sin b. \sin c. \cos \hat{A}$$

$$\cos \hat{A} = -\cos .\hat{B}. \cos \hat{C} + \sin \hat{C}/ \cos a$$

$$\frac{\sin a}{\sin \hat{A}} = \frac{\sin b}{\sin \hat{B}} = \frac{\sin c}{\sin \hat{C}}$$

Reciprocal lattice relations

$$A^* = \frac{b \wedge c}{V}, \; B^* = \frac{c \wedge a}{V}, \; C^* = \frac{a \wedge b}{V}$$

$$\cos \alpha^* = \frac{\cos \beta. \cos \gamma - \cos \alpha}{\sin \beta. \sin \gamma} \Leftrightarrow \cos \alpha = \frac{\cos \beta^*. \cos \gamma^* - \cos \alpha^*}{\sin \beta^*. \sin \gamma^*}$$

$$\|A^*\| = \frac{1}{a. \sin \beta^*. \sin \gamma} = \frac{1}{a. \sin \beta. \sin \gamma^*}$$

$$N^*_{hkl} \perp (hkl) \qquad \|N^*_{hkl}\|.d_{hkl} = 1$$

The axis $[uvw]$ of a zone containing the plan (hkl) is such that:

$$h.u + k.v + l.w = 0$$

Appendix E
Using the Programs

The programs on the disk which accompanies the book are designed to help you through the more difficult parts of the course. All the programs have graphic displays, many of which were used in the numerous illustrations. This appendix consists of a brief description of what you can do using each of the programs. I have designed all the programs to run under Windows, now that this has become a standard. When I first started writing the programs the only available user-friendly interface was Microsoft's 'VisualBasic', which explains my choice of programming language. Many PCs are now equipped with Windows 95, but to ensure compatibility with Windows 3.1x, the programs are all in 16-bit versions.

Thus the applications all follow Windows standards and conventions.

—In certain menus, you have to supply information into a dialogue box.
—All these dialogue boxes are 'modal', i.e. you have to close them before the program can continue. If a value you enter would lead to a 'fatal' error the program waits for you to enter an allowed value before continuing. If the error would not be 'fatal', the erroneous value is simply replaced by zero.
—You must fill in all the data fields before closing the dialogue box. You can pass from one field to another by using the tabulation [Tab] key. You can go back to the previous field by pressing [Shift][Tab].
—Some of the programs have an external help file. This file has to be in the same folder as the application. [F1] gives the Help summary page. When a menu has been opened, [F1] gives direct access to the relevant page in the Help file. These help files contain both information on using the application and theoretical notes on the topic involved.
—Several applications can be used at once; to switch from one to another, use the Windows task bar (Windows 95) or [Alt][Tab] (Windows 3.1x).

Minimum system requirements:
A 386 processor with 8 Mo of RAM. For reasonable response times I recommend as a minimum configuration a Pentium 100 or 486DX4-100. The programs have been tested with VGA and SVGA graphic cards. To ensure compatibility, I have only used 16 colours. You must of course have a mouse. The graphics have been successfully tested on practically all the Hewlett-Packard 'Laserjet' and 'Deskjet' printers.

As well as printing, you can create HP-GL format files which are recognised by most popular word processor and graphics applications.

FORMS

This program enables you to visualise convex solids and crystals. To display an object, use the menu 'File/Load' to select the folder and the name of the file in the list (double click on the name); the files have the extension '.FOR'. The commands available are displayed in a window. [+] and [−] keys are those on the number pad. The difference of effect between [+] and [PgUp] is only apparent when you are in perspective mode. You can rotate the object, change its size and display it in perspective or isometric projection. In transparent mode the hidden faces are displayed dotted.

You can create your own objects. You are advised to print a list (Menu 'Print /Print data') of the data relating to a few of the objects supplied (starting with simple examples such as the cube and the octahedron) so as to understand the system of representation. Next, make a rough sketch of your object, and determine the Miller indices of its faces. By using fractional indices you can give the faces different developments. Finally, you use the Gp program (with the correct parameters) to calculate the coordinates of the vertices. You can then create your data file (Menu 'File/Create a file'). It is essential to enter the indices of the face vertices in trigonometrical sense for the test of face visibility to function correctly. Several examples of actual crystals are included together with all the crystalline forms in the atlas.

GP: Point Groups

See chapters 2, 3 and 6.

1—Lattice calculations:
First, you must select a group of the system to which the compound you are studying belongs (e.g. mmm for an orthorhombic compound)), then enter the lattice parameters (Menu: 'Options/Choice of parameters'). You can then use the various options in the 'Calculations' menu: calculate the modulus of a reciprocal or direct row, the angle between rows, the coordinates of a vertex, the indices of an edge between two planes (Zones option) and so on.

2—Stereographic projection:
With this program you can display the stereographic projections of the 32 symmetry point groups together with projections of the poles (representing the normals to the crystal faces). You can work either with angles or Miller indices. The program automatically generates all the equivalent directions in terms of the symmetry operations of the group. Simultaneous display of several forms is possible, together with various operations such as zone tracing and rotation of the diagram.

With the Menu 'Net' you can print out a Wulff or polar net; these nets enable certain functions in the program to be carried out manually.

SYM: Symmetry Elements

See chapters 4, 6 and 7.

This program has been designed to enable you to display the elementary symmetry operations and to show the essential difference between the internal symmetry of an object and the symmetry elements which can be applied to that object.

Using the menu 'Selection/Symmetry element', choose one of the symmetry elements from the list. You can adjust the viewing angle, the size of the objects and the mode of projection using keyboard commands. The latter are identical to those in the FORMS program. With the command 'Selection/Helix' you can display screw axes. The menu 'Objects' allows you to choose different objects with internal symmetries 1, m, mm2 or mmm.

LAUE: Simulation of Laue Diagrams

See chapters 10 and 11.

Initially, this program was written as an aid to crystal orientation by the Laue method. A photograph is first produced and, in the program, the orientation of the crystal is modified so as to make the calculated pattern and the pattern actually obtained coincide. In this way the difference between the actual position and the required position can be determined. It then only remains to adjust the crystal orientation accordingly to obtain almost perfect adjustment.

In the 'File' menu, choose the option 'Input data' and enter the lattice parameters (a, b, c, α, β, γ). Modify the minimum and maximum wavelengths if necessary (the minimum wavelength depends on the X-ray tube supply voltage). Choose the film-specimen distance. Select the systematic extinctions for the space group of the compound (consult the 'International Tables' if you are not sure).

In the menu 'Drawing/Rotation', select the mode of determination of the crystal orientation; the default setting is 'Row' it is also possible to work with angles. In the same menu select either the transmission ('Direct spectrum') or reflection ('Back spectrum') method. The theoretical diagram is then calculated and displayed. If you select 'Gnomonic' the gnomonic projection is calculated instead of the spectrum. You can also determine all the spots which are in zone with a given row.

The default values in the dialogue boxes will display the diffraction patterns of cubic structures. Values of the 'Upper indices' in the dialogue boxes are calculated values; these can be decreased. The program has a Help file with a few notes on the theory, and in particular a page describing orientation of the crystal in the laboratory frame.

BRAGG: SIMULATION OF ROTATING CRYSTAL PATTERNS

See chapters 10, 12 and 13.

This program simulates the common methods of monochromatic X-ray diffraction. Only spot *positions* are calculated, not *intensities* which would require the use of much more sophisticated programs.

In the menu 'File' select the option 'Input data' The dialogue box is practically the same as in the Laue program. Enter the lattice parameters (a, b, c, α, β, γ). Choose the X-ray wavelength; a list of the common anodes is given, but you can set a different value if you wish. Select the systematic extinctions for the space group of the compound by clicking in the list displayed. (Again, consult the 'International Tables' if you are not sure).

In the menu 'Choice', select the technique used. For single crystal methods, choose the rotation row. For the Weissenberg method, you must also specify the reciprocal row chosen as origin for the rotation of the crystal. For the three methods using a single crystal, print-outs are given in formats of the most common cameras (12×12 cm for Buerger photographs, 180 mm circumference Bragg and Weissenberg photographs). This enables experimental results to be compared with those calculated by the program.

In the Bragg method, the rotation is assumed to be complete, and there are therefore superimposed spots. One option enables you to eliminate from the list of spots all those which correspond to the same angle of diffraction.

This same option is offered for powder diagrams. In this mode, the program displays a sorted list of lattice spacings and the corresponding indices.

CRYS: Studying crystal structures

See chapter 16.

This program displays the internal structure of crystals. Select the menu 'File/Open' and in the dialogue box open the folder containing files with the extension '.CRI'.

Double click on the chosen file to load it. The data for the compound selected are displayed and on clicking on 'OK' the unit cell is displayed. The default mode is 'transparent', which gives the greatest clarity; the atoms are shown as transparent circles whose radius is a quarter of the real radius. In compact mode (menu 'Structure/ Compact mode'), the atoms are represented by opaque discs with a radius equal to that of the atom. By selecting the menu 'Structure/Increments', you can change the direction of observation of the structure to suit the aspect you want to study, using either the keyboard arrows or the mouse (click on the on-screen arrows). You can also set a particular row or direction along which you wish to look, using the menu Structure/ Along a row.

It is not always easy to understand the internal pattern in a crystal through a single unit-cell. Using the option 'Structure/Multiple' you can increase the volume of space depicted; the maximum number of atoms is limited to 400.

The display can be modified with various other options: display the base vectors, the unit-cell, a caption, numbering of the atoms to enable them to be identified. The radii of the atoms and their colour can both be modified. If you wish you can save any modifications in a new data file.

The commands in the menu 'Coordinations' enable you to examine the the neighbourhood of an atom. With the command 'Lengths' you can find the distance between two atoms, and the command 'Angles' is used to find the angle between two bonds. 'Normal mode' enables you to select those atoms surrounding a given atom which fulfil certain conditions (type of atom and distance from the given atom). Finally, using the 'Automatic' command you can draw the coordination polyhedra of the structure.

With the command 'Structure/Lines' you can draw lines anywhere on the structure, for example to show bonds; these can be printed but any rotation of the structure will caused them to be erased irreversibly.

Several printing options are included in the menu 'Print'. Before using the option 'Print/HP-GL file', it is advisable to check that the filter in the importing program used recognizes such files properly. If the 'Colour Print' option is selected when using a black and white printer, the results will, depending on the printer configuration, either be a range of greys or pure black and white.

You can create your own data files. Using a text editor such as 'NotePad', you should study the structure of the data files included with the program and read the associated Help file. If you are considering using homogeneous matrices, you should carefully read the corresponding page in the Help file.

GRAPH: A Graphic Calculator

This is not a crystallography program but a simulation of a numeric/graphic calculator. You can draw curves $Y = f(X, K)$ where X is a variable and K a parameter. The function to be studied, the range of X and the step are written in a dialogue box. If the step is equal to zero, the value of the function is calculated for the value of X which corresponds to the lower bound. If the step is different from zero, the program plots the curve $Y = f(X)$ in the selected range. By modifying the values of the lower bound and the step, it is possible to avoid values that would lead to a division by zero. You can work in cartesian coordinates, polar coordinates or semi-logarithmic coordinates (Bode diagram). In this system of coordinates, the lower bound must of course be positive. The program uses the usual priority rules for operators. In case of doubt, you can use brackets. The program detects any syntax errors in the functions but does not indicate where they are.

Bibliography

International Tables for Crystallography
volume A: Space-group symmetry, Kluwer Academic Publishers, Dordrecht (1989)
volume B: Reciprocal lattice, Kluwer Academic Publishers, Dordrecht (1993)
volume C: Mathematical, Physical and Chemical Tables, Kluwer Academic Publishers, Dordrecht (1992)

AZAROFF, L. V. and BUERGER, M. J. *The powder method in X-ray crystallography.* McGraw-Hill, New York (1958).

BACON G. E. *Neutron Diffraction.* Oxford University Press, New York (1975).

BORCHARD T- OTT W. *Crystallography.* Springer-Verlag, Berlin (1993).

BOUASSE H. *Cristallographie géométrique.* Delagrove, Paris (1929).

BUERGER M. J. *The precession method.* Wiley, New York (1964).

BUERGER M. J. *Introduction to crystal geometry.* McGraw-Hill, New York (1964).

EBERHART, J.-P. *Analyse structurale et chimique des matériaux.* Dunod, Paris (1989).

GIACOVAZZO C. *Fundamentals of Crystallography.* International Union of Crystallography. Oxford University Press (1992).

GUINIER A. *X-ray diffraction.* Freeman, San Francisco (1983).

HAYMANN P. *Théorie dynamique de la microscope et diffraction éléctronique.* PUF, Paris (1974).

HERMANN C. *Internationale Tabellen zur Bestimmung von Kristallstrukturen.* Gebrüder Borntraeger, Berlin (1935).

HULIN M. and BETBEDER O. *Théorie des groupes appliqué à la physique.* Editions de Physique, Les Ulis (1991).

JANAOT C. *Quasicrystals.* Clarendon Press, Oxford (1992).

JEFFERY J. W. *Methods in X-ray crystallography.* Academic Press, London (1971).

MEGAW, H. D. *Crystal structures: a working approach.* Saunders, Philadelphia (1973).

NYE J. F. *Physical properties of crystals.* Clarendon Press, Oxford (1985).

WYCKOFF R. W. G. *Crystal structures*, Vol 1–6. J.Wiley, new York (1962–71).

Index